Essentials of Signals and Systems

Essentials of Signals and Systems

EMILIANO R. MARTINS

Department of Electrical and Computer Engineering
University of São Paulo, Brazil

Registered Offices
John Wiley & Sons, Inc., 111 River Street, Hoboken, NJ 07030, USA
John Wiley & Sons Ltd, The Atrium, Southern Gate, Chichester, West Sussex, PO19 8SQ, UK

Editorial Office
The Atrium, Southern Gate, Chichester, West Sussex, PO19 8SQ, UK

For details of our global editorial offices, customer services, and more information about Wiley products visit us at www.wiley.com.

Library of Congress Cataloging-in-Publication Data

Names: Martins, Emiliano R., author. | John Wiley & Sons, publisher.
Title: Essentials of signals and systems / Emiliano R. Martins.
Description: Hoboken, NJ : Wiley, 2023. | Includes bibliographical
    references and index.
Identifiers: LCCN 2022049576 (print) | LCCN 2022049577 (ebook) | ISBN
    9781119909217 (paperback) | ISBN 9781119909224 (adobe pdf) | ISBN
    9781119909231 (epub)
Subjects: LCSH: Signal theory (Telecommunication). | Signals and signaling.
    | Signal processing–Data processing.
Classification: LCC TK5102.5 .M29225 2023 (print) | LCC TK5102.5 (ebook)
    | DDC 621.382/23–dc23/eng/20221107
LC record available at https://lccn.loc.gov/2022049576
LC ebook record available at https://lccn.loc.gov/2022049577

Cover Design: Wiley
Cover Image: © Andriy Onufriyenko/Getty Images

Set in 11/13pt Computer Modern by Straive, Chennai, India
Printed and bound by CPI Group (UK) Ltd, Croydon, CR0 4YY

C9781119909217_300123

Contents

Preface

Like Achilles played by Brad Pitt, signals and systems are a powerful and beautiful theory. Have you ever wondered why is it the case that, if any waveform passes through an electric circuit, the circuit modifies the waveform, except if it is a sinusoidal wave? For example, if the circuit input is a square wave, then the output is not a square wave. If the input is triangular, then the output is not triangular. But if the input is sinusoidal, then the output is also sinusoidal, only with a different amplitude and phase. Why is that? And why does the same thing happen in so many different systems? For instance, if you apply a square wave force to the mass of a spring–mass system, the mass does not move like a square wave. But, if the force is sinusoidal, and only if it is sinusoidal, then the mass movement is also sinusoidal. What is so special about sinusoidal waves that their form is maintained in so many different systems?

As you will learn in this textbook, this is a consequence of conservation of frequency in linear and time invariant systems. This is an extremely useful concept, with applications in a wide range of topics in engineering and physics. For example, by changing time for space, the same theory explains Snell's law of reflection and refraction, and many other conservation properties of electromagnetic, optical, and quantum systems. And this is just a taste of the power of this theory.

However, signals and systems enjoy an unfair reputation of being a tough nut to crack. It is not. In fact, if you already know elementary notions of linear algebra – like vectors, change of basis, eigenvectors, and eigenvalues – then you already know the most important bits of the theory of signals and systems. It is only a matter of spelling them out. So, as Socrates, in history's most cringing epistemological simile, put it: you are already pregnant with signals and systems; I am only the midwife. But if you have never taken a linear algebra course, no need to worry, we will introduce the main ideas from scratch.

So you, reader, are in for a treat. With a little bit of effort, you will learn one of the most useful mathematical theories in engineering and physics. And, yes, this is a mathematical theory, but I will make sure to show you the physical meaning of the crucial concepts, like Fourier transforms and the frequency response. You will learn not only their meaning, but also how these mathematical objects can be measured in real life.

Now let me tell you a bit about the lay of the land. We begin by stretching and warming up in Chapter 1, where we review the basic concepts of linear algebra that are pertinent to signals and systems. Though I assume that you have already taken at least one course in linear algebra, the review introduces all ideas from scratch, so that a reader unfamiliar with linear algebra can follow it without difficulty. The only caveat is that the review is not in the rigorous format of a traditional exposition of linear algebra. Indeed, it takes some liberties that are only acceptable in a book that is not about linear algebra per se, as it is the case here. In short: if you have never taken a course in linear algebra, you should have no difficulties in following the review, but you should not forget that this is only a review, not a rigorous exposition of the theory. Finally, I will introduce a notation that may be new even to students familiar with linear algebra. It is the famous Dirac notation, after Paul Dirac, one of the greats of the twentieth century. The main motivation to introduce the Dirac notation is that it helps to highlight a crucial aspect of the theory of signals and systems, namely, the difference between the representation of a signal and the signal itself.

The course per se begins in Chapter 2, which is a foundational chapter, where we will learn the theory of representation of signals. We will see that the same ideas of linear algebra can also be applied to signals. In linear algebra, the same vector can be represented in different bases. In signals and systems, signals play the role of vectors, and as such can also be represented in different bases. This notion leads to two central representations: the time domain representation and the frequency domain representation. The latter is the representation involving Fourier transforms, so a big chunk of Chapter 2 is about Fourier transforms. We will learn what their algebraic and physical meanings are, and that a Fourier transform is a scalar product.

In Chapter 3, we deal with the theory of representation of systems. We are mainly interested in a particular but quite common class of systems, the so-called linear and time invariant systems. We will learn that they play the same role that matrices play in linear algebra: matrices are objects that transform one vector into another; likewise, systems are objects that transform one signal (so, the vector) into another. Then, we will learn that the time representation of a system is a differential equation. You will probably agree that differential equations are not the most straightforward things to solve in the universe. But here we will learn that they are not straightforward because they are like matrices that are not diagonal. In algebra, however, a general matrix can be made diagonal if a basis of eigenvectors is used. Likewise, differential equations can also be made diagonal by choosing a basis of eigenvectors. That leads to a simple representation of systems: the frequency domain representation, which is a diagonal representation. So, along the way, we will also learn the easiest way to solve differential equations. In this chapter, we will also learn why frequency is conserved in linear and time invariant systems, and so we will finally find out what is so special about sinusoidal waves. Then we will catch a glimpse of applications of this theory in other fields, like optics and quantum mechanics.

In Chapters 4 and 5, we will reap what we have sown and apply the freshly learned concepts to some examples, namely, analysis of electric circuits and filter design. We will learn what signals and systems have to do with phasors and impedances (they are intimately connected) and the basics of filter design.

In Chapter 6, we will learn that sometimes systems go bananas and require a more general representation. That will lead to the Laplace transform, which is a generalization of the Fourier transform.

We change gears in Chapters 7–10, which deal with discrete signals and systems. In Chapter 7 and 8, we will learn how to obtain Fourier transforms using computers. Maybe you have already heard about the fast Fourier transform (*fft*), which is a widely used algorithm in engineering and physics. We will learn that this algorithm implements a Discrete Fourier Transform (DFT), which is the topic of Chapter 8. So, we will learn what the *fft* is about and how to use it in practice.

Chapter 9 presents the basic ideas of the representation of discrete systems. The theories of discrete and continuous systems (the latter is the subject of Chapter 3) are quite similar, so we focus attention on their differences. Finally, in Chapter 10, we will learn the z-transform, which is a version of the Laplace transform for discrete signals, and it is widely used in control engineering.

Now it is time to brew the coffee and play the focusing/meditation playlist. I hope you will enjoy the ride.

First note to instructors: in my experience, I found that students tend to prefer an exposition of the continuous time theory in one block, without interweaving it with the discrete time theory. Such a preference guided the organization of content in this textbook. If, however, you prefer to teach continuous and discrete time in parallel, then I suggest the following sequence of exposition: Chapters 1, 2, 7, 8, 3, 4, 5, 9, 6, and 10.

Second note to instructors: to bring out the difference between a signal and its representation, I introduce the Dirac notation for vectors and scalar products in Section 1.10 and use it in a few places in Chapters 2 and Chapter 3. But I keep it simple, and do not discuss the notion of 'duals', since it is not necessary for our purposes. The Dirac notation is used to represent scalar products between signals, and to distinguish between a signal itself and its representation.

About the Author

Emiliano R. Martins majored in electrical engineering at the University of São Paulo (Brazil), obtained a master's degree in electrical engineering from the same university, another master's degree in photonics from the Erasmus Mundus Master in Photonics (European consortium), and a PhD in physics from the University of St. Andrews (UK). He has been teaching signals and systems in the Department of Electrical and Computer Engineering of the University of São Paulo (Brazil) since 2016. He is also the author of *Essentials of Semiconductor Device Physics*.

Acknowledgments

The raison d'être of all my endeavours lies within three persons: my wife Andrea, and my two children Tom and Manu. Without you, no book would be written, no passion would be meaningful.

I am grateful to my parents Elcio and Ana for teaching me what really matters, and to my sister Barbara for showing me the easiest way to learn what really matters.

I also thank all my students who contributed with invaluable feedback over all these years. You were the provers guiding the recipe of the pudding. If I have lived up to your expectations, then I have done my job well.

I am grateful to the Wiley team who brought the manuscript to life.

Last, but certainly not least, I thank my editors Martin Preuss and Jenny Cossham for believing in this project.

About the Companion Website

This book is accompanied by a companion website:

www.wiley.com/go/martins/essentialsofsignalsandsystems

This website includes:

- Solutions answers

1

Review of Linear Algebra

Learning Objectives

We review the concepts of linear algebra that are relevant to signals and systems. There are two key ideas of particular importance to subsequent chapters: (i) the idea that the coordinate representation of a vector depends on the choice of basis vectors and (ii) the idea that a basis of eigenvectors simplifies a matrix equation. In Chapters 2 and 3, these two key ideas will be extended to a new algebraic space: the function space.

1.1 Introduction

In this chapter, we review concepts of linear algebra pertinent to signals and systems. I recommend that you read it even if you are already familiar and comfortable with linear algebra because, besides defining the notation and terminology, we will also make some important connections between linear algebra and signals and systems. This is certainly not a rigorous exposition of linear algebra, but we build it roughly from scratch, so that a reader without any previous knowledge of algebra can follow it.

The theory of signals and systems is essentially a generalization of linear algebra. Thus, the core ideas in the exposition of signals and systems will be illustrated by drawing analogies with a Euclidean space. In Chapter 2, we will learn that some intimidating integrals are only linear combinations of vectors, and some others are only scalar products.

Linear algebra deals with two mathematical objects: vectors and operators. Operators are objects that act on vectors, turning them into another vector. The set of all vectors upon which the operators can act is called a space, and if an operator acts on a vector of a space, then the resulting vector also belongs to the

Essentials of Signals and Systems, First Edition. Emiliano R. Martins.
© 2023 John Wiley & Sons Ltd. Published 2023 by John Wiley & Sons Ltd.
Companion website: www.wiley.com/go/martins/essentialsofsignalsandsystems

space. Furthermore, if we add two vectors from a given space, the result is also a vector from the same space. The most common space, which we will deal with in this chapter, is the Euclidean space.

We will use two types of notations for vectors. In the beginning, we will use the more standard notation, denoting a vector by placing a little arrow on the top of a letter, like this: $\vec{v}$. In Section 1.10, we will introduce a more elegant notation, namely, the Dirac notation. For now, though, we stick to the little arrows.

We will also use two types of notations for operators. First, we will denote an operator by a capital letter followed by curly brackets, like this: $T\{\ \}$. Thus, we denote the action of the operator $T\{\ \}$ on a vector $\vec{v}$ resulting in another vector $\vec{u}$, by:

$$\vec{u} = T\{\vec{v}\} \tag{1.1}$$

In Section 1.10, we will also introduce the Dirac notation for operators.

Pay attention to a subtlety of the terminology: even though we can define many kinds of operations involving vectors, we reserve the term 'operator' for an object that turns a vector into another vector. But not all operations turn a vector into another. For example, in Section 1.2, we will define an operation that maps two vectors into a scalar (scalars are defined in Section 1.2), so we cannot call the object that does this operation an 'operator', because it does not turn a vector into another vector. Keep in mind this distinction between 'operation' and 'operator'.

1.2   Vectors, Scalars, and Bases

In linear algebra, we are allowed to multiply vectors by numbers, and the result is another vector. These numbers are also called 'scalars'. A more precise definition of a scalar involves the notion of invariance, but we do not need to go there, so we will make do with the notion of a scalar as being just a number.

Now, let us define an operation (mind you: I said an 'operation', not an 'operator') that maps two vectors into a scalar. Think of this operation as a kind of mathematical machine, in which there are two input slots and one output slot. So, you put one vector in one of the input slots, another vector in the other input slot, and out comes a scalar in the output slot. To denote this operation, we will use the symbol ■, and the rule is: if we put one vector on each side of ■, out comes a scalar, like this:

DEFINITION 1

$$\vec{u}\ ■\ \vec{v} = c \tag{1.2}$$

In **Definition 1**, c is a number (a scalar). So, **Definition 1** is only specifying that the object $\vec{u}\ ■\ \vec{v}$ is a scalar: it is the output of the mathematical machine. Now, one feature that may be novel even to students familiar with linear algebra is that, here, we will allow $\vec{u}\blacksquare\vec{v}$ to be a complex number. So, in **Definition 1**, c may be a complex number.

Let us define other two properties of our mapping. The first one is that the mapping is linear, and as such obeys **Definition 2**:

DEFINITION 2

$$\vec{u}\blacksquare(a\vec{v} + b\vec{w}) = a(\vec{u}\blacksquare\vec{v}) + b(\vec{u}\blacksquare\vec{w}) \tag{1.3}$$

As implied in **Definition 2**, the object $(a\vec{v} + b\vec{w})$ is itself a vector. We say that this vector was formed by a 'linear combination' of the vectors $\vec{v}$ and $\vec{w}$.

Notice that we have defined linearity in terms of linear combinations on the right-hand side of the $\blacksquare$, but we have said nothing about linear combinations on the left-hand side, such as $(a\vec{v} + b\vec{w})\blacksquare\vec{u}$. To check what happens with operations involving linear combinations on the left-hand side, we need our third definition:

DEFINITION 3

$$\vec{u}\blacksquare\vec{v} = (\vec{v}\blacksquare\vec{u})^* \tag{1.4}$$

where $(\vec{v}\blacksquare\vec{u})^*$ is the complex conjugate of $(\vec{v}\blacksquare\vec{u})$.

We call the operation obeying **Definitions 1.1--1.3** a 'scalar product'. The terminology reflects the fact that the operation associates two vectors with a scalar.

Worked Exercise: Linear Combinations on the Left-hand Side of the Scalar Product

As an exercise in logic, let us check how the linear combination in parenthesis in the expression $(a\vec{v} + b\vec{w})\blacksquare\vec{u}$ can be extracted from the parenthesis.

Solution:

First, recall that $(a\vec{v} + b\vec{w})$ is itself a vector. Call it $\vec{p}$:

$$(a\vec{v} + b\vec{w}) = \vec{p}$$

Thus:

$$(a\vec{v} + b\vec{w})\blacksquare\vec{u} = \vec{p}\blacksquare\vec{u} \tag{1.5}$$

But, according to Definition 3:

$$\vec{p}\blacksquare\vec{u} = (\vec{u}\blacksquare\vec{p})^* \tag{1.6}$$

Recall that $\vec{u}\,\blacksquare\,\vec{p} = \vec{u}\,\blacksquare\,(a\vec{v} + b\vec{w})$. Moreover, according to Definition 2:

$$\vec{u}\,\blacksquare\,(a\vec{v} + b\vec{w}) = a(\vec{u}\,\blacksquare\,\vec{v}) + b(\vec{u}\,\blacksquare\,\vec{w}) \tag{1.7}$$

Substituting Equation (1.7) into Equation (1.6):

$$\vec{p}\,\blacksquare\,\vec{u} = (\vec{u}\,\blacksquare\,\vec{p})^* = [a(\vec{u}\,\blacksquare\,\vec{v}) + b(\vec{u}\,\blacksquare\,\vec{w})]^* = a^*(\vec{u}\,\blacksquare\,\vec{v})^* + b^*(\vec{u}\,\blacksquare\,\vec{w})^* \tag{1.8}$$

But, according to Definition 3:

$$(\vec{u}\,\blacksquare\,\vec{v})^* = (\vec{v}\,\blacksquare\,\vec{u}) \text{ and } (\vec{u}\,\blacksquare\,\vec{w})^* = (\vec{w}\,\blacksquare\,\vec{u}) \tag{1.9}$$

Substituting Equation (1.9) into Equation (1.8):

$$\vec{p}\,\blacksquare\,\vec{u} = a^*(\vec{v}\,\blacksquare\,\vec{u}) + b^*(\vec{w}\,\blacksquare\,\vec{u})$$

Thus, with the help of Equation (1.5), we conclude that:

CONCLUSION

$$(a\vec{v} + b\vec{w})\,\blacksquare\,\vec{u} = a^*(\vec{v}\,\blacksquare\,\vec{u}) + b^*(\vec{w}\,\blacksquare\,\vec{u}) \tag{1.10}$$

Pay close attention to Equation (1.10), and compare it to Equation (1.3): in the way we defined the scalar product, to extract the scalars inside the parenthesis on the left-hand side of the $\blacksquare$, we have to complex conjugate them (as in Equation (1.10)). But, if the parenthesis is on the right-hand side of the $\blacksquare$, then we do not need to conjugate them (as in Equation (1.3)). Some textbooks define the scalar product in the opposite way (not conjugate if it is on the left part, but conjugate if it is on the right part), so we need to pay attention to the way it is defined.

We also need to define some terminology. If I take the scalar product of a vector with itself, then the result is obviously a scalar. It is also a real scalar, as you will prove in the exercise list. We call the square root of this scalar 'the magnitude of the vector'. So, the magnitude is a real number associated with a vector. Using the symbol $|\vec{u}|$ to denote the magnitude of the vector $\vec{u}$, then, by definition:

$$|\vec{u}| = \sqrt{\vec{u}\,\blacksquare\,\vec{u}} \tag{1.11}$$

A vector with magnitude one is called a 'unit vector'. To specify that a vector is a unit vector, we will replace the arrow with a funny hat, like this: $\hat{u}$. Thus, again by definition, a vector only deserves a hat if its magnitude is one, that is:

$$|\hat{u}| = 1 \tag{1.12}$$

One key notion in linear algebra is the notion of orthogonality. By definition, two vectors are orthogonal if the scalar product between them is zero.

Now, suppose that two unit vectors $\widehat{x}$ and $\widehat{y}$ are orthogonal (that is, suppose that $\widehat{x}\blacksquare\widehat{y} = 0$). Now suppose that a vector $\overrightarrow{u}$ is formed by a linear combination of $\widehat{x}$ and $\widehat{y}$:

$$\overrightarrow{u} = a\widehat{x} + b\widehat{y}$$

If we apply the scalar product between $\widehat{x}$ and $\overrightarrow{u}$, then, with the help of Definition 2, we find:

$$\widehat{x}\blacksquare\overrightarrow{u} = a(\widehat{x}\blacksquare\widehat{x}) + b(\widehat{x}\blacksquare\widehat{y})$$

But, by definition, $\widehat{x}$ is a unit vector (that is, $\widehat{x}\blacksquare\widehat{x} = 1$) and orthogonal to $\widehat{y}$ (that is: $\widehat{x}\blacksquare\widehat{y} = 0$). Therefore:

$$\widehat{x}\blacksquare\overrightarrow{u} = a \cdot 1 + b \cdot 0 = a$$

Likewise:

$$\widehat{y}\blacksquare\overrightarrow{u} = b$$

The fact that $\widehat{x}\blacksquare\overrightarrow{u} = a$ and $\widehat{y}\blacksquare\overrightarrow{u} = b$ when $\overrightarrow{u} = a\widehat{x} + b\widehat{y}$ entails that a vector $\overrightarrow{u}$ can be formed by a linear combination of two vectors $\widehat{x}$ and $\widehat{y}$ only if $\overrightarrow{u}$ is not orthogonal to both $\widehat{x}$ and $\widehat{y}$. After all, if $\overrightarrow{u}$ was orthogonal to $\widehat{x}$ and $\widehat{y}$, then a and b would be zero. Of course, this conclusion also holds for linear combinations involving more than two vectors.

Now let us do a thought mathematical experiment. Imagine you have a bag full of vectors. This bag is your vector space. Pick a vector from the bag. Next, pick a second vector, but this time you must choose one that is orthogonal to the first. Now you have two orthogonal vectors. But suppose you want a third one, and it needs to be orthogonal to both vectors you already picked. If you **cannot** find a third one that is orthogonal to both vectors you already picked, then your bag is a two-dimensional Euclidean space. If, on the other hand, you can pick a third one, that means your bag is a higher dimensional space. But how higher? Well, that depends. If, for example, you can find three vectors that are orthogonal to each other (each one must be orthogonal to the other two), but you **cannot** find a fourth one that is orthogonal to all the other three, that means your bag is a three-dimensional Euclidean space. And so on.

So, the largest possible group of orthogonal vectors defines the dimension of the space. We call this group a '**basis of the space**'. Thus, any pair of orthogonal vectors forms a basis of a two-dimensional Euclidean space. Any three orthogonal vectors form a basis of a three-dimensional Euclidean space. And so on. If, furthermore, the basis is formed only by unit vectors, then we say that it is an **orthonormal basis**. Unless explicitly stated otherwise, **from now on, when I use the term 'basis', I mean an orthonormal basis**.

Suppose you have a space of a certain dimension, for example, a three-dimensional Euclidean space. We have just seen that, to assert that the space is three-dimensional is tantamount to asserting that a basis has three vectors, and that it is not possible to find a fourth vector that is orthogonal to all these three basis vectors. So, that means any vector of this space can be formed by a linear combination of these basis vectors. If it could not, then it would have to be orthogonal to all these three basis vectors, which would entail that your space is not three-dimensional after all, but at least four-dimensional. **Therefore, any vector can be formed by a linear combination of the basis vectors**. This is one of the foundational concepts in our course in signals and systems: we will be constantly expressing signals in terms of linear combinations of other signals, the latter forming a basis of the space where the signals live (I defer more information about this new space until Chapter 2).

For the sake of simplicity, from now on we assume a two-dimensional Euclidean space, but the concepts we will be reviewing can be straightforwardly extended to higher dimensional Euclidean spaces.

Let us suppose again that vectors $\hat{x}$ and $\hat{y}$ are orthogonal. If they are orthogonal, then necessarily they form a basis of the two-dimensional Euclidean space. Thus, any vector of this space can be formed by a linear combination of $\hat{x}$ and $\hat{y}$. For example, a vector $\vec{u}$ can be expressed as:

$$\vec{u} = u_x\hat{x} + u_y\hat{y} \tag{1.13}$$

The numbers u_x and u_y are called the 'coordinates' of vector $\vec{u}$ with respect to the basis $\hat{x}$ and $\hat{y}$. Once a basis has been specified, then a vector is uniquely specified by giving its coordinates. Thus, we can specify the vector $\vec{u}$ by specifying its coordinates u_x and u_y.

But, if the coordinates uniquely specify a vector, then any information about the vector can be expressed in terms of its coordinates. In particular, the scalar product between two vectors can be expressed in terms of their coordinates. Thus, suppose we have a vector $\vec{u}$, as specified in Equation (1.13), and another vector $\vec{v}$, specified as:

$$\vec{v} = v_x\hat{x} + v_y\hat{y} \tag{1.14}$$

Let us find an expression for the scalar product $\vec{v} \blacksquare \vec{u}$ in terms of their coordinates. We begin by expressing $\vec{u}$ in terms of a linear combination of basis vectors:

$$\vec{v} \blacksquare \vec{u} = \vec{v} \blacksquare (u_x\hat{x} + u_y\hat{y})$$

Now we use Definition 2:

$$\vec{v} \blacksquare \vec{u} = \vec{v} \blacksquare (u_x\hat{x} + u_y\hat{y}) = (\vec{v} \blacksquare \hat{x})u_x + (\vec{v} \blacksquare \hat{y})u_y$$

Now we express $\vec{v}$ in terms of a linear combination of basis vectors:

$$\vec{v} \blacksquare \vec{u} = (\vec{v} \blacksquare \hat{x})u_x + (\vec{v} \blacksquare \hat{y})u_y = [(v_x\hat{x} + v_y\hat{y}) \blacksquare \hat{x}]u_x + [(v_x\hat{x} + v_y\hat{y}) \blacksquare \hat{y}]u_y$$

And now we use the **Conclusion** (Equation (1.10)) to extract the scalars from the parenthesis. Thus:

$$\vec{v} \blacksquare \vec{u} = [(v_x\hat{x} + v_y\hat{y}) \blacksquare \hat{x}]u_x + [(v_x\hat{x} + v_y\hat{y}) \blacksquare \hat{y}]u_y = [v_x^*(\hat{x} \blacksquare \hat{x}) + v_y^*(\hat{y} \blacksquare \hat{x})]u_x$$
$$+ [v_x^*(\hat{x} \blacksquare \hat{y}) + v_y^*(\hat{y} \blacksquare \hat{y})]u_y$$

But if $\hat{x}$ and $\hat{y}$ form a basis, then $\hat{x} \blacksquare \hat{y} = \hat{y} \blacksquare \hat{x} = 0$. Furthermore, the basis vectors are unit vectors, that is: $\hat{x} \blacksquare \hat{x} = \hat{y} \blacksquare \hat{y} = 1$. Thus, we conclude that:

$$\vec{v} \blacksquare \vec{u} = v_x^* u_x + v_y^* u_y \tag{1.15}$$

Notice that the coordinates of the vector $\vec{v}$ are conjugated in the scalar product. This happened because $\vec{v}$ is on the left-hand side of the $\blacksquare$. So, do not forget that the coordinates of the left-hand side vector are conjugated in the scalar product.

We can also express the magnitude of a vector in terms of its coordinates. According to Equation (1.11):

$$|\vec{u}| = \sqrt{\vec{u} \blacksquare \vec{u}}$$

Using Equation (1.15) in Equation (1.11):

$$|\vec{u}| = \sqrt{\vec{u} \blacksquare \vec{u}} = \sqrt{u_x^* u_x + u_y^* u_y} \tag{1.16}$$

The root square of the product of a complex number with its conjugate is called the 'magnitude of the complex number'. Thus, $|c|$ denotes the magnitude of the complex number c:

$$|c| = \sqrt{c^* c} \tag{1.17}$$

Notice that both $|\vec{u}|$ and $|c|$ are called magnitudes, but they are not the magnitudes of the same object ($|\vec{u}|$ is the magnitude of a vector and $|c|$ is the magnitude of a complex number). Using Equation (1.17) in Equation (1.16), we find:

$$|\vec{u}| = \sqrt{|u_x|^2 + |u_y|^2} \tag{1.18}$$

Are you tired? If you are, please rest a little bit before reading Section 1.3. It is not difficult, but it is crucial that you understand it well.

1.3 Vector Representation in Different Bases

In Section 1.2, I mentioned that the idea of expressing a vector as a linear combination of basis vectors is a foundational idea in our course. Students usually have no problem with this notion. What tends to confuse them is the fact that because there are lots of bases (in fact, an infinite number of bases) of a space, so there are lots of possible representations (actually, an infinite number of possible representations). Each group of coordinates of a vector is a representation of this vector in a given basis. **Thus, the very same vector can be represented with different coordinates, using different bases.** The confusion arises when we do not take care to distinguish between the vector itself and its representation in a given basis. In other words, beware not to confuse a vector with its coordinates. As I have just said, the coordinates uniquely specify a vector once a basis is chosen. But, if we change the basis, then the very same vector is specified by different coordinates. Let us spell out this difference. Suppose that we picked the basis $\widehat{x}$ and $\widehat{y}$, and specified a vector $\overrightarrow{u}$ by specifying its coordinates u_x and u_y, like in Equation (1.13), which I repeat below for convenience.

Equation (1.13)

$$\overrightarrow{u} = u_x\widehat{x} + u_y\widehat{y}$$

Now, suppose that we specify a unit vector, which I call $\widehat{a_1}$. As always, to specify a vector, we must specify its coordinates. Thus, say that we chose the coordinates a_{1x} and a_{1y} to specify $\widehat{a_1}$:

$$\widehat{a_1} = a_{1x}\widehat{x} + a_{1y}\widehat{y} \tag{1.19}$$

Notice that, if, by construction, we want $\widehat{a_1}$ to be a unit vector, then according to Equation (1.18) its coordinates must satisfy the condition:

$$\sqrt{|a_{1x}|^2 + |a_{1y}|^2} = 1 \tag{1.20}$$

Ok, so let us say that we picked a pair a_{1x}, a_{1y} that satisfies Equation (1.20). Now let us specify another unit vector $\widehat{a_2}$ with the coordinates a_{2x} and a_{2y}:

$$\widehat{a_2} = a_{2x}\widehat{x} + a_{2y}\widehat{y} \tag{1.21}$$

Of course, if we want $\widehat{a_2}$ to be a unit vector, then we must guarantee that:

$$\sqrt{|a_{2x}|^2 + |a_{2y}|^2} = 1 \tag{1.22}$$

But, besides being unit vectors, now we also require $\widehat{a_2}$ to be orthogonal to $\widehat{a_1}$:

$$\widehat{a_2}\blacksquare\widehat{a_1} = 0 \tag{1.23}$$

In other words, besides satisfying Equation (1.22), the coordinates of $\widehat{a_2}$ must also satisfy:

$$a_{2x}^*a_{1x} + a_{2y}^*a_{1y} = 0 \tag{1.24}$$

To obtain Equation (1.24), I used Equation (1.15) in Equation (1.23).

Ok, so let us suppose we have done this job: we picked a pair of coordinates a_{1x}, a_{1y} that satisfies Equation (1.20), and then we found a new pair a_{2x}, a_{2y} that satisfies Equation (1.22) and Equation (1.24). So, we have two new unit vectors that are orthogonal to each other. In other words, we have a new basis: the vectors $\widehat{a_1}$ and $\widehat{a_2}$ also form a basis. Thus, any vector in the space can be expressed as a linear combination of $\widehat{a_1}$ and $\widehat{a_2}$. In particular, the very same $\overrightarrow{u}$ of Equation (1.13) can also be expressed in the basis formed by $\hat{a}_1$ and $\hat{a}_2$. But the coordinates will be different. We denoted the coordinates of $\vec{u}$ with respect to the basis formed by $\hat{x}$ and $\hat{y}$ by u_x and u_y, but in the new basis we have new coordinates. Let us call the coordinates with respect to the new basis U_1 and U_2. Then:

$$\vec{u} = U_1\hat{a}_1 + U_2\hat{a}_2 \tag{1.25}$$

But, since this is the very same $\vec{u}$ of Equation (1.13), then:

$$\vec{u} = u_x\hat{x} + u_y\hat{y} = U_1\hat{a}_1 + U_2\hat{a}_2 \tag{1.26}$$

Suppose you chose to specify $\vec{u}$ in the basis $\hat{x}$ and $\hat{y}$. That means you specified u_x and u_y. Now, suppose you also chose a new basis $\hat{a}_1$ and $\hat{a}_2$. Recall that you specified these vectors $\hat{a}_1$ and $\hat{a}_2$ by specifying their coordinates in the basis $\hat{x}$ and $\hat{y}$. So, at this moment you have u_x and u_y specified, and you also have a_{1x}, a_{1y} specified and a_{2x}, a_{2y} specified. Thus, these six numbers have been given. Now, I ask you: if these six numbers have been specified, then are you free to pick whatever U_1 and U_2 you want? Of course not. If U_1 and U_2 specify the vector $\vec{u}$, and this vector has already been specified by the coordinates u_x and u_y, then U_1 and U_2 have also been already specified. Does that mean that we can express U_1 and U_2 solely in terms of u_x and u_y? Of course not, because we also need information about the new basis. In other words, we must be able to express U_1 and U_2 in terms of u_x, u_y, a_{1x}, a_{1y}, a_{2x}, and a_{2y}.

Let us do it! Let us express: U_1 and U_2 in terms of u_x, u_y, a_{1x}, a_{1y}, a_{2x}, and a_{2y}. It is not difficult at all. All we need to do is to recall that, to extract a coordinate, we need to take the scalar product between the vector and the unit vector, with the unit vector on the left-hand side of the scalar product. For example:

$$\hat{a}_1 \blacksquare \vec{u} = \hat{a}_1 \blacksquare (U_1\hat{a}_1 + U_2\hat{a}_2) = U_1(\hat{a}_1 \blacksquare \hat{a}_1) + U_2(\hat{a}_1 \blacksquare \hat{a}_2)$$

where Definition 2 was used. But $\hat{a}_1$ and $\hat{a}_2$ form an orthonormal basis, so $\hat{a}_1 \blacksquare \hat{a}_1 = 1$ and $\hat{a}_1 \blacksquare \hat{a}_2 = 0$. Therefore:

$$\hat{a}_1 \blacksquare \vec{u} = U_1 \tag{1.27}$$

Likewise:

$$\hat{a}_2 \blacksquare \vec{u} = U_2 \tag{1.28}$$

Thus, to find U_1, all we need to do is to find $\hat{a}_1 \blacksquare \vec{u}$. But we know the coordinates of both vectors in the basis formed by $\hat{x}$ and $\hat{y}$. Thus, with the help of Equation (1.15):

$$U_1 = \hat{a}_1 \blacksquare \vec{u} = a_{1x}^* u_x + a_{1y}^* u_y \tag{1.29}$$

In the same spirit, we have:

$$U_2 = \hat{a}_2 \blacksquare \vec{u} = a_{2x}^* u_x + a_{2y}^* u_y \tag{1.30}$$

So, if we have specified $\vec{u}$ in one basis, its coordinates in the new basis are uniquely specified in terms of Equation (1.29) and Equation (1.30).

In Equation (1.29) and Equation (1.30), we chose to calculate the scalar product using the coordinates in the basis $\hat{x}$ and $\hat{y}$. We chose to use this basis because, by assumption, we already knew the coordinates with respect to this basis. But we could have chosen to calculate it in the new basis (or, for that matter, in any basis we want). For example, say that we have the coordinates of $\vec{u}$ and $\vec{v}$ in the basis formed by $\hat{a}_1$ and $\hat{a}_2$:

$$\vec{u} = U_1\hat{a}_1 + U_2\hat{a}_2$$

and:

$$\vec{v} = V_1\hat{a}_1 + V_2\hat{a}_2 \tag{1.31}$$

Then, following the same procedure that led to Equation (1.15), we find that:

$$\vec{v} \blacksquare \vec{u} = V_1^* U_1 + V_2^* U_2 \tag{1.32}$$

Recall that the scalar product is a mapping of two vectors into a scalar. When we change the basis, we change only the representation, that is, only the coordinates, but we do not change the vectors themselves. **Thus, the scalar product does not depend on the representation: it does not matter which basis we use; the scalar product must be the same.** For example, the scalar product of Equation (1.32) is the same of Equation (1.15). In other words:

$$\vec{v} \blacksquare \vec{u} = v_x^* u_x + v_y^* u_y = V_1^* U_1 + V_2^* U_2 \tag{1.33}$$

In the exercise list, you will be asked to check that the last equality in Equation (1.33) is indeed true.

Often, the basis formed by the vectors $\hat{x}$ and $\hat{y}$ is called a 'canonical basis'. Here, we will also adopt this terminology, because it will come in quite handy when we compare vectors with signals. But I need to warn you that this terminology is a bit inadequate, because the term 'canonical' has an overtone of 'special'. For example, the Christian texts that made it into the Bible are called 'canonical', with obvious connotations of legitimacy and uniqueness. But, here, I want you to forget about the religious connotation, and think of the term 'canonical basis' as just a name for this basis. We could just as well have called it 'the apocryphal basis', and it would not have made any difference. Indeed, the canonical basis is as significant and as meaningful as any other basis.

For example, in Equation (1.29) and Equation (1.30), we expressed the coordinates of the new basis vectors in terms of the coordinates with respect to the canonical basis. But we did not do that because the canonical basis is special. We only did that because we had assumed that the vectors had been specified in the canonical basis. But we could have specified the vector in the new basis. If we had done that, then we would have needed to find u_x and u_y in terms of U_1 and U_2 and the new basis, following the same logic as before.

To emphasize that all bases are on an equal footing, let us express u_x and u_y in terms of U_1 and U_2 and the new basis. Thus, recalling that:

$$u_x = \widehat{x} \blacksquare \overrightarrow{u} \tag{1.34}$$

we need to compute $\widehat{x} \blacksquare \overrightarrow{u}$. By assumption, now we have the coordinates of $\overrightarrow{u}$ with respect to the new basis, so this is the natural choice to compute the scalar product. We also need the coordinates of $\widehat{x}$ with respect to the new basis. Calling them X_1 and X_2:

$$\widehat{x} = X_1 \widehat{a_1} + X_2 \widehat{a_2} \tag{1.35}$$

It then follows that:

$$X_1 = \widehat{a_1} \blacksquare \widehat{x}$$

But Equation (1.19) entails that $\widehat{x} \blacksquare \widehat{a_1} = a_{1x}$. Furthermore, according to Equation (1.4), $\widehat{a_1} \blacksquare \widehat{x} = a_{1x}$ entails that $\widehat{a_1} \blacksquare \widehat{x} = a_{1x}^*$. Therefore:

$$X_1 = \widehat{a_1} \blacksquare \widehat{x} = a_{1x}^* \tag{1.36}$$

Likewise:

$$X_2 = \widehat{a_2} \blacksquare \widehat{x} = a_{2x}^* \tag{1.37}$$

Thus:

$$u_x = \widehat{x} \blacksquare \overrightarrow{u} = X_1^* U_1 + X_2^* U_2 = a_{1x} U_1 + a_{2x} U_2 \tag{1.38}$$

In the same spirit:

$$u_y = \widehat{y} \blacksquare \overrightarrow{u} = a_{1y} U_1 + a_{2y} U_2 \tag{1.39}$$

So, it does not matter which basis you use to specify your vector.

To close this section, one last bit of notation. Often, the coordinates of a vector are specified in a column, like this:

$$\begin{pmatrix} u_x \\ u_y \end{pmatrix}$$

We need to be careful with this notation, as it often leads to confusion. The problem is that it is easy to think that $\begin{pmatrix} u_x \\ u_y \end{pmatrix}$ by itself specifies a vector. It does not. To be specified, a vector needs both the coordinates AND the basis. Furthermore, completely different coordinates can specify the same vector. So, for example, the same vector $\overrightarrow{u}$ can be specified by the coordinates $\begin{pmatrix} u_x \\ u_y \end{pmatrix}$ with respect to the canonical basis, and, also, by the coordinates $\begin{pmatrix} U_1 \\ U_2 \end{pmatrix}$ with respect to the new basis. If I only give you a pair of coordinates, but do not tell you the basis, then you cannot know what the vector is. You need to know the basis!

Now, there is one sense in which a canonical basis is indeed special, but it is a bit silly. It is the sense that, if a pair of coordinates is specified in a column, and there is no mention whatsoever of what basis has been used, then it is implied that it is the canonical basis. But we really need to keep in mind that the same vector can be specified by completely different coordinates, like $\begin{pmatrix} u_x \\ u_y \end{pmatrix}$ and $\begin{pmatrix} U_1 \\ U_2 \end{pmatrix}$. I will highlight this issue again and again, and it will become more evident when we consider the representation of operators, which is the subject of Section 1.4.

1.4   Linear Operators

In signals and systems, operators are the mathematical objects describing systems. Recall that an operator is an object that transforms one vector into another vector. We are interested in studying a particular class of operators, namely, linear operators. By definition, a linear operator obeys the following property:

DEFINITION OF LINEARITY

An operator $T\{\ \}$ is linear if, and only if, the following equality is true:

$$T\{a\vec{u} + b\vec{v}\} = aT\{\vec{u}\} + bT\{\vec{v}\} \tag{1.40}$$

Let us pause to reflect on the meaning of a linear operation. Recall from Equation (1.1) that the vector inside the curly brackets is the vector on which the operator is acting. So, in a relationship of the form $T\{\vec{v}\} = \vec{w}$, $\vec{v}$ is the input of the operator and $\vec{w}$ is the output. Thus, the object $T\{\vec{v}\}$ is itself the output of the operator when the input is $\vec{v}$. According to Equation (1.40), to know how a linear operator acts on a vector formed by a linear combination of other vectors (left-hand side of Equation (1.40)), all we need to know is how the operator acts on each vector of the linear combination separately (right-hand side of Equation (1.40)). Using a more formal language, if the output of the linear combination (left-hand side of Equation (1.40)) is identical to the linear combination of the outputs (right-hand side of Equation (1.40)), then the operator is linear.

Since this is a crucial property (and we will soon spell out why), to gain a solid intuition about what it means, let us see two examples of systems, one described by a linear operator, and the other by a nonlinear operator.

I have already mentioned that in Chapter 2 we will learn that signals are vectors in a different space. Suppose that you have a certain song, and that it is represented by the signal g. So, the song g is a vector in this example. Now, suppose you want to modify your song. Thus, you pass it through a system (recall that the system is described by an operator) and out comes a new signal, call it h. The system could be, for example, an equalizer. So, g is your input, the equalizer is the system (the operator), and h is the output. Now, imagine a second situation, where you have one part of the same song g stored in another signal, call it $a1$, and the other part stored in yet another signal, call it $a2$. For example, $a1$ could be the vocal and $a2$ could be the instruments. If you play $a1$ and $a2$ together, then you get your song, that is, $g = a1 + a2$. In this new situation, instead of passing g through the equalizer, you decided to pass $a1$ and $a2$ separately. Call $b1$ the output of $a1$ and $b2$ the output of $a2$. So, first you passed $a1$, and obtained $b1$. Then you passed $a2$ and obtained $b2$. If your equalizer is linear, then, if you combine the output $b1$ with the output $b2$, the result will be h, that is, $h = b1 + b2$. The experiment where you obtained h directly, by plugging g in the input, is akin to the left-hand side of Equation (1.40), while the experiment where you first obtained $b1$, and then $b2$, and then combined them, is akin to the right-hand side of Equation (1.40). If both ways give the same result, then the system is linear.

The equalizer is an example of a linear system. Can you think of an example of a nonlinear system? I like the example of an oven: it is not very algebraic, but it is quite intuitive. Think of the ingredients of a cake as being the vectors, and the oven as being the system. If you get all the ingredients, mixes them up like granny taught you, and then put them all together in the oven, out comes a cake. This procedure is akin to the left-hand side of Equation (1.40): you put all the ingredients (the vectors) together in the input of the operator (the operator is the oven), and the output is your cake. Now, to test if the oven is linear, you need to perform a second experiment, this time following the recipe of the right-hand side of Equation (1.40). According to the right-hand side, you need to find out what happens if you put each ingredient separately in the oven, bake them, and then mix the already baked ingredients. So, first, in go the eggs, bake them, take them out. Then, in goes the flour, bake it, take it out. Then the sugar, and so on. After all the ingredients have been baked, and only then, you mix them up. Of course, the result is not a cake, so, in this example, the right-hand side of Equation (1.40) is not the same as the left-hand side of Equation (1.40): the left-hand side is a cake, but the right-hand side is a disgrace to your coffee break. Conclusion, an oven is a nonlinear system.

Coming back to the question of why linearity is so important.

A key idea in engineering and physics is intimately related with the linearity of systems. The idea is that, since a vector (or signal) can be described as a linear combination of basis vectors (or basis signals), then we know how a linear operator acts on any vector (or any signal) if we know how it acts on the basis vectors (or basis signals). As we will see in Chapter 3, the laws of physics can be treated as operators, so you can imagine how powerful and general this idea is. We will come back to this notion again and again in this course, but for now I want to give you at least one example of how it works.

Suppose we have an operator $T\{\ \}$ and three vectors $\overrightarrow{u}$, $\overrightarrow{v}$, and $\overrightarrow{w}$, specified in the canonical basis:

$$\overrightarrow{u} = u_x \widehat{x} + u_y \widehat{y}$$

$$\overrightarrow{v} = v_x \widehat{x} + v_y \widehat{y}$$

$$\overrightarrow{w} = w_x \widehat{x} + w_y \widehat{y}$$

Suppose we need to find the vectors resulting from the action of $T\{\}$ on $\vec{u}$, $\vec{v}$, $\vec{w}$. If $T\{\}$ is nonlinear, then we need to compute the action of $T\{\}$ on each of these vectors separately, that is, we need to find $T\{\vec{u}\}$, $T\{\vec{v}\}$, and $T\{\vec{w}\}$. So, we have three jobs to do. But, if $T\{\}$ is linear, then all we need to know is its action on $\hat{x}$ and $\hat{y}$. Indeed, if $T\{\}$ is linear, then:

$$T\{\vec{u}\} = T\{u_x\hat{x} + u_y\hat{y}\} = u_xT\{\hat{x}\} + u_yT\{\hat{y}\}$$

$$T\{\vec{v}\} = T\{v_x\hat{x} + v_y\hat{y}\} = v_xT\{\hat{x}\} + v_yT\{\hat{y}\}$$

$$T\{\vec{w}\} = T\{w_x\hat{x} + w_y\hat{y}\} = w_xT\{\hat{x}\} + w_yT\{\hat{y}\}$$

Notice that the property of linearity was used in the last equality of each equation. So, when $T\{\}$ is linear, if we know its action on the basis vectors, that is, if we know $T\{\hat{x}\}$ and $T\{\hat{y}\}$, then we know its action on $\vec{u}$, $\vec{v}$, and $\vec{w}$. Thus, we reduced three jobs (the computation of $T\{\vec{u}\}$, $T\{\vec{v}\}$, and $T\{\vec{w}\}$) to two jobs (the computation of $T\{\hat{x}\}$ and $T\{\hat{y}\}$). But, of course, $T\{\hat{x}\}$ and $T\{\hat{y}\}$ specify the action not only on $\vec{u}$, $\vec{v}$, and $\vec{w}$, but on any vector whatsoever of the vector space. For example, consider the case of a vector field. A vector field associates each point of space with a vector. Since there are an infinite number of points of space, a vector field is a collection of infinitely many vectors. So, if we want to know the action of a nonlinear operator on a vector field, we need to calculate the action on each vector individually. But we have infinitely many vectors, so, like Sisyphus, we have one darn long job to do: it will take nothing short of eternity to finish this job. If the operator is linear, however, our eternal job is reduced to two jobs (or three if the space is three-dimensional): all we need to know is the action of the operator on the basis vectors. A theory that reduces an eternal job to two or three jobs must be a powerful theory indeed. Notice that vector fields appear all the time in engineering and physics (for example, an electric field is a vector field). As I mentioned earlier, this idea of analyzing a system in terms of its action on basis vectors is central to signals and systems, in particular, and to engineering and physics, in general.

1.5 Representation of Linear Operators

In Section 1.3 (Vector Representation in Different Bases), we saw that the representation of vectors, that is, their coordinates, depends on the basis. In this section, we will follow a similar line of reasoning to obtain the representation of linear operators.

An operator is an object that transforms one vector into another. Thus, the representation of an operator must be closely connected with how it acts on vectors. So, suppose again that we have a linear operator $T\{\}$, and like in Equation (1.1) – which I repeat below for convenience – let us suppose that it acts on a vector $\vec{v}$ (the input) and the result is a vector $\vec{u}$ (the output), that is:

Equation (1.1)

$$\vec{u} = T\{\vec{v}\}$$

A representation of vectors is nothing else than their coordinates with respect to a given basis. Furthermore, a vector is fully specified when its coordinates are specified (again assuming that a basis has been chosen). Since an operator is an object that transforms a vector into another vector, and a vector is fully specified by its coordinates, then an operator must be an object that transforms one set of coordinates into another set of coordinates. Thus, the representation of an operator must be a mathematical object that specifies how the operator transforms the coordinates of the input vector into the coordinates of the output vector. To find what object this is, let us expand the vectors $\vec{u}$ and $\vec{v}$ of Equation (1.1) into a linear combination of canonical basis vectors. With this expansion, Equation (1.1) reads:

$$u_x\widehat{x} + u_y\widehat{y} = T\{v_x\widehat{x} + v_y\widehat{y}\} \tag{1.41}$$

Since we are dealing with linear operators, then, according to Equation (1.40), $T\{v_x\widehat{x} + v_y\widehat{y}\} = v_x T\{\widehat{x}\} + v_y T\{\widehat{y}\}$. Thus:

$$u_x\widehat{x} + u_y\widehat{y} = T\{\widehat{x}\}v_x + T\{\widehat{y}\}v_y \tag{1.42}$$

Recall that we want to find how the operator transform input coordinates into output coordinates. To do it, we can isolate each output coordinate on the left-hand side of Equation (1.42). First, to extract u_x, we take the scalar product between $\widehat{x}$ and both sides of Equation (1.42). Thus:

$$\widehat{x}\blacksquare[u_x\widehat{x} + u_y\widehat{y}] = \widehat{x}\blacksquare[T\{\widehat{x}\}v_x + T\{\widehat{y}\}v_y]$$

Since, according to Definition 2, the scalar product is itself a linear operation (mind you, I said linear 'operation', not linear 'operator'), we get:

$$u_x(\widehat{x}\blacksquare\widehat{x}) + u_y(\widehat{x}\blacksquare\widehat{y}) = (\widehat{x}\blacksquare T\{\widehat{x}\})v_x + (\widehat{x}\blacksquare T\{\widehat{y}\})v_y$$

But $(\widehat{x}\blacksquare\widehat{x}) = 1$ and $(\widehat{x}\blacksquare\widehat{y}) = 0$, so we conclude that:

$$u_x = (\widehat{x}\blacksquare T\{\widehat{x}\})v_x + (\widehat{x}\blacksquare T\{\widehat{y}\})v_y$$

Recall that $T\{\widehat{x}\}$ and $T\{\widehat{y}\}$ are vectors: they are the output vectors when the operator acts on each basis vector. Furthermore, $(\widehat{x}\blacksquare T\{\widehat{x}\})$ and $(\widehat{x}\blacksquare T\{\widehat{y}\})$ are scalars: they are the scalar products between the basis vector $\widehat{x}$ and the vectors resulting from the action of the operator on each basis vectors.

Following an analogous procedure, we extract the coordinate u_y by taking the scalar product between $\widehat{y}$ and both sides of Equation (1.42). Thus:

$$\widehat{y}\blacksquare[u_x\widehat{x} + u_y\widehat{y}] = \widehat{y}\blacksquare[T\{\widehat{x}\}v_x + T\{\widehat{y}\}v_y]$$

which leads to:

$$u_y = (\widehat{y}\blacksquare T\{\widehat{x}\})v_x + (\widehat{y}\blacksquare T\{\widehat{y}\})v_y$$

Let us collect these two equations in a single equation number:

$$u_x = (\hat{x}\cdot T\{\hat{x}\})v_x + (\hat{x}\cdot T\{\hat{y}\})v_y$$

$$u_y = (\hat{y}\cdot T\{\hat{x}\})v_x + (\hat{y}\cdot T\{\hat{y}\})v_y \tag{1.43}$$

According to Equation (1.43), once a basis has been chosen (and, for now, we chose the canonical basis), then the action of the operator $T\{\}$ is fully specified by four scalars: $(\hat{x}\cdot T\{\hat{x}\})$, $(\hat{x}\cdot T\{\hat{y}\})$, $(\hat{y}\cdot T\{\hat{x}\})$, and $(\hat{y}\cdot T\{\hat{y}\})$. Indeed, if we know these four scalars, then we know how the operator transforms the coordinates of the input $\vec{v}$ into the coordinates of the output $\vec{u}$. Thus, an operator $T\{\}$ is fully specified by determining the scalar products between each basis vector, and the vectors resulting from the action of the operator on these basis vectors. Notice that this specification is tantamount to specifying the vectors $T\{\hat{x}\}$ and $T\{\hat{y}\}$. Indeed, $\hat{x}\cdot T\{\hat{x}\}$ and $\hat{y}\cdot T\{\hat{x}\}$ are, respectively, the x and y coordinates of the vector $T\{\hat{x}\}$. Likewise, $\hat{x}\cdot T\{\hat{y}\}$ and $\hat{y}\cdot T\{\hat{y}\}$ are, respectively, the x and y coordinates of the vector $T\{\hat{y}\}$.

Recall that, to specify a vector in a basis of a two-dimensional space, we need to specify two coordinates, that is, two numbers (and they can be complex numbers). In the same spirit, Equation (1.43) is teaching that, to specify a linear operator in a basis of a two-dimensional space, we need to specify four numbers. So, in a two-dimensional space, we need two numbers to specify a vector and four numbers to specify a linear operator. If we had a three-dimensional space, we would need three numbers to specify a vector and nine numbers to specify an operator. And so on for higher dimensions.

To clean up the notation, we can denote each of these numbers specifying the operator by:

$$t_{mn} = \hat{m}\cdot T\{\hat{n}\} \tag{1.44}$$

In Equation (1.44), m and n stand for x and/or y. Thus, Equation (1.44) encapsulates the four possibilities of a two-dimensional space: $t_{xx} = \hat{x}\cdot T\{\hat{x}\}$, $t_{xy} = \hat{x}\cdot T\{\hat{y}\}$, $t_{yx} = \hat{y}\cdot T\{\hat{x}\}$, and $t_{yy} = \hat{y}\cdot T\{\hat{y}\}$. With the help of this notation, we can rewrite Equation (1.43) as:

$$u_x = t_{xx}v_x + t_{xy}v_y$$

$$u_y = t_{yx}v_x + t_{yy}v_y \tag{1.45}$$

Probably you have already guessed that the most usual form of expressing the two equations above is by means of a matrix equation, as in:

$$\begin{pmatrix} u_x \\ u_y \end{pmatrix} = \begin{pmatrix} t_{xx} & t_{xy} \\ t_{yx} & t_{yy} \end{pmatrix} \begin{pmatrix} v_x \\ v_y \end{pmatrix} \tag{1.46}$$

Thus, the representation of a linear operator is a matrix. In Section 1.3 (Vector Representation in Different Bases), I pleaded with you not to forget that the representations of vectors depend on the basis: if we know only the coordinates, but not the basis, then we know nothing about the vectors. In the same spirit, now I will beg, implore, and entreat that you do not forget that a matrix on its own does not specify a linear operator: we also need to know the basis. Indeed, if we had chosen a different basis, then we would have found a relation in the same form of Equation (1.46), but involving a completely different matrix and completely different vector coordinates. Mind you, if we had chosen a different basis, then the matrix and coordinates would be completely different, but they would still be a representation of the very same operator and vectors, that is, it would still be a representation of Equation (1.1), involving the same vectors, and the same operator.

Let us see one example to drive home this issue of different representations of the same object. Let us begin once again with Equation (1.1), involving the very same operator $T\{\ \}$, and the very same vectors $\vec{u}$ and $\vec{v}$. For convenience, I repeat the relation below:

Equation (1.1)

$$\vec{u} = T\{\vec{v}\}$$

But now, instead of expanding the vectors in the canonical basis, as we did in Equation (1.41), let us expand them in a different basis. For convenience, we use again the basis $\hat{a}_1$ and $\hat{a}_2$. To emphasize that we are not altering the vectors (and we are not altering the operator either), I will write them explicitly in both bases (as we had done in Equation (1.26)):

$$\vec{u} = u_x \hat{x} + u_y \hat{y} = U_1 \hat{a}_1 + U_2 \hat{a}_2$$

$$\vec{v} = v_x \hat{x} + v_y \hat{y} = V_1 \hat{a}_1 + V_2 \hat{a}_2$$

So, instead of expanding $\vec{u}$ and $\vec{v}$ in the canonical basis, as we did in Equation (1.41), now we expand them in the new basis. With the expansion in the new basis, Equation (1.1) reads:

$$U_1 \hat{a}_1 + U_2 \hat{a}_2 = T\{V_1 \hat{a}_1 + V_2 \hat{a}_2\}$$

Since $T\{\ \}$ is the same operator as before, it is still linear. Thus, the equation above reduces to:

$$U_1 \hat{a}_1 + U_2 \hat{a}_2 = T\{\hat{a}_1\} V_1 + T\{\hat{a}_1\} V_2 \tag{1.47}$$

To extract coordinates U_1 and U_2, we take the scalar product between both sides of Equation (1.47), and each basis vector ($\hat{a}_1$ and $\hat{a}_2$). These scalar products lead to the two equations below:

$$U_1 = (\hat{a}_1 \blacksquare T\{\hat{a}_1\}) V_1 + (\hat{a}_1 \blacksquare T\{\hat{a}_2\}) V_2$$

$$U_2 = (\hat{a}_2 \blacksquare T\{\hat{a}_1\}) V_1 + (\hat{a}_2 \blacksquare T\{\hat{a}_2\}) V_2 \tag{1.48}$$

Once again, we can define a more compact notation: in the same spirit of Equation (1.44), we define:

$$t_{mn} = \hat{a}_m \blacksquare T\{\hat{a}_n\} \tag{1.49}$$

In Equation (1.49), instead of standing for x and y as before, now m and n stand for 1 and 2; and the four coefficients encapsulated by Equation (1.49) are: $t_{11} = \hat{a}_1 \blacksquare T\{\hat{a}_1\}$, $t_{12} = \hat{a}_1 \blacksquare T\{\hat{a}_2\}$, $t_{21} = \hat{a}_2 \blacksquare T\{\hat{a}_1\}$, and $t_{22} = \hat{a}_2 \blacksquare T\{\hat{a}_2\}$. Thus, Equation (1.48) can be written as:

$$U_1 = t_{11}V_1 + t_{12}V_2$$

$$U_2 = t_{21}V_1 + t_{22}V_2 \tag{1.50}$$

Finally, its matrix form is:

$$\begin{pmatrix} U_1 \\ U_2 \end{pmatrix} = \begin{pmatrix} t_{11} & t_{12} \\ t_{21} & t_{22} \end{pmatrix} \begin{pmatrix} V_1 \\ V_2 \end{pmatrix} \tag{1.51}$$

How does Equation (1.51) compare with Equation (1.46)? If you look only at the numbers, you may think that they are completely different equations. After all, U_1 is **not** the same number as u_x, t_{12} is **not** the same number as t_{xy}, and so on. The equations involve completely different numbers. But they refer to exactly the same objects: they are different representations of the same equality. In other words, they are different, but completely equivalent, representations of the same Equation (1.1).

Why does this matter? Suppose that you are given a problem where the numbers t_{xx}, t_{xy}, t_{yx}, t_{yy}, u_x, and u_y are given, and the problem asks you to find v_x and v_y. The problem just gives you the numbers, says nothing about bases and how awesome linear algebra is. Then your job is essentially to invert the matrix, which would not take too long to do, because this example assumes a two-dimensional space. But if it was in a higher dimensional space, then your heart may faint at the job of inverting a big matrix. The best practice, which we will be constantly exploring in this course, is to change the representation. So, we treat these numbers as coordinates of a vector $\vec{u}$, the coordinates of a vector $\vec{v}$, and the coefficients of an operator $T\{\ \}$, all in the canonical basis. Then, we look for a new convenient basis in which the matrix has a simpler form, and solve the problem in this new basis. Once the problem has been solved in the convenient basis, then we go back to the original basis. That is signals and systems in a nutshell. As we will learn in Chapter 3, the matrix inversion problem is analogous to the solution of a differential equation representing a system; and the strategy is the same: go to a more convenient basis, solve the problem there, and go back to the original basis.

I hope you are happy to find out that we are sailing charted waters. But you may be wondering what exactly a 'convenient basis' is. This is the subject of Section 1.6.

1.6 Eigenvectors and Eigenvalues

A convenient basis is a basis of eigenvectors, because in a basis of eigenvectors the matrix becomes diagonal. It is easy to prove this with the tools we developed in the previous sections, but first let us first define these two terms, 'eigenvectors' and 'eigenvalues', explicitly.

An eigenvector $\vec{a}$ of an operator $A\{\}$ is a vector that, when acted upon by the operator, results in the same vector multiplied by a number. Thus, if $\vec{a}$ is an eigenvector of $A\{\}$, then:

$$A\{\vec{a}\} = \lambda \vec{a} \tag{1.52}$$

The multiplying number λ is the eigenvalue. Notice that, if a vector is an eigenvector of a certain operator, then it will probably not be an eigenvector of another operator. Thus, if you are told that a certain vector is an eigenvector, you must always ask 'of which operator?'.

If an operator in a two-dimensional Euclidean space has two orthogonal eigenvectors, then it is possible to form a (orthonormal) basis with these eigenvectors. If the space is three-dimensional, then we need three orthogonal eigenvectors, and so on. Not all operators of a Euclidean space can provide a (orthonormal) basis with its eigenvectors. But, if the matrix representing the operator is Hermitian, then it is guaranteed that we can find a (orthonormal) basis with its eigenvectors. A Hermitian matrix is one that is equal to its self-adjoint, and the self-adjoint is the complex conjugate of the transpose of the matrix. That is a bucketload of definitions, but they are not tremendously relevant to our purposes because, as we will see in Chapters 2 and Chapter 3, it is always possible to find a basis of eigenvectors for the operators we are interested in. So, do not worry too much about these definitions. Let me just give you some information about Hermitian matrixes, and then we move on.

We denote the self-adjoint of a matrix A by a little cross: $\dagger$. Thus, to denote that B is the self-adjoint of A, we write:

$$B = A^{\dagger} \tag{1.53}$$

Since the self-adjoint is the transpose conjugate, then Equation (1.53) entails that the matrix coefficients of B and A are related as:

$$B_{mn} = A_{nm}^{*} \tag{1.54}$$

For a matrix to be Hermitian, it must be equal to its self-adjoint, so a Hermitian matrix obeys the symmetry relation:

$$A_{mn} = A_{nm}^{*} \tag{1.55}$$

Of course, not every matrix is Hermitian, so it is not always guaranteed that we can form a (orthonormal) basis with their eigenvectors in a Euclidean space. But, again, that will not be a problem in this course. Our main interest here is to illustrate the idea that a basis of eigenvectors greatly facilitates the analysis, and when we deal with signals and systems, we will have this basis ready for us. Thus, to illustrate the idea in the Euclidean space, I will assume that the operator of interest is Hermitian, to guarantee that its eigenvectors form a (orthonormal) basis.

So, suppose that we have a certain operator $A\{\ \}$ in a two-dimensional Euclidean space, and that its two eigenvectors form a (orthonormal) basis. Adopting the same notation we used earlier for the 'new basis', we denote these eigenvectors by $\widehat{a_1}$ and $\widehat{a_2}$, and their eigenvalues by λ_1 and λ_2. Thus:

$$A\{\widehat{a_1}\ \} = \lambda_1 \widehat{a_1} \tag{1.56}$$

$$A\{\widehat{a_2}\ \} = \lambda_2 \widehat{a_2} \tag{1.57}$$

Recall that we can use Equation (1.49) to find the representation of any operator in any basis. This time, since we have specified a certain operator $A\{\ \}$, I will denote its coefficients by a_{mn}, instead of t_{mn}. Thus, according to Equation (1.49):

$$a_{mn} = \widehat{a_m} \blacksquare A\{\widehat{a_n}\} \tag{1.58}$$

So, let us evaluate these coefficients one by one. We begin with a_{11}:

$$a_{11} = \widehat{a_1} \blacksquare A\{\widehat{a_1}\}$$

Using Equation (1.56):

$$a_{11} = \widehat{a_1} \blacksquare A\{\widehat{a_1}\} = \widehat{a_1} \blacksquare (\lambda_1 \widehat{a_1}) = \lambda_1 (\widehat{a_1} \blacksquare \widehat{a_1})$$

where Definition 2 (Equation (1.3)) was used in the last equality.
But $\widehat{a_1} \blacksquare \widehat{a_1} = 1$, therefore:

$$a_{11} = \lambda_1$$

Now we evaluate a_{12}:

$$a_{12} = \widehat{a_1} \blacksquare A\{\widehat{a_2}\}$$

Using Equation (1.57):

$$a_{12} = \widehat{a_1} \blacksquare A\{\widehat{a_2}\} = \widehat{a_1} \blacksquare (\lambda_2 \widehat{a_2}) = \lambda_2 (\widehat{a_1} \blacksquare \widehat{a_2})$$

But, by assumption, $\widehat{a_1}$ and $\widehat{a_2}$ are orthogonal, that is, $\widehat{a_1} \blacksquare \widehat{a_2} = 0$. Therefore:

$$a_{12} = 0$$

The other two coefficients can be found in an analogous way:

$$a_{21} = \widehat{a_2} \blacksquare A\{\widehat{a_1}\} = \widehat{a_2} \blacksquare (\lambda_1 \widehat{a_1}) = \lambda_1 (\widehat{a_2} \blacksquare \widehat{a_1}) = 0$$

$$a_{22} = \widehat{a_2} \blacksquare A\{\widehat{a_2}\} = \widehat{a_2} \blacksquare (\lambda_2 \widehat{a_2}) = \lambda_2 (\widehat{a_2} \blacksquare \widehat{a_2}) = \lambda_2$$

Thus, we conclude that the matrix representing the operator $A\{\ \}$ in the basis of its eigenvectors is of the form:

$$\begin{pmatrix} a_{11} & a_{12} \\ a_{21} & a_{22} \end{pmatrix} = \begin{pmatrix} \lambda_1 & 0 \\ 0 & \lambda_2 \end{pmatrix}$$

It is a diagonal matrix, and its diagonal elements are its eigenvalues. Now, let us spell out how we can use this special basis to solve matrix equations.

1.7 General Method of Solution of a Matrix Equation

In this section, I want to show you step by step the method of solution of a matrix equation in terms of its diagonal representation. These steps are not difficult, and it is likely that you already know how to do it. My main interest in showing these steps explicitly is that I will refer to them in Chapter 3, where we deal with the general method of finding the output of linear and time invariant systems.

Suppose we are given a matrix equation of the form:

$$\begin{pmatrix} u_x \\ u_y \end{pmatrix} = \begin{pmatrix} a_{xx} & a_{xy} \\ a_{yx} & a_{yy} \end{pmatrix} \begin{pmatrix} v_x \\ v_y \end{pmatrix} \tag{1.59}$$

Our job is to find v_x and v_y in terms of all the other parameters. As I mentioned earlier, this problem is analogous to the solution of a differential equation.

We begin by interpreting Equation (1.59) as being a particular representation of Equation (1.60):

$$\vec{u} = A\{\vec{v}\} \tag{1.60}$$

By convention, we specify that Equation (1.59) is a representation of Equation (1.60) in the canonical basis. Thus, we write:

$$\vec{u} = u_x \hat{x} + u_y \hat{y}$$

$$\vec{v} = v_x \hat{x} + v_y \hat{y}$$

$$a_{mn} = \hat{m} \blacksquare A\{\hat{n}\} \tag{1.61}$$

where m and n stand for x and/or y.

The next step is to find the representation in the most convenient basis, that is, the basis of eigenvectors. Again, we denote the basis of eigenvectors of $A\{\}$ by $\hat{a}_1$ and $\hat{a}_2$, and write them explicitly in the canonical basis:

$$\hat{a}_1 = a_{1x} \hat{x} + a_{1y} \hat{y}$$

$$\hat{a}_2 = a_{2x} \hat{x} + a_{2y} \hat{y} \tag{1.62}$$

There are standard techniques for finding the eigenvalues and eigenvectors coordinates, and if you have already taken a course in linear algebra, then probably you have already done lots of exercises about this. These techniques are not crucial to our course, so I will not review them (but a few problems in the exercise list at the end of this chapter involve finding these coordinates). I will just assume that, from the knowledge of the matrix coefficients a_{mn}, we have already obtained the coordinates a_{1x}, a_{1y}, a_{2x}, and a_{2y}.

Now we write the vectors $\vec{u}$ and $\vec{v}$ explicitly in terms of this new basis:

$$\vec{u} = u_x\hat{x} + u_y\hat{y} = U_1\hat{a}_1 + U_2\hat{a}_2$$

$$\vec{v} = v_x\hat{x} + v_y\hat{y} = V_1\hat{a}_1 + V_2\hat{a}_2 \tag{1.63}$$

By assumption, we know u_x and u_y. So, we can find U_1 and U_2 in terms of u_x and u_y. Thus, as in Equation (1.29) and Equation (1.30):

$$U_1 = \hat{a}_1\blacksquare\vec{u} = a_{1x}^*u_x + a_{1y}^*u_y$$

$$U_2 = \hat{a}_2\blacksquare\vec{u} = a_{2x}^*u_x + a_{2y}^*u_y \tag{1.64}$$

Next, we write the representation of Equation (1.60) in the basis of eigenvectors of $A\{\}$. We have seen that this representation reads:

$$\begin{pmatrix} U_1 \\ U_2 \end{pmatrix} = \begin{pmatrix} \lambda_1 & 0 \\ 0 & \lambda_2 \end{pmatrix} \begin{pmatrix} V_1 \\ V_2 \end{pmatrix} \tag{1.65}$$

Of course, in this representation, the matrix is diagonal. Consequently, there is no mixing of coordinates: U_1 depends only on V_1 and U_2 depends only on V_2. Thus, it becomes a trivial matter to find the coordinates of $\vec{v}$ in terms of the coordinates of $\vec{u}$. They are:

$$V_1 = \frac{U_1}{\lambda_1}, \quad V_2 = \frac{U_2}{\lambda_2} \tag{1.66}$$

In a sense, we have already finished our job, because we have already found the vector $\vec{v}$:

$$\vec{v} = \frac{U_1}{\lambda_1}\hat{a}_1 + \frac{U_2}{\lambda_2}\hat{a}_2 \tag{1.67}$$

But we might want to know explicitly what the coordinates of $\vec{v}$ with respect to the canonical basis are. We could find them by taking the scalar product with $\hat{x}$ and $\hat{y}$, just like we did in Equation (1.38) and Equation (1.39). But, for our purposes, I find it more illustrative to obtain the coordinates by expanding the eigenvectors in the canonical basis. Thus:

$$\vec{v} = \frac{U_1}{\lambda_1}\hat{a}_1 + \frac{U_2}{\lambda_2}\hat{a}_2 = \frac{U_1}{\lambda_1}(a_{1x}\hat{x} + a_{1y}\hat{y}) + \frac{U_2}{\lambda_2}(a_{2x}\hat{x} + a_{2y}\hat{y})$$

Reorganizing:

$$\vec{v} = \left(\frac{U_1}{\lambda_1}a_{1x} + \frac{U_2}{\lambda_2}a_{2x}\right)\hat{x} + \left(\frac{U_1}{\lambda_1}a_{1y} + \frac{U_2}{\lambda_2}a_{2y}\right)\hat{y} \tag{1.68}$$

From which we conclude that:

$$v_x = \frac{U_1}{\lambda_1}a_{1x} + \frac{U_2}{\lambda_2}a_{2x}$$
$$v_y = \frac{U_1}{\lambda_1}a_{1y} + \frac{U_2}{\lambda_2}a_{2y}$$

(1.69)

And we have completed our job.

To conclude this section, consider again a vector $\vec{v}$ (it can be any vector) expressed in the basis of eigenvectors:

$$\vec{v} = V_1\hat{a}_1 + V_2\hat{a}_2$$

The column vector representation of this equality in the canonical basis is:

$$\begin{pmatrix} v_x \\ v_y \end{pmatrix} = V_1 \begin{pmatrix} a_{1x} \\ a_{1y} \end{pmatrix} + V_2 \begin{pmatrix} a_{2x} \\ a_{2y} \end{pmatrix}$$

(1.70)

This form is analogous to the inverse Fourier transform, which we will study in Chapter 2. Notice that Equation (1.70) entails that $v_x = V_1a_{1x} + V_2a_{2x}$ and $v_y = V_1a_{1y} + V_2a_{2y}$, which are the same expressions in Equation (1.69) when V_1 and V_2 are given by Equation (1.66). In Chapter 2, we will refer to Equation (1.70) to illustrate the algebraic meaning of the inverse Fourier transform.

1.8 The Closure Relation

The closure relation is a useful tool to change a representation from one basis to another. It is an expression of the feature that the identity operator can be built from basis vectors. By definition, the identity operator $I\{\ \}$ transforms any vector $\vec{u}$ into itself:

$$\vec{u} = I\{\vec{u}\}$$

(1.71)

To find how we can build the identity operator from basis vectors, consider the basis formed by $\hat{a}_1$ and $\hat{a}_2$ (it can be any basis, and the only reason I am not using the canonical basis is that I do not want to run the risk of you thinking that it is special).

Expanding the vector $\vec{u}$ in this basis:

$$\vec{u} = U_1\hat{a}_1 + U_2\hat{a}_2$$

We have seen that $U_1 = \hat{a}_1 \blacksquare \vec{u}$ and $U_2 = \hat{a}_2 \blacksquare \vec{u}$. Thus:

$$\vec{u} = (\hat{a}_1 \blacksquare \vec{u})\hat{a}_1 + (\hat{a}_2 \blacksquare \vec{u})\hat{a}_2$$

Interchanging the order of scalars and vectors, we can express the relation above as:

$$\vec{u} = \hat{a}_1(\hat{a}_1 \blacksquare \vec{u}) + \hat{a}_2(\hat{a}_2 \blacksquare \vec{u})$$

I left the parenthesis there to emphasize that we have scalars multiplied by vectors, but we do not actually need them because there is no ambiguity without them. Thus, removing the parenthesis, we obtain:

$$\vec{u} = \hat{a}_1 \hat{a}_1 \blacksquare \vec{u} + \hat{a}_2 \hat{a}_2 \blacksquare \vec{u} \tag{1.72}$$

Now notice that the object $\hat{a}_1 \hat{a}_1 \blacksquare + \hat{a}_2 \hat{a}_2 \blacksquare$ is an operator: it receives a vector (which is plugged on the right-hand side of the squares) and spits out a vector. But it is also the identity operator, because the vector it spits out is the same that has been plugged in. Thus, we can express the identity operator as $I\{\ \}$:

$$I\{\ \} = \hat{a}_1 \hat{a}_1 \blacksquare + \hat{a}_2 \hat{a}_2 \blacksquare \tag{1.73}$$

In Equation (1.73), the vector going inside the curly brackets is attached to the right-hand side of the $\blacksquare$.

Recall that we could have done the same thing in any other basis. For example, in the canonical basis:

$$I\{\ \} = \widehat{x}\widehat{x} \blacksquare + \widehat{y}\widehat{y} \blacksquare \tag{1.74}$$

I emphasize that the operator of Equation (1.74) is exactly the same operator as in Equation (1.73): both are the identity operator. In other words: $\hat{a}_1 \hat{a}_1 \blacksquare + \hat{a}_2 \hat{a}_2 \blacksquare = \widehat{x}\widehat{x} \blacksquare + \widehat{y}\widehat{y} \blacksquare$.

Equation (1.73) and Equation (1.74) are closure relations. One way we can find out if a collection of unit vectors forms a basis is to test if they satisfy the closure relation. In other words, we construct an operator by placing each basis vector next to each other and next to the symbol of the scalar product (like $\widehat{x}\widehat{x} \blacksquare + \widehat{y}\widehat{y} \blacksquare$). If the operator we constructed in this way turns out to be the identity operator, then our collection of unit vectors forms a basis.

1.9   Representation of Linear Operators in Terms of Eigenvectors and Eigenvalues

Another relevant notion of algebra that we are going to explore in Chapter 3 is the notion that, if the eigenvectors of a linear operator forms a (orthonormal) basis, then the operator is fully characterized by its eigenvalues and eigenvectors (this notion is connected with the concept of a 'frequency response'). If we know the eigenvectors and eigenvalues, then we know everything we need to know about the operator. That makes sense, right? After all, if we know the eigenvectors and eigenvalues, then we know the representation of the operator in a specific basis, which in this case is the basis of eigenvectors (recall that the representation is a diagonal matrix of eigenvalues). And we have seen that the representation of an operator in any basis fully specifies the operator.

In this section, we will find an explicit expression for a linear operator in terms of its eigenvectors and eigenvalues. It is quite straightforward: consider again a linear operator $A\{\ \}$ whose eigenvectors are $\widehat{a}_1$ and $\widehat{a}_2$, and that they form a basis. Now, consider the action of $A\{\ \}$ on a given vector $\overrightarrow{u}$, whose coordinates in the basis of eigenvectors of $A\{\ \}$ are U_1 and U_2:

$$A\{\overrightarrow{u}\} = A\{U_1\widehat{a}_1 + U_2\widehat{a}_2\} = A\{\widehat{a}_1\}U_1 + A\{\widehat{a}_2\}U_2$$

But $A\{\widehat{a}_1\} = \lambda_1\widehat{a}_1$ and $A\{\widehat{a}_2\} = \lambda_2\widehat{a}_2$. Therefore:

$$A\{\overrightarrow{u}\} = \lambda_1\widehat{a}_1 U_1 + \lambda_2\widehat{a}_2 U_2$$

But $U_1 = \widehat{a}_1\blacksquare\overrightarrow{u}$ and $U_2 = \widehat{a}_2\blacksquare\overrightarrow{u}$. Thus:

$$A\{\overrightarrow{u}\} = \lambda_1\widehat{a}_1\widehat{a}_1\blacksquare\overrightarrow{u} + \lambda_2\widehat{a}_2\widehat{a}_2\blacksquare\overrightarrow{u}$$

which entails that:

$$A\{\ \} = \lambda_1\widehat{a}_1\widehat{a}_1\blacksquare + \lambda_2\widehat{a}_2\widehat{a}_2\blacksquare \tag{1.75}$$

Notice that, even though we have derived Equation (1.75) in the basis of eigenvectors of $A\{\ \}$, this equation is in fact general; after all, the scalar product does not depend on the basis. For example, suppose we already know the eigenvalues and the coordinates of the eigenvectors with respect to the canonical basis, but we do not know the representation of $A\{\ \}$ in the canonical basis. We can easily find this representation using the last expression in Equation (1.61) together with Equation (1.75). Thus:

$$a_{mn} = \widehat{m}\blacksquare A\{\widehat{n}\} = \widehat{m}\blacksquare(\lambda_1\widehat{a}_1\widehat{a}_1\blacksquare\widehat{n} + \lambda_2\widehat{a}_2\widehat{a}_2\blacksquare\widehat{n})$$
$$= \lambda_1(\widehat{m}\blacksquare\widehat{a}_1)(\widehat{a}_1\blacksquare\widehat{n}) + \lambda_2(\widehat{m}\blacksquare\widehat{a}_2)(\widehat{a}_2\blacksquare\widehat{n}) \tag{1.76}$$

where m and n stand for x and/or y.

1.10 The Dirac Notation

It is time to doff old-fashioned arrow-shaped hats and don the swanky Dirac notation.

The Dirac notation was introduced by the awesome Paul Dirac to handle operations with quantum states, and it has been adopted in most fields dealing with vectors and algebraic operations. The main reason I want to use it in this textbook is that it helps to differentiate between a signal and its representation.

In the Dirac notation, a vector $\overrightarrow{u}$ is denoted by:

$$|u\rangle$$

This is just a change of notation: instead of using $\vec{u}$, we use $|u\rangle$. The latter symbol is called a 'ket', so you can read the symbol $|u\rangle$ as 'the ket u'.

The only disadvantage of the Dirac notation is that it does not differentiate between vectors and unit vectors. For example, the unit vector $\hat{x}$ is denoted by $|x\rangle$, and we have to infer either from context or by explicit specification whether the ket is a unit ket or not.

The expansions of $|u\rangle$ and $|v\rangle$ in the canonical basis are written as:

$$|u\rangle = u_x|x\rangle + u_y|y\rangle$$

$$|v\rangle = v_x|x\rangle + v_y|y\rangle \tag{1.77}$$

A major advantage of the Dirac notation is its way of expressing scalar products. We have been denoting the scalar product with a ■, and we have seen that it matters on which side we plug in the vectors. For example, in the scalar product $\vec{v}\,■\,\vec{u}$, the coordinates of the vector $\vec{v}$ are conjugated and of the vector $\vec{u}$ are not conjugated (see Equation (1.33)). In the Dirac notation, a ket replaces the vector that goes on the right-hand side of the ■ (that is, the vector whose coordinates are not conjugated). Meanwhile, the vector that goes on the left-hand side of the ■ is replaced with a 'dual' of the ket, which we call a 'bra' and denote by:

$$\langle v|$$

The symbol $\langle v|$ is read: 'the bra v'. Each ket has an associated bra. You can think of a bra as a mathematical object that, when combined with a ket, returns a scalar. Thus, our previous scalar product $\vec{v}\,■\,\vec{u}$ is now denoted by:

$$\langle v|u\rangle = v_x^* u_x + v_y^* u_y \tag{1.78}$$

So, if you combine a bra $\langle v|$ with a ket $|u\rangle$, you get a bra-c-ket $\langle v|u\rangle$. That is pithy terminology indeed.

Notice that Equation (1.78) entails that:

$$\langle v|u\rangle = \langle u|v\rangle^* \tag{1.79}$$

No surprise at all in Equation (1.79): it is just Definition 3 (Equation (1.4)) expressed in the Dirac notation.

As a little exercise, let us see how Equation (1.10) (which is our **Conclusion**, showing that we must conjugate the scalars when extracting them from the linear combination on the left side of the scalar product) is expressed in the Dirac notation. Suppose we define a ket p as:

$$|p\rangle = a|v\rangle \tag{1.80}$$

Our job is to express the bra p in terms of the bra v. We follow the same logic that led to Equation (1.10). We begin by evaluating the scalar product involving an arbitrary bra u and the ket p:

$$\langle u|p \rangle = \langle u \mid (a|v)\rangle \tag{1.81}$$

According to Definition 2 (Equation (1.3)), we can extract a from the scalar product. Thus:

$$\langle u|p \rangle = \langle u \mid (a|v)\rangle = a\langle u|v\rangle \tag{1.82}$$

According to Equation (1.79), Equation (1.82) entails that:

$$\langle p|u \rangle = \langle u|p\rangle^* = [a\langle u|v\rangle]^*$$

Therefore:

$$\langle p|u \rangle = a^*\langle u|v\rangle^* = a^*\langle v|u\rangle \tag{1.83}$$

Thus, we have concluded that:

If:

$$|p\rangle = a|v\rangle$$

then:

$$\langle p| = a^*\langle v| \tag{1.84}$$

Equation (1.84) is just a particular case of Equation (1.10).

In the Dirac notation, an operator is denoted by a capital letter, but without curly brackets. Thus, in the Dirac notation, the relation $\vec{u} = T\{\vec{v}\}$ reads:

$$|u\rangle = T|v\rangle \tag{1.85}$$

The Dirac notation for the matrix coefficients of a linear operator is quite cute. Recall that, in the old notation, the coefficients of the matrix representation of a linear operator $T\{\}$ in a basis formed by $\widehat{a_1}$ and $\widehat{a_2}$ were: $t_{mn} = \widehat{a_m} \blacksquare T\{\widehat{a_n}\}$, where m and n stood for 1 and/or 2. In the Dirac notation, the scalar product $\widehat{a_m} \blacksquare T\{\widehat{a_n}\}$ is expressed as $\langle a_m|T|a_n\rangle$, where the command is: first find the ket $T|a_n\rangle$ and then form the scalar product of this ket with the bra a_m.

Thus, in the Dirac notation, the coefficients of the matrix representation of the operator T in the basis formed by $|a_1\rangle$ and $|a_2\rangle$ are:

$$t_{mn} = \langle a_m|T|a_n\rangle \tag{1.86}$$

Much more elegant than $t_{mn} = \widehat{a_m} \blacksquare T\{\widehat{a_n}\}$.

Worked Exercise: The Bra of the Action of an Operator on a Ket

A ket u is defined as:

$$|u\rangle = T|v\rangle \tag{1.87}$$

Find the bra u.

Solution:

If the ket u depends on the operator T, then its bra must also depend on some operator related to T. We still do not know what operator that is, so we call it B. The bra u must depend on B and on the bra v. But in which order? Can we write $\langle u| = B\langle v|$? No, we cannot. To see why not, plug a ket p on both sides of the equation: $\langle u|p\rangle = B\langle v|p\rangle$. This is completely inconsistent: $\langle u|p\rangle$ is a scalar, but $B\langle v|p\rangle$ is a scalar times an operator, which is another operator. So, we cannot write $\langle u| = B\langle v|$. But we can write $\langle u| = \langle v|B$. Let us test if that makes sense by plugging the ket p again on both sides: $\langle u|p\rangle = \langle v|B|p\rangle$. That makes sense: we have scalars on both sides of the identity. Thus, the bra u must be of the form:

$$\langle u| = \langle v|B \tag{1.88}$$

We need to find the relationship between B and T. The most intuitive way is to compare the representation of Equation (1.87) with the representation of Equation (1.88). This time, for the sake of convenience, we denote our basis kets by numbers, so $|1\rangle$ is one basis ket and $|2\rangle$ is the other basis ket. First, we write the kets and matrix coefficients explicitly in terms of the basis vectors:

$$|u\rangle = u_1|1\rangle + u_2|2\rangle$$

$$|v\rangle = v_1|1\rangle + v_2|2\rangle$$

$$t_{mn} = \langle m|T|n\rangle$$

$$b_{mn} = \langle m|B|n\rangle \tag{1.89}$$

Equation (1.89) is just a bit of bookkeeping to assist us along the way. To find the representation of Equation (1.87), we need to 'project it on the basis vectors', that is, we need to take the scalar product with the basis vectors (just like we did in Section 1.3 [Vector Representation in Different Bases]). Thus:

$$\langle 1|u\rangle = \langle 1|T|v\rangle$$

$$\langle 2|u\rangle = \langle 2|T|v\rangle \tag{1.90}$$

Recall that $\langle 1|u \rangle = u_1$ and $\langle 2|u \rangle = u_2$. Using these identities and expanding the ket v:

$$u_1 = \langle 1|T\left[v_1|1\rangle + v_2|2\rangle\right] = \langle 1|T|1\rangle v_1 + \langle 1|T|2\rangle v_2 = t_{11}v_1 + t_{12}v_2$$

$$u_2 = \langle 2|T\left[v_1|1\rangle + v_2|2\rangle\right] = \langle 2|T|1\rangle v_1 + \langle 2|T|2\rangle v_2 = t_{21}v_1 + t_{22}v_2 \tag{1.91}$$

Not surprisingly, Equation (1.91) can be expressed in terms of a matrix equation:

$$\begin{pmatrix} u_1 \\ u_2 \end{pmatrix} = \begin{pmatrix} t_{11} & t_{12} \\ t_{21} & t_{22} \end{pmatrix} \begin{pmatrix} v_1 \\ v_2 \end{pmatrix} \tag{1.92}$$

Thus, we have found the representation of Equation (1.87).

Now we obtain the matrix equation representing Equation (1.88). To obtain it, we just plug in the basis kets on the right side of the bras, forming scalar products. Watch out for the order of the symbols:

$$\langle u|1 \rangle = \langle v|B|1 \rangle$$

$$\langle u|2 \rangle = \langle v|B|2 \rangle \tag{1.93}$$

Recall that $\langle u|1 \rangle = \langle 1|u \rangle^* = u_1^*$ and $\langle u|2 \rangle = \langle 2|u \rangle^* = u_2^*$. We also need to expand the bra v. Using Equation (1.84) and the second line of Equation (1.89), we find:

$$\langle v| = v_1^* \langle 1| + v_2^* \langle 2| \tag{1.94}$$

With the help of Equation (1.94), Equation (1.93) can be recast as:

$$u_1^* = [v_1^* \langle 1| + v_2^* \langle 2|]B|1\rangle = \langle 1|B|1\rangle v_1^* + \langle 2|B|1\rangle v_2^* = b_{11}v_1^* + b_{21}v_2^*$$

$$u_2^* = [v_1^* \langle 1| + v_2^* \langle 2|]B|2\rangle = \langle 1|B|2\rangle v_1^* + \langle 2|B|2\rangle v_2^* = b_{12}v_1^* + b_{22}v_2^* \tag{1.95}$$

Thus, we have found the matrix representation of Equation (1.88):

$$\begin{pmatrix} u_1^* \\ u_2^* \end{pmatrix} = \begin{pmatrix} b_{11} & b_{21} \\ b_{12} & b_{22} \end{pmatrix} \begin{pmatrix} v_1^* \\ v_2^* \end{pmatrix} \tag{1.96}$$

We want to compare Equation (1.96) with Equation (1.92), so it is more convenient to conjugate both sides of Equation (1.96), which results in:

$$\begin{pmatrix} u_1 \\ u_2 \end{pmatrix} = \begin{pmatrix} b_{11}^* & b_{21}^* \\ b_{12}^* & b_{22}^* \end{pmatrix} \begin{pmatrix} v_1 \\ v_2 \end{pmatrix} \tag{1.97}$$

Equation (1.92) and Equation (1.97) together entail that:

$$\begin{pmatrix} b_{11}^* & b_{21}^* \\ b_{12}^* & b_{22}^* \end{pmatrix} = \begin{pmatrix} t_{11} & t_{12} \\ t_{21} & t_{22} \end{pmatrix} \tag{1.98}$$

Notice that the off-diagonal indices are swapped. For example, $b_{12}^* = t_{21}$. In general, according to Equation (1.98), $b_{mn}^* = t_{nm}$. In other words, the matrix representation of B is the conjugated transpose of the matrix representation of T (mind you, the matrix on the left-hand side of Equation (1.98) is NOT the representation of B, but it contains the coefficients of the representation of B). In short, B is the self-adjoint of T:

$$B = T^\dagger$$

We have thus concluded that:

If:

$$|u\rangle = T|v\rangle$$

then:

$$\langle u| = \langle v|T^\dagger \tag{1.99}$$

I find this result quite cool. Do you agree? Or do you think I should try paragliding?

To close this section, let us write explicitly the closure relation and the representation of an operator in terms of its eigenvectors and eigenvalues. If $|1\rangle$ and $|2\rangle$ are basis kets, then their closure relation reads:

$$I = |1\rangle\langle 1| + |2\rangle\langle 2| \tag{1.100}$$

Compare Equation (1.100) with Equation (1.73) or Equation (1.74). It is the same algorithm: take the scalar product with a basis vector, and multiply the result by the same basis vector, then do repeat the process with the other basis vector and then add them up.

In the same spirit, the representation of an operator A in terms of its eigenvectors $|a_1\rangle$ and $|a_2\rangle$, whose eigenvalues are λ_1 and λ_2, respectively, is:

$$A = \lambda_1|a_1\rangle\langle a_1| + \lambda_2|a_2\rangle\langle a_2| \tag{1.101}$$

Equation (1.101) is the version of Equation (1.75) in Dirac notation.

1.11 Exercises

Exercise 1

Prove that the scalar product between a vector and itself (like $\vec{u} \blacksquare \vec{u}$) is a real number.

Exercise 2

Considering a three-dimensional space, express the scalar product between two vectors in terms of their coordinates

Exercise 3

Assuming a two-dimensional space, prove that the scalar product does not depend on the basis.

Hint: expand two vectors explicitly in two different bases, as in Equation (1.63). Then express the coordinates in one basis in terms of the coordinates in the other basis. For example: $u_x = \widehat{x} \blacksquare \vec{u} = \widehat{x} \blacksquare (U_1 \widehat{a}_1 + U_2 \widehat{a}_2) = U_1 a_{1x} + U_2 a_{2x}$. Do that for all coordinates and then combine them together to prove that $v_x^* u_x + v_y^* u_y = V_1^* U_1 + V_2^* U_2$.

Exercise 4

Suppose the matrix representation of a linear operator in the basis $\widehat{a}_1$, $\widehat{a}_2$ is $\begin{pmatrix} a_{11} & a_{12} \\ a_{21} & a_{22} \end{pmatrix}$, and the matrix representation of the very same linear operator in the basis $\widehat{b}_1$, $\widehat{b}_2$ is $\begin{pmatrix} b_{11} & b_{12} \\ b_{21} & b_{22} \end{pmatrix}$. Express the matrix coefficients b_{mn} in terms of the matrix coefficients a_{mn} and scalar products between basis vectors.

Exercise 5

(a) In the canonical basis, a certain operator has the following matrix representation:

$$\begin{pmatrix} 0 & 1 \\ 1 & 0 \end{pmatrix}$$

Find the column representation of the eigenvectors of this operator in the canonical basis. Write these eigenvectors explicitly as linear combinations of the canonical basis vectors $\widehat{x}$ and $\widehat{y}$.

(b) Now consider the following vector representation in the canonical basis:

$$\begin{pmatrix} p \\ q \end{pmatrix}$$

Express this vector as a linear combination of the basis formed by eigenvectors of the operator of part (a).

(c) Find the column representation of all vectors (that is, of the vector of part (b), and of the eigenvectors themselves) in the basis of eigenvectors.

Exercise 6

Consider the matrix below:

$$\begin{pmatrix} 3 & \sqrt{3} \\ \sqrt{3} & 5 \end{pmatrix}$$

(a) Prove that this matrix is Hermitian.

(b) Find its eigenvalues and eigenvectors.

(c) Prove that the eigenvectors are orthogonal.

(d) If a vector representation in the 'original' basis is $\begin{pmatrix} 2 \\ 4 \end{pmatrix}$, find the representation of this same vector in the basis of eigenvectors.

(e) In the matrix equation below, find v_1 and v_2 by changing the representation to the basis of eigenvectors.

$$\begin{pmatrix} 10 \\ 20 \end{pmatrix} = \begin{pmatrix} 3 & \sqrt{3} \\ \sqrt{3} & 5 \end{pmatrix} \begin{pmatrix} v_1 \\ v_2 \end{pmatrix}$$

Exercise 7

Once upon a time, there were three matrices more famous than the three little pigs. They were called Pauli matrices and were denoted by the symbols σ_x, σ_y, and σ_z. Their eigenvectors and eigenvalues were:

$$\sigma_x \Rightarrow \frac{1}{\sqrt{2}} \begin{pmatrix} 1 \\ 1 \end{pmatrix}, \lambda = 1; \quad and \quad \frac{1}{\sqrt{2}} \begin{pmatrix} 1 \\ -1 \end{pmatrix}, \lambda = -1$$

$$\sigma_y \Rightarrow \frac{1}{\sqrt{2}} \begin{pmatrix} 1 \\ i \end{pmatrix}, \lambda = 1; \quad and \quad \frac{1}{\sqrt{2}} \begin{pmatrix} 1 \\ -i \end{pmatrix}, \lambda = -1$$

$$\sigma_z \Rightarrow \begin{pmatrix} 1 \\ 0 \end{pmatrix}, \lambda = 1; \quad and \quad \begin{pmatrix} 0 \\ 1 \end{pmatrix}, \lambda = -1$$

Use Equation (1.75) to find Pauli matrices.

Exercise 8

Assuming a two-dimensional space, consider the following relation:

$$|u\rangle = A|v\rangle$$

The representation of the relation above in a certain basis is:

$$\begin{pmatrix} u_1 \\ u_2 \end{pmatrix} = \begin{pmatrix} a_{11} & a_{12} \\ a_{21} & a_{22} \end{pmatrix} \begin{pmatrix} v_1 \\ v_2 \end{pmatrix}$$

In this basis, the scalar product between an arbitrary ket c and the bra u is:

$$\langle u|c\rangle = (u_1^* \ u_2^*) \begin{pmatrix} c_1 \\ c_2 \end{pmatrix}$$

where c_1 and c_2 are the coordinates of ket $|c\rangle$ in this basis.
  Prove that:

$$\langle u|c\rangle = (v_1^* \ v_2^*) \begin{pmatrix} a_{11}^* & a_{21}^* \\ a_{12}^* & a_{22}^* \end{pmatrix} \begin{pmatrix} c_1 \\ c_2 \end{pmatrix}$$

Pay attention to the order of the indices. By proving the relation above, you have proved that $\langle u| = \langle v|A^\dagger$. In other words, you have proved Equation (1.99).

Exercise 9

The closure relation is a useful tool to choose the basis in which a scalar product is done. Consider the scalar product $\langle v|u \rangle$. Prove that, if you write the identity operator I as the closure relation in a certain basis of your choice, then $\langle v|I|u \rangle$ returns the scalar product $\langle v|u \rangle$ evaluated in the basis of your choice.

Exercise 10

Consider the following expansion of kets $|u\rangle$ and $|v\rangle$ in the canonical basis:

$$|u\rangle = \cos\theta|x\rangle + i\sin\theta|y\rangle$$

$$|v\rangle = \sin\theta|x\rangle - i\cos\theta|y\rangle$$

where i is the imaginary number.

(a) Do $|u\rangle$ and $|v\rangle$ form an orthonormal basis?

(b) If $|u\rangle$ is an eigenvector of the operator T with eigenvalue λ_1, and $|v\rangle$ is an eigenvector of the same operator T with eigenvalue λ_2, then what is the matrix representation of T in the **canonical** basis?

(c) Still using the representation in the canonical basis, check whether $|u\rangle$ e $|v\rangle$ are indeed eigenvectors of T with their corresponding eigenvalues.

Interlude: Signals and Systems: What is it About?

We have warmed, we have stretched. Now, Jack, it is time to hit the road. So, what is it about?

In a broad sense, a signal is anything that carries information, while a system is anything that modifies a signal.

Signals can be quite sophisticated, like the ultra-short electromagnetic pulses travelling through optical fibres and connecting computers all around the world; or sophisticated but mundane, like speech and writing; or rudimentary, like puffs of smoke.

The actual information carried by a signal is coded. For example, a puff of smoke every five seconds may mean *'we have chicken for dinner'*, or this same signal may mean *'we are under attack, run for your lives'*. The code itself is thus extraneous to the signal, and as such it is not the subject of this textbook.

Signals are also 'embodied' in physical entities. For example, signals may be embodied in electromagnetic waves, or in electric currents, or in ink, etc. Furthermore, they can propagate through a medium. For instance, signals embodied in electromagnetic waves can propagate through air or through optical fibres. There are disciplines fully devoted to study the physical entities carrying signals. Electromagnetism is a paradigmatic example.

As it propagates through a medium, a signal is modified. For example, the shapes of electromagnetic pulses are gradually distorted as they propagate through optical fibres. Thus, these signals require treatment, literally to be get back in shape, which is a major source of limitation of the Internet speed. If the medium of propagation modifies a signal, and a system is anything that modifies a signal, then the medium itself is a system.

Systems can deteriorate or improve signals. An optical fibre is an example of a system deteriorating signals. Devices and circuits designed to treat signals are examples of systems improving signals. Again, there are disciplines fully devoted to study systems, for example, electric and electronic circuits. Electromagnetism is again another example, as it deals with the physical laws describing propagation of electromagnetic waves.

Essentials of Signals and Systems, First Edition. Emiliano R. Martins.
© 2023 John Wiley & Sons Ltd. Published 2023 by John Wiley & Sons Ltd.
Companion website: www.wiley.com/go/martins/essentialsofsignalsandsystems

A signal is a pattern: it is something that changes with respect to some other thing, usually time or space. For example, think of an audio signal coming out of a speaker. The membrane of the speaker is moving, which means that its amplitude is changing with time. So, the amplitude of the membrane constitutes a signal. We represent this signal by a generic function of time, like $g(t)$: the value of $g(t)$ at a certain time t denotes the amplitude of the membrane at time t. So, if the speaker is playing your favourite song, then $g(t)$ is a mathematical object representing your favourite song: a plot of $g(t)$ against time is a graphical representation of how a speaker membrane must move to create your favourite song. But just before being embodied in the membrane vibration, your song must have been embodied in an electric current. Thus, for example, the same $g(t)$ could also denote the value of the current in a resistor. What I want to convey with this example is that, for this course, it does not matter how a signal is embodied: we are interested in the pattern, that is, on $g(t)$. Thus, to this course, it is immaterial whether $g(t)$ is the amplitude of a speaker membrane or air vibration, or an electric current, or an electromagnetic wave or whatever.

A signal could also be something that changes with respect to space. For example, the Mona Lisa is a well-defined spatial distribution of colours. So, suppose that you have a colour palette, and that you ascribe a number to designate each colour of your palette. Now suppose you paint the Mona Lisa using this palette. After painting it, you attach coordinate axes to the canvas. If you do that, then there will be a signal $g(x, y)$ representing the Mona Lisa, with the value of $g(x, y)$ denoting the colour in the position x, y. Thus, if you get a new canvas, and paint each point x, y with the colour $g(x, y)$, out comes the Mona Lisa.

Even though signals can be either time or space dependent (or both), we will build the theory on the assumption that the signals are time dependent. The same theory, however, also applies to space-dependent signals. In fact, in Chapter 3, we will see some examples of application of the theory to spatial dependent signals. Signals can also be discrete, and we will deal with discrete signals and systems in Chapters 7–10.

The main objective of this course is to learn a handful of general and extremely useful properties that signals obey when they are acted on by a special class of systems: the so-called linear and time-invariant systems. There is an enormous range of systems falling into this category, so the concepts we will learn here have a wide range of applications, like electromagnetism, principles of communications, electronic circuits, sensors, medical instrumentation, and quantum mechanics, to name only a few. The core of the theory is covered in Chapters 2 and Chapter 3. We begin with the theory of representation of signals, which is the subject of Chapter 2.

2

Representation of Signals

Learning Objectives

Chapters 2 and 3 introduce the backbone concepts of signals and systems. Chapter 2 introduces the theory of representation of signals: signals are like vectors of a continuous algebraic space: the function space. As such, they admit different coordinate representations, depending on the choice of basis. We will learn the two most important representations: the time domain representation and the frequency domain representation. And we will learn how to change back and forth between these two representations by means of the famous Fourier transforms. Later, in Chapter 3, we will learn that one of these representations – the frequency domain representation – is the representation in the basis of eigenvectors; as such it is the representation that diagonalizes the operators.

2.1   Introduction

In the Interlude, I said that even though signals can be either time or space dependent, we will build the theory assuming time dependence, like $g(t)$. But a signal is not any function of time. A signal must belong to the set of square-integrable functions. A function $g(t)$ is square-integrable if the integral:

$$\int_{-\infty}^{+\infty} |g(t)|^2 dt = A \qquad (2.1)$$

does not blow up, that is, if A is finite.

Of course, it is easy to conceive of functions that are not square-integrable. For example, $g(t) = 1$ is not square-integrable. But if $g(t)$ represents a physical signal, something that can be embodied in physical quantities, then $g(t)$ is square-integrable. Notice that $g(t) = 1$ cannot represent a physical signal because

Essentials of Signals and Systems, First Edition. Emiliano R. Martins.
© 2023 John Wiley & Sons Ltd. Published 2023 by John Wiley & Sons Ltd.
Companion website: www.wiley.com/go/martins/essentialsofsignalsandsystems

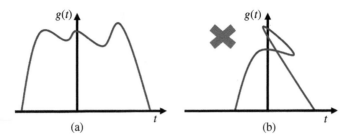

Figure 2.1 Illustration of a single-valued function (a) and non-single-valued function (b). Only (a) can represent a signal.

that would be a signal that never begun and will never end; a signal that has been equal to one for all eternity for ever and ever and ever and ever.

As I have already mentioned several times, we will learn in this chapter that signals are like vectors, so they also belong to an algebraic space. The space signals belong to is often loosely called 'function space', or Hilbert space, but I do not want to go into a formal definition of this space. It suffices to say that it is the set of all square-integrable functions. What really matters here is that signals are like vectors, and as such the vector operations and notions we are familiar with from linear algebra also apply to signals.

A signal must also be single valued: if you pick a specific time t_0, then there is only one value of the signal at t_0. In other words, $g(t_0)$ is a well-defined number. This property is illustrated in Figure 2.1: only $g(t)$ of Figure 2.1a is a single-valued function, so only Figure 2.1a can represent a signal.

Before we move on to the actual business of this chapter, which is the theory of representation of signals, we need to learn an operation that appears often in this theory: the operation of convolution.

2.2 The Convolution

This section introduces the operation of convolution. Our main interest is not in learning how to evaluate an integral of convolution, though we will see two examples of calculation in the worked exercises. Our main interest is in understanding what the convolution does: in other words, in understanding its algebraic meaning.

A convolution is an operation involving two functions, and the result of the operation is another function. By definition, if the function $g(t)$ is the convolution between the functions $a(t)$ and $b(t)$, then:

$$g(t) = \int_{-\infty}^{+\infty} a(x)b(t-x)dx \tag{2.2}$$

It is convenient to denote the integral of convolution without having to write it explicitly. For that, we use the symbol $*$, so that $a(t) * b(t)$ denotes the convolution between $a(t)$ and $b(t)$. Thus, by definition:

$$a(t) * b(t) = \int_{-\infty}^{+\infty} a(x)b(t - x)dx \qquad (2.3)$$

Inside the integral, the time t is treated as a constant. But, if we change t, then we change $b(t - x)$, and, consequently, the result of the integral changes. Thus, the result of the integral depends on t. In other words, the result of the integral is a function of t.

So, what exactly is the integral in Equation (2.2) doing? Obviously, it is building a function $g(t)$ out of two functions $a(t)$ and $b(t)$. And it is doing that using an infinite set of functions, where each function of this set is a shifted version of $b(t)$. In other words, each value of x specifies a different function $b(t - x)$. For a fixed value of x, say $x = 3$, $b(t - x)$ is a time-shifted version of $b(t)$, in this case $b(t - 3)$. This time-shifted function is multiplied by a number specified by $a(x)$. For example, the function $b(t - 3)$ is multiplied by the number $a(3)$.

Equation (2.2) is a recipe to obtain a new function $g(t)$ by adding up these time-shifted versions of $b(t)$, each one of them with a 'weight' set by $a(x)$. To better visualize what is happening, we can express the integral of convolution as a Riemann sum. Thus:

$$g(t) = \int_{-\infty}^{+\infty} a(x)b(t - x)dx \approx \sum_{n=-\infty}^{n=\infty} a_n b(t - n\Delta x)\Delta x \qquad (2.4)$$

where $a_n = a(n\Delta x)$, and the sum is identical to the integral in the limit of Δx going to zero.

The Riemann sum in Equation (2.4) is a sum over infinitely many functions of time, each one being a time-shifted version of $b(t)$. To help visualizing it, three terms of the sum are explicitly shown below:

$$g(t) \approx \Delta x \sum_{n=-\infty}^{n=\infty} a_n b(t - n\Delta x) = \Delta x [\ldots + a_0 b(t) + a_1 b(t - \Delta x) + a_2 b(t - 2\Delta x) + \ldots]$$

$$(2.5)$$

One term explicitly shown is just the function $b(t)$ itself, multiplied by $a_0 = a(0)$. Another term is the function $b(t - \Delta x)$, which is the original function $b(t)$, but shifted a little bit to the right (more precisely, shifted by an amount Δx). And this shifted function is multiplied by $a_1 = a(\Delta x)$. Thus, how much $b(t)$ has been shifted determines the instant at which the function $a(t)$ is evaluated to multiply the shifted $b(t)$: if we shifted $b(t)$ by an amount Δx, then we must multiply it by $a(t)$ evaluated at the instant $t = \Delta x$, that is, by $a(\Delta x)$. Another term shown explicitly is $b(t - 2\Delta x)$, which is multiplied by $a_2 = a(2\Delta x)$. The contributions of these three terms are shown in Figure 2.2: we start with $b(t)$ (blue function) and multiply it by $a(0)$ (blue point). Then we get $b(t - \Delta x)$ (green function) and multiply it by $a(\Delta x)$ (green point). Then we add up these two functions (the weighted versions of the blue and green curves). Then we get $b(t - 2\Delta x)$ (red function) and multiply it by

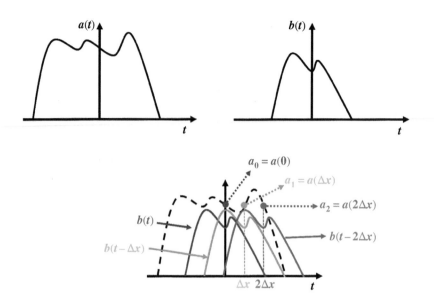

Figure 2.2 Illustration of how the convolution builds up a new function by adding up shifted versions of the function $b(t)$, each one of them multiplied by $a(t)$ evaluated at the instant defined by how much $b(t)$ has been shifted.

$a(2\Delta x)$ (red point), and add it to the previous two, and so on. In Equation (2.5) and Figure 2.2, I showed only functions shifted to the right (that is, positive n), but, of course, the sum also involves functions shifted to the left (that is, negative n). The function $g(t)$ is what we get when we do this process taking infinitesimal time-shifting steps Δx.

One key feature to notice is that, **in the integral of convolution, only $b(t)$ and its shifted versions are functions of time: the function $a(t)$ has been turned into a set of coefficients.** We will often take advantage of this feature.

Another significant feature of the convolution is that it is a symmetric function, that is:

$$a(t) * b(t) = b(t) * a(t) \qquad (2.6)$$

Probably you are a bit surprised at this symmetry property since it is not evident at all that the convolution is a symmetric operation. After all, one of the functions does not even play the role of a function, but of a list of coefficients. But we will prove later in this chapter, when we deal with properties of Fourier transforms, that it is indeed symmetric.

As said earlier, it is most relevant to understand what the convolution does: it picks two functions, treats one as a list of coefficients, creates lots of (actually, infinitely many) shifted versions of the other, multiplies each of them according to the list of coefficients, and adds them all up to build a third one. And since one of the functions became just a list of coefficients, only the shifted versions of the other function are actual functions of time. That is what it does. Please take time to let this sink in, look again at Figure 2.2 and make sure you understood the algorithm of the convolution.

Assuming that now you are comfortable with what the convolution does, we can take a look at how to evaluate it, that is, how we can find $g(t)$ in Equation (2.2).

First notice that Equation (2.2) involves infinitely many integrals: each value of t specifies a different integral. For example, for $t = 0$, Equation (2.2) reduces to:

$$g(0) = \int_{-\infty}^{+\infty} a(x)b(-x)dx$$

To find $g(0)$, first we turn $a(t)$ into $a(x)$: this is just a change of symbols, as illustrated in Figure 2.3a. We also need to do this for $b(t)$, but now we have $b(-x)$ instead of $b(x)$, which means that we also need to flip $b(x)$ around the vertical axis, as also illustrated in Figure 2.3a. Then we multiply the two functions $a(x)$ and $b(-x)$ and integrate them.

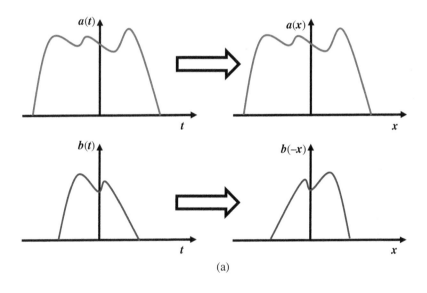

(a)

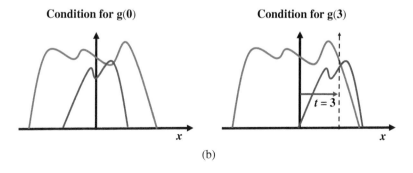

(b)

Figure 2.3 Illustration of how to evaluate the integral of convolution.

Now suppose we want to evaluate $g(3)$:

$$g(3) = \int_{-\infty}^{+\infty} a(x)b(3-x)dx$$

In the evaluation of $g(3)$, the process involving $a(t)$ is the same as before: just change the horizontal axis from t to x. In respect of $b(t)$, however, there is a difference: what enters the integral is no longer $b(-x)$, but $b(3-x)$. That means that we must shift $b(-x)$ to the right, by the amount set by the instant of time at which we are evaluating $g(t)$, which, in this case, is $t = 3$. The functions $a(x)$ and $b(3-x)$ are illustrated in Figure 2.3b. Notice that, if t is a sufficiently large number so that there is no overlap between $a(x)$ and $b(t-x)$, then their product is zero, which results in $g(t) = 0$.

To practise evaluating integrals of convolution, two worked examples are shown below. It is important to understand them well, but do not get too caught up in evaluating these integrals: remember that we are mostly interested in interpreting what the convolution does.

Worked Exercise: First Example of Convolution

Evaluate the convolution between $a(t) = t^2$ and a rectangular function.

Solution:

The rectangular function is defined as:

$$rect(t) = 1 \quad for - c \leq t \leq c$$

$$rect(t) = 0 \; oustide \; the \; interval - c \leq t \leq c \tag{2.7}$$

Calling $g(t)$ the function resulting from the convolution between $a(t)$ and $rect(t)$, we have:

$$g(t) = a(t) * rect(t)$$

To evaluate the convolution, we first turn $a(t) = t^2$ into $a(x) = x^2$. Next, we turn $rect(t)$ into $rect(x)$, and then we turn $rect(x)$ into $rect(-x)$. Notice, however, that $rect(x)$ is an even function, which means that $rect(-x) = rect(x)$. Furthermore, if $rect(-x) = rect(x)$, then necessarily $rect(t-x) = rect(x-t)$. Therefore, we can work with $rect(x-t)$ instead of $rect(t-x)$, which is more convenient.

From the definition of the rectangular function (Equation (2.7)), we find:

$$rect(x) = 1 \; for - c \leq x \leq c$$

$$rect(x) = 0 \; oustide \; the \; interval - c \leq x \leq c$$

which entails that:

$$rect(x - t) = 1 \; for - c \leq x - t \leq c$$

$$rect(x - t) = 0 \; oustide \; the \; interval - c \leq x - t \leq c$$

From which we conclude that:

$$rect(x - t) = 1 \; for - c + t \leq x \leq c + t$$

$$rect(x - t) = 0 \; oustide \; the \; interval - c + t \leq x \leq c + t \tag{2.8}$$

The functions $a(x) = x^2$ and $rect(x - t)$ are illustrated in Figure 2.4, the latter for two different values of t, namely, $t = 0$ and $t = 5$.

Next, we write the integral of convolution explicitly:

$$g(t) = \int_{-\infty}^{+\infty} x^2 rect(t - x) dx = \int_{-\infty}^{+\infty} x^2 rect(x - t) dx$$

Since $rect(x - t) = 0$ outside the interval $-c + t \leq x \leq c + t$, the integral above reduces to:

$$g(t) = \int_{-\infty}^{+\infty} x^2 rect(x - t) dx = \int_{(-c+t)}^{(c+t)} x^2 rect(x - t) dx$$

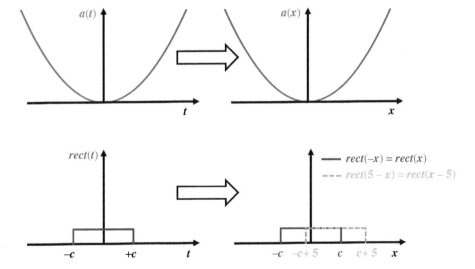

Figure 2.4 Illustration of modifications in the quadratic and rectangular functions. The rectangular function is even, so $rect(x) = rect(-x)$. Likewise, $rect(t - x) = rect(x - t)$. The green dashed rectangle illustrates the function $rect(x - 5)$.

Furthermore, $rect(x - t) = 1$ inside this interval. Thus:

$$g(t) = \int_{(-c+t)}^{(c+t)} x^2 rect(x - t)dx = \int_{(-c+t)}^{(c+t)} x^2 dx = \frac{x^3}{3}\bigg|_{(-c+t)}^{c+t} = \frac{(c+t)^3 - (-c+t)^3}{3}$$

$$(2.9)$$

Notice that the result is indeed a function of time t. It is as if we had calculated infinitely many integrals, each one for a different value of t, and then grouped the results together into Equation (2.9).

Worked Exercise: Second Example of Convolution

Evaluate the convolution between the rectangular function and itself.

Solution:

From Equation (2.7) and Equation (2.8), we have:

$$rect(x) = 1 \; for - c \leq x \leq c$$

$$rect(x) = 0 \; oustide \; the \; interval - c \leq x \leq c$$

and:

$$rect(x - t) = 1 \quad for - c + t \leq x \leq c + t$$

$$rect(x - t) = 0 \; oustide \; the \; interval - c + t \leq x \leq c + t$$

As illustrated in Figure 2.5, we can distinguish four different situations involved in the integral of convolution of the rectangular function with itself.

First, for $t \leq -2c$, there is no overlap between the rectangular functions, so their product must be zero; consequently, the integral is also zero. Thus, we immediately conclude that:

$$g(t) = rect(t) * rect(t) = 0 \; for \; t \leq -2c \tag{2.10}$$

Second, for $-2c \leq t \leq 0$, there is overlap between $x = -c$ (which is the tail of the red rectangle – the horizontal coordinate was omitted in Figure 2.5 to avoid clutter) and $x = c + t$ (which is the tip of the blue-dashed rectangle). Thus:

$$g(t) = rect(t) * rect(t) = \int_{(-c)}^{(c+t)} dx = c + t - (-c) = 2c + t$$

Therefore:

$$g(t) = 2c + t \; for \; -2c \leq t \leq 0 \tag{2.11}$$

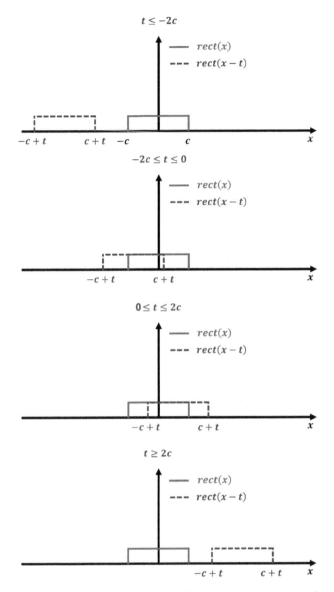

Figure 2.5 Illustration of the four domains of integration involved in the convolution of a rectangular function with itself.

Third, for $0 \leq t \leq 2c$, there is overlap between $x = -c + t$ (the tail of the blue dashed rectangle) and $x = c$ (the tip of the red rectangle). Thus:

$$g(t) = rect(t) * rect(t) = \int_{(-c+t)}^{(c)} dx = c - (-c + t) = 2c - t$$

Therefore:

$$g(t) = 2c - t \; for \; 0 \leq t \leq 2c \tag{2.12}$$

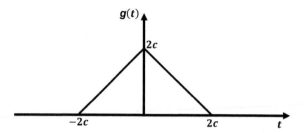

Figure 2.6 The convolution of a rectangular function with itself results in a triangular function.

Fourth, and finally, there is no overlap for $t \geq 2c$. Thus:

$$g(t) = 0 \ for \ t \geq 2c \tag{2.13}$$

According to Equation (2.10)–Equation (2.13), the convolution between a rectangular function and itself is a triangular function, as illustrated in Figure 2.6.

2.3 The Impulse Function, or Dirac Delta

The last mathematical tool we need to acquire before embarking on the theory of representations is the impulse function, also known as Dirac Delta (the most remarkable feat of Paul Dirac was his contribution in merging quantum mechanics with special relativity; nevertheless, this is the second time his name appears in our course, which has little to do with merging quantum mechanics with special relativity; some people seem to have been born with eight brains).

There are many ways of defining the impulse function. Here, I will define it in terms of a normalized rectangular function (the rectangular function itself was defined in Equation (2.7)). Thus, we define the function $q(t)$ as:

$$q(t) = \frac{1}{T} \ for - \frac{T}{2} \leq t \leq \frac{T}{2}$$

$$q(t) = 0 \ oustide \ the \ interval - \frac{T}{2} \leq t \leq \frac{T}{2} \tag{2.14}$$

Examples of $q(t)$ for different values of T are shown in Figure 2.7. Notice that, regardless of the interval T, the 'area' of $q(t)$ is always one, that is:

$$\int_{-\infty}^{\infty} q(t)dt = \int_{-\frac{T}{2}}^{\frac{T}{2}} \frac{1}{T}dt = \frac{T}{T} = 1 \tag{2.15}$$

A mathematician may frown upon the choice of $q(t)$, because $q(t)$ has discontinuities. In fact, $q(t)$ cannot even represent a physical signal, because no physical signal can go from zero to $1/T$ instantaneously (physical signals cannot have discontinuities). But $q(t)$ is a convenient idealization of a signal that rises quickly,

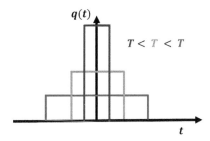

Figure 2.7 Examples of normalized rectangular functions $q(t)$ for different values of T.

stays constant for some time, then goes back to zero quickly. So, we stick to $q(t)$ in defining the impulse function, because its simplicity more than compensates for its lack of rigour. A more rigorous treatment, however, can be found in G. Barton's 'Elements of Green's function and Propagation' (see references section).

We denote the impulse function (or Dirac delta) by $\delta(t)$, and define it as a normalized rectangular function in the limit of vanishing T:

$$\delta(t) = \lim_{T \to 0} q(t) \tag{2.16}$$

You can think of the impulse function as a super thin and super tall rectangular function (something like 'super model meets basketball player'). But it is still a rectangular function, so it still has unit area. Thus:

$$\int_{-\infty}^{\infty} \delta(t)dt = \lim_{T \to 0} \left[\int_{-\infty}^{\infty} q(t)dt \right] = \lim_{T \to 0} \left[\int_{-\frac{T}{2}}^{\frac{T}{2}} \frac{1}{T}dt \right] = \lim_{T \to 0} \left[\frac{T}{T} \right] = 1 \tag{2.17}$$

So, $\delta(t)$ is kind of funny. It has unit area, yet it is zero everywhere, except at the origin (because T is vanishingly small). Thus, to compensate for the fact that its width is infinitesimal, its height must go to infinity. So, it is a function that is infinite at the origin, zero outside the origin, and has unit area, as summarized below:

$$\delta(t) = 0, \, for \, t \neq 0$$

$$\delta(0) = \infty$$

$$\int_{-\infty}^{\infty} \delta(t)dt = 1 \tag{2.18}$$

These properties are represented graphically by an upwards pointing arrow placed at the origin, as illustrated in Figure 2.8. The arrow pointing upwards is a reminder that the impulse function goes to infinity at the origin. But the arrow is placed only at the origin, so it is zero outside the origin, as is the impulse function. Furthermore, the height of the arrow represents the area of the impulse function. Since the impulse function has unit area, the height of the arrow is one.

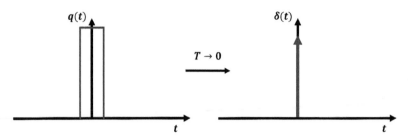

Figure 2.8 Graphical representation of the impulse function as an upwards pointing arrow placed at the origin. The arrow is a reminder that the impulse function goes to infinity at the origin, but it is zero outside the origin. The height of the arrow indicates the area. Since the impulse function has unit area, the height of the arrow is one.

Now consider what happens when we integrate the product of a signal $g(t)$ and the impulse function:

$$\int_{-\infty}^{\infty} g(t)\delta(t)dt =?$$

To evaluate this integral, we can use the definition of $\delta(t)$, that is, we replace it with a normalized rectangular function, and check what happens when T goes to zero:

$$\int_{-\infty}^{\infty} g(t)\delta(t)dt = \lim_{T\to 0}\int_{-\infty}^{+\infty} g(t)q(t)dt = \lim_{T\to 0}\int_{-\frac{T}{2}}^{+\frac{T}{2}} g(t)\frac{1}{T}dt \qquad (2.19)$$

Look at the last integral in Equation (2.19): it is evaluated only between $-T/2$ and $T/2$. But, as illustrated in Figure 2.9a, if we make this interval vanishingly small, then $g(t)$ can be treated as a constant inside this interval. Furthermore, since this is a vanishingly small interval around the origin, the value of $g(t)$ inside this interval is just its value at the origin. Thus:

$$\int_{-\infty}^{\infty} g(t)\delta(t)dt = \lim_{T\to 0}\int_{-\frac{T}{2}}^{+\frac{T}{2}} g(t)\frac{1}{T}dt = \lim_{T\to 0}\int_{-\frac{T}{2}}^{+\frac{T}{2}} g(0)\frac{1}{T}dt$$

But $g(0)$ is just a constant, so we can remove it from the integral:

$$\int_{-\infty}^{\infty} g(t)\delta(t)dt = \lim_{T\to 0}\int_{-\frac{T}{2}}^{+\frac{T}{2}} g(0)\frac{1}{T}dt = g(0)\lim_{T\to 0}\int_{-\frac{T}{2}}^{+\frac{T}{2}} \frac{1}{T}dt$$

Now the integral is only the 'area' of the normalized rectangular function, which is always one, for any value of T. So, we conclude that:

$$\int_{-\infty}^{\infty} g(t)\delta(t)dt = g(0) \qquad (2.20)$$

Equation (2.20) is an example of the 'sifting property' of the impulse function: when we integrate the product of a signal with the impulse function, the integral

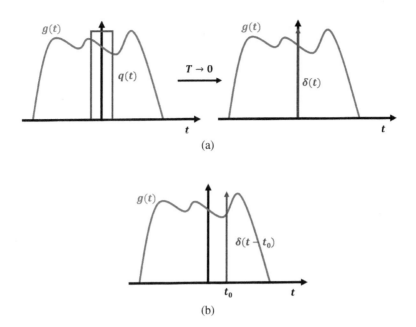

Figure 2.9 Representation of the sifting property of the impulse function. (a) The signal $g(t)$ can be treated as a constant inside a vanishingly small interval around the origin. (b) Sifting property of a time-shifted impulse function.

selects the value of the signal at the 'position' of the impulse function. As illustrated in Figure 2.9b, we could repeat the same reasoning for a time-shifted impulse function $\delta(t - t_0)$, which is just the impulse function 'located' at time $t = t_0$. Thus:

$$\int_{-\infty}^{+\infty} g(t)\delta(t - t_0)dt = \lim_{T \to 0} \int_{-\infty}^{+\infty} g(t)q(t - t_0)dt = \lim_{T \to 0} \int_{-\frac{T}{2}+t_0}^{\frac{T}{2}+t_0} g(t)\frac{1}{T}dt$$

$$= g(t_0) \lim_{T \to 0} \int_{-\frac{T}{2}+t_0}^{\frac{T}{2}+t_0} \frac{1}{T}dt = g(t_0)$$

Therefore:

$$\int_{-\infty}^{+\infty} g(t)\delta(t - t_0)dt = g(t_0) \tag{2.21}$$

According to Equation (2.21), if we integrate the product of a signal $g(t)$ with an impulse function located at $t = t_0$, then the result of the integral is the value of the signal at $t = t_0$.

We will be constantly referring to Equation (2.21) in this textbook.

Finally, notice that, since a rectangular function is an even function (that is, $q(t) = q(-t)$), so is the impulse function:

$$\delta(t) = \delta(-t) \tag{2.22}$$

2.4   Convolutions with Impulse Functions

The sifting property of the impulse function entails a funny consequence: if we convolve a signal $g(t)$ with the impulse function, the result is the signal $g(t)$ itself. In this section, we spell out this property as a bridge to the time domain representation of signals.

Consider Equation (2.21) again, which I repeat below for convenience:

Equation (2.21)

$$\int_{-\infty}^{+\infty} g(t)\delta(t - t_0)dt = g(t_0)$$

The result of this integral is not a signal: it is just a number. A signal is a function of time t, but the result of the integral above is just the number $g(t_0)$, that is, the value of $g(t)$ at the specific time $t = t_0$. Indeed, t_0 is just a constant, just a number, and so is $g(t_0)$. But we can turn the result of this integral into a signal by inverting the roles of t and t_0. After all, if it is true that $\int_{-\infty}^{+\infty} g(t)\delta(t - t_0)dt = g(t_0)$, then, necessarily, it must also be true that:

$$\int_{-\infty}^{+\infty} g(t_0)\delta(t_0 - t)dt_0 = g(t) \tag{2.23}$$

From a purely mathematical point of view, Equation (2.23) and Equation (2.21) are identical; after all, Equation (2.23) was obtained from Equation (2.21) by only swapping t for t_0. But though they are mathematically identical, their interpretations are completely different: while the result of the integral in Equation (2.21) is just a number ($g(t_0)$ is just a number), the result of the integral in Equation (2.23) is a signal: it is the signal $g(t)$. Furthermore, on the one hand, the integral in Equation (2.21) involves the product of two functions of t ($g(t)$ and $\delta(t - t_0)$), and the product is integrated over t, thus returning the number $g(t_0)$. The integral in Equation (2.23), on the other hand, does not involve the product of functions of t. Instead, it is a sum of infinitely many impulse functions (each value of t_0 specifies a different impulse function, that is, a different arrow 'located' at a different instant $t = t_0$). And each one of these impulse functions is multiplied by a 'weight': the number $g(t_0)$. Thus, contrary to Equation (2.21), there is no product of functions of time in Equation (2.23), but a sum over weighted time-shifted versions of the impulse function, resulting in the signal $g(t)$.

A weighted sum over time-shifted versions of a function returning another function is reminiscent of the convolution. It is indeed a convolution, and we can make it more explicit by rewriting the integral in Equation (2.23) in a slightly different form. Notice that Equation (2.22) entails that $\delta(t_0 - t) = \delta(t - t_0)$. Thus, we can recast Equation (2.23) as:

$$\int_{-\infty}^{+\infty} g(t_0)\delta(t - t_0)dt_0 = g(t) \tag{2.24}$$

Compare Equation (2.24) with Equation (2.3), which I repeat below for convenience

Equation (2.3)

$$a(t) * b(t) = \int_{-\infty}^{+\infty} a(x)b(t-x)dx$$

In Equation (2.24), the integral is over t_0, so t_0 is a dummy index: we can replace it with any symbol we want. For example, if you are fond of complicated looking Greek letters, you can write Equation (2.24) as $\int_{-\infty}^{+\infty} g(\xi)\delta(t-\xi)d\xi = g(t)$. But if you are fond of the good old Latin alphabet, you can write it as $\int_{-\infty}^{+\infty} g(x)\delta(t-x)dx = g(t)$. By comparing this latter form with Equation (2.3), we can immediately recognize that the integral in Equation (2.24) is the convolution between $g(t)$ and $\delta(t)$. Thus, we have concluded that:

$$g(t) * \delta(t) = g(t) \tag{2.25}$$

This is a major result, and it will lead to the notion of a basis of the function space (to be discussed in Section 2.5).

To conclude this section, let us make sure Equation (2.24) makes good sense by evaluating its integral for two values of t. First, let us evaluate the integral for $t = 0$:

$$\int_{-\infty}^{+\infty} g(t_0)\delta(-t_0)dt_0$$

Since the impulse function is even, $\delta(-t_0) = \delta(t_0)$ (see Equation (2.22)), so:

$$\int_{-\infty}^{+\infty} g(t_0)\delta(-t_0)dt_0 = \int_{-\infty}^{+\infty} g(t_0)\delta(t_0)dt_0$$

This last integral is the same as Equation (2.20), but the dummy index in Equation (2.20) is t and in the integral above it is t_0. Since it is immaterial what symbol is used for a dummy index, Equation (2.20) entails that:

$$\int_{-\infty}^{+\infty} g(t_0)\delta(-t_0)dt_0 = \int_{-\infty}^{+\infty} g(t_0)\delta(t_0)dt_0 = g(0)$$

as expected.

Now, let us evaluate the integral in Equation (2.24) for $t = 3$:

$$\int_{-\infty}^{+\infty} g(t_0)\delta(3-t_0)dt_0$$

Again, using the fact that the impulse function is an even function (Equation (2.22)):

$$\int_{-\infty}^{+\infty} g(t_0)\delta(3-t_0)dt_0 = \int_{-\infty}^{+\infty} g(t_0)\delta(t_0-3)dt_0$$

This last integral has the same form of Equation (2.21), but again with another dummy index. For convenience, just change the dummy index for t:

$$\int_{-\infty}^{+\infty} g(t_0)\delta(t_0 - 3)dt_0 = \int_{-\infty}^{+\infty} g(t)\delta(t - 3)dt$$

Comparing the integral above with Equation (2.21), we conclude that:

$$\int_{-\infty}^{+\infty} g(t_0)\delta(t_0 - 3)dt_0 = \int_{-\infty}^{+\infty} g(t)\delta(t - 3)dt = g(3)$$

as expected.

Worked Exercise: The Convolution with $\delta(t - a)$

Evaluate $g(t) * \delta(t - a)$

Solution:

Define: $y(t) = \delta(t - a)$. Thus:

$$g(t) * \delta(t - a) = g(t) * y(t) = \int_{-\infty}^{+\infty} g(t_0)y(t - t_0)dt_0$$

If $y(t) = \delta(t - a)$, then $y(t - t_0) = \delta(t - a - t_0) = \delta(-t_0 + (t - a))$. Furthermore, the impulse function is even (see Equation (2.22)), so $\delta(-t_0 + (t - a)) = \delta(t_0 - (t - a))$, from which we conclude that $y(t - t_0) = \delta(t_0 - (t - a))$.
Thus:

$$g(t) * \delta(t - a) = \int_{-\infty}^{+\infty} g(t_0)y(t - t_0)dt_0 = \int_{-\infty}^{+\infty} g(t_0)\delta(t_0 - (t - a))dt_0$$

According to the sifting property of the impulse function (Equation (2.21)), the last integral above results in:

$$\int_{-\infty}^{+\infty} g(t_0)\delta(t_0 - (t - a))dt_0 = g(t - a)$$

Therefore:

$$g(t) * \delta(t - a) = g(t - a) \tag{2.26}$$

Thus, according to Equation (2.26), if we convolve a function with an impulse function 'located' at $t = a$, the result is the function shifted to $t = a$.

2.5   Impulse Functions as a Basis: The Time Domain Representation of Signals

We have acquired the mathematical tools required to introduce the theory of representation of signals. We begin with the time domain representation, which is analogous to the representation of a vector in the canonical basis. In Section 2.8, we will see another crucial representation: the frequency domain representation.

In Section 2.2 (The Convolution), we learned that the meaning of $g(t) = a(t) * b(t)$ is that $g(t)$ is built up by adding up the functions $b(t - x)$, each one with a 'weight' $a(x)$. This process is analogous to a linear combination of basis vectors, with the functions $b(t - x)$ playing the role of basis vectors, and $a(x)$ playing the role of coordinates. The only significant difference is that, in a linear combination of vectors, we have a discrete number of vectors, whereas in the integral of convolution we have a continuum of functions $b(t - x)$. We have a continuum because x can be any real number, and each value of x indexes a different function $b(t - x)$. Furthermore, if we have a continuum of functions, then we need an integral to add them up. Therefore, the sum in a linear combination of vectors in a discrete space becomes an integral of convolution in a continuous space (mind you, I am not saying that only an integral of convolution is a linear combination; indeed, in Section 2.8, we will see a linear combination that is not an integral of convolution). In short, the integral of convolution is a linear combination in the function space.

A linear combination involving a continuum of functions $b(t - x)$ is related to the most significant difference between the function space and the Euclidean space: a Euclidean space is discrete, whereas the function space is continuous. In a discrete space, we can count the basis vectors, like 'the first basis vector is $\widehat{x}$, and the second is $\widehat{y}$ and there is nothing between $\widehat{x}$ and $\widehat{y}$'. But we cannot do that in a continuous space. For, suppose that the functions $b(t - x)$ form a basis. Then we pick one of them, say $b(t - 7.3)$. Which one comes next? This question does not make sense because, since x is any real number, and there are always infinitely many real numbers between 7.3 and any other real number, then there are always infinitely many functions between $b(t - 7.3)$ and any other function of the set $b(t - x)$. Thus, the function space has an infinite dimension, because a basis of this space is a set with infinitely many functions; and the basis has infinitely many functions because there is a continuum of basis functions, in the same sense that the functions $b(t - x)$ form a continuum, with each different function indexed by a different value of x.

But what on earth could these basis functions of the function space be? Well, what is a basis of a space? It is a collection of vectors (functions) that can be used to construct any vector (function) of the space from a linear combination of these basis vectors (functions). Can you think of a group of functions that can be used to construct any other function? (Actually, not any other function, but a signal – it must belong to the Hilbert space). I bet you can; after all, it is in the title of this section: the set of all impulse functions is a basis of the function space (the 'set of all impulse functions' is the collection of all impulse functions, where each impulse function is 'located' at a different time $t = t_0$).

Maybe you are asking yourself why the impulse functions form a basis. But we have already proved that they do form a basis, because we have already proved that any signal $g(t)$ can be constructed from a linear combination of impulse functions: this is precisely the message of Equation (2.24).

Since this is such a crucial notion in signals and system, I will emphasize it by breaking the convolution of Equation (2.24) into a Riemann sum. I want to emphasize that the integral in Equation (2.24) is just a sum over many impulse functions, each one with a different weight. In other words, that the integral in Equation (2.24) is a linear combination of impulse functions. Thus, breaking it into the Reimann sum, Equation (2.24) reads:

$$g(t) = \int_{-\infty}^{+\infty} g(t_0)\delta(t - t_0)dt_0 \approx \Delta t_0 \sum_{n=-\infty}^{+\infty} g(n\Delta t_0)\delta(t - n\Delta t_0) \qquad (2.27)$$

To clean up the notation a bit, we define the 'coefficient' (or, if you prefer to use proper algebraic language straight away, the 'coordinate') $g_n = g(n\Delta t_0)$. We also write a few terms of the sum explicitly:

$$g(t) = \int_{-\infty}^{+\infty} g(t_0)\delta(t - t_0)dt_0 \approx \Delta t_0 \sum_{n=-\infty}^{+\infty} g_n\delta(t - n\Delta t_0)$$
$$= \Delta t_0[\ldots + g_0\delta(t) + g_1\delta(t - \Delta t_0) + g_2\delta(t - 2\Delta t_0) + \ldots] \qquad (2.28)$$

As illustrated by Equation (2.28), $g(t)$ is the result of a linear combination of many (actually, infinitely many) impulse functions, each one 'located' at a different time $t = n\Delta t_0$, and each one multiplied by its coordinate g_n. For example, $\delta(t - 2\Delta t_0)$ is an impulse function 'located' at $t = 2\Delta t_0$, and it is multiplied by the coordinate $g_2 = g(2\Delta t_0)$. I emphasize again that there is only one type of function of time t in the sum (and integral) of Equation (2.28): only the impulse functions are functions of time t, whereas $g(t_0)$ is only a group of coordinates indexed by t_0.

The terms of the sum in Equation (2.28) are illustrated in Figure 2.10. The heights of the arrows depend on the coordinates of the impulse functions. Thus, if $g(t)$ is represented by the red curve in Figure 2.10, its value at a specific time t_0 sets the height of the impulse function 'located' at this specific time t_0. The heights of the arrows still represent the 'area' of the function, but now it is the area of the impulse function times its coordinate, that is, it is the area of $g(t_0)\delta(t - t_0)$; but, since $\delta(t - t_0)$ has unit area, then the area of $g(t_0)\delta(t - t_0)$ coincides with $g(t_0)$. Furthermore, as the time step Δt_0 goes to zero, the arrows come together, approaching a continuum, and the Riemann sum converges to the convolution integral.

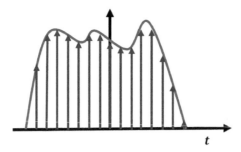

Figure 2.10 Graphical representation of the terms of the Riemann sum in Equation (2.28); $g(t)$ is represented by the red curve.

Thus, we have a basis of our new space: it is a basis formed by infinitely many impulse functions. We need infinitely many impulse functions because we need impulse functions 'located' at all possible times $t = t_0$, and t_0 is any real number. Therefore, a linear combination in a continuous space must be represented by an integral (in this case, the convolution integral), instead of a sum, as in a discrete space.

To gain a solid ground about this notion of signals as vectors of a continuous space, let us compare the convolution integral in Equation (2.27) with its analogue in a three-dimensional Euclidean space.

To begin with, think of the meaning of a function $g(t)$. What does $g(t)$ signify? It is an implicit denotation of infinitely many numbers: each value of t indexes a different number $g(t)$. So, we can think of $g(t)$ as an implicit denotation of a list of numbers. Furthermore, we have just seen that, in Equation (2.27), $g(t_0)$ is interpreted as a coordinate multiplying the specific basis function $\delta(t - t_0)$. For example, $g(3.85)$ is the coordinate multiplying the basis function $\delta(t - 3.85)$. Therefore, $g(t)$ is a list of numbers with a specific algebraic meaning: it is the list of coordinates of the signal with respect to the basis of impulse functions. Thus, by specifying $g(t)$ (for example, by specifying the plot of $g(t)$, like the red curve in Figure 2.10), we are specifying the signal (the vector) by giving its coordinates in the basis of impulse functions.

But, if $g(t)$ is a list of coordinates specifying a signal (vector) in a basis, it must be analogous to the column vector representation in the Euclidean space, like this:

$$g(t) \leftrightarrow \begin{pmatrix} g_x \\ g_y \\ g_z \end{pmatrix} \tag{2.29}$$

In Equation (2.29), the symbol $\leftrightarrow$ reads 'is analogous to'. We can make a closer connection with this analogy by 'discretizing' $g(t)$. This discretization allows the representation of $g(t)$ as a column vector with infinitely many slots. For the sake of clarity, let us assume a specific discretization step of 0.1 seconds. Thus, the representation of the discrete version of $g(t)$ is a column vector like this:

$$
\begin{pmatrix} \vdots \\ g(-0.2) \\ g(-0.1) \\ g(0) \\ g(0.1) \\ g(0.2) \\ \vdots \end{pmatrix} \leftrightarrow \begin{pmatrix} g_x \\ g_y \\ g_z \end{pmatrix} \tag{2.30}
$$

If we are working with a discrete space like the Euclidean space, then only a handful of coordinates specify a vector. For example, in the three-dimensional space, we need only three coordinates to specify a vector. Thus, it is not a big deal to write these three coordinates explicitly in a column, like it is done on the right-hand side of Equation (2.30). But, if we are working with a continuous space, we cannot write the column vector coordinates explicitly, because there are infinitely many coordinates. We cannot even write a section of it because there are always infinitely many coordinates between any two coordinates (since t is a continuous variable). And, even if we decide to discretize it to approximate the representation, as we have done in Equation (2.30), we still end up with a vector with infinitely many coordinates, because t is allowed to range from $-\infty$ to ∞. So, we just go 'oh bother', and write $g(t)$ as an implicit denotation of this collection of infinitely many coordinates specifying our signal. This is the spirit of Equation (2.29). So, do not forget that $g(t)$ is the list of coordinates with respect to the basis of impulse functions (which is akin to the canonical basis).

Let us spell out the role of t in the algebraic interpretation of signals. So, suppose once again that the red curve of Figure 2.10 is a graphical representation of $g(t)$. Say you come from left to right in the plot (towards increasing t), writing down a few values of $g(t)$. For example, you write down $g(-0.1)$, $g(0)$, $g(0.1)$, $g(0.2)$, and so on. This process is analogous to moving through the slots of the column representation of Equation (2.30), and writing down g_x, then g_y, and so on. Therefore, t can be interpreted as indexing the slot of the column. This is precisely what the analogy of Equation (2.30) is showing: $t = 0$ indexes the middle slot, $t = 0.1$ indexes the one below it, and so on like this:

$$
\begin{array}{rll}
 & \vdots & \\
slot & t = -0.2 & \rightarrow \\
slot & t = -0.1 & \rightarrow \\
slot & t = 0 & \rightarrow \\
slot & t = 0.1 & \rightarrow \\
slot & t = 0.2 & \rightarrow \\
 & \vdots &
\end{array}
\begin{pmatrix} \vdots \\ g(-0.2) \\ g(-0.1) \\ g(0) \\ g(0.1) \\ g(0.2) \\ \vdots \end{pmatrix} \tag{2.31}
$$

Thus, in the slot $t = 0$ goes the coordinate $g(0)$, in the slot $t = 0.1$ goes the coordinate $g(0.1)$, and so on.

We have seen that $g(t)$ in Equation (2.27) is analogous to the column representation of a vector. What about the convolution integral? What is its analogous expression in a discrete space? Let us begin with the elements involved in the convolution integral, that is, with the impulse functions. What is the analogue of the impulse function in the Euclidean space?

To answer this question, notice that the impulse function is just a particular function of time. So, we could ascribe a symbol for it, like $b(t) = \delta(t - t_0)$. Let me be more specific and choose a particular t_0 as an example, say $t_0 = 5$. So, in our example $b(t) = \delta(t - 5)$. This is just a function of time t, so it is still just a collection of coordinates, in the same spirit of $g(t)$. Thus, $b(t) = \delta(t - 5)$ must also be analogous to a column vector representation in the Euclidean space. But it is a special column because it is a representation of one of the functions that are part of the basis. How does the column representation of a vector that is part of the basis look like in Euclidean space? For example, say we have the canonical basis $\widehat{x}, \widehat{y}, \widehat{z}$. What is the column representation of, say, vector $\widehat{y}$, in this same basis? Well, you know it, it is:

$$\begin{pmatrix} 0 \\ 1 \\ 0 \end{pmatrix}$$

After all, $\widehat{y} = 0\widehat{x} + 1\widehat{y} + 0\widehat{z}$. The column vector representation of a basis vector in the same basis is always a column filled with 0, but having a single slot with the number 1 (mind you, this is not peculiar to the canonical basis: any vector from any basis that is represented in the same basis of which it is part has a column representation with a single slot with the number 1, and all other slots with 0; but if a basis vector is represented in another basis, this is no longer true – for example, in Section 1.3 (Vector Representation in Different Bases), the coordinates of vector $\widehat{x}$ with respect to the basis $\widehat{a}_1, \widehat{a}_2$ were a^*_{1x} and a^*_{2x}).

Now, let us compare the column vector representing a basis vector in the Euclidean space, like:

$$\begin{pmatrix} 0 \\ 1 \\ 0 \end{pmatrix}$$

with the 'column vector' of a single impulse function, like $b(t) = \delta(t - 5)$.

As shown by its graphical representation in Figure 2.11, the function $b(t) = \delta(t - 5)$ is zero everywhere, except at the instant $t = 5$, where it goes to infinity. Thus, the coordinate representation of $b(t) = \delta(t - 5)$ is akin to the column vector coordinate representation of a basis vector in Euclidean space (with respect to its own basis): whereas in the Euclidean space we have 0 in all slots, except for a single slot, that gets the number 1, the coordinate representation of $b(t) = \delta(t - 5)$ is zero everywhere, except at the slot $t = 5$, where it goes to infinity. But it has unit area, so it is the area that plays the role of the number 1. As we will see soon, the unit area is related to the orthonormalization of the basis – as opposed to only orthogonalization (just like, in a discrete space, the number 1 in the slot of the column guarantees orthonormalization – as opposed to only orthogonalization).

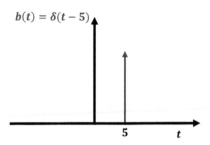

Figure 2.11 Graphical representation of the function $b(t) = \delta(t - 5)$.

Now we can make a full comparison between Equation (2.27) and its analogue in the Euclidean space. The integral in Equation (2.27) is a linear combination of 'column vectors' $\delta(t - t_0)$, each one located at a different position t_0, which is akin to a linear combination of the column vectors representing the basis vectors in the Euclidean space. Thus:

$$\int_{-\infty}^{+\infty} g(t_0)\delta(t - t_0)dt_0 \leftrightarrow g_x \begin{pmatrix} 1 \\ 0 \\ 0 \end{pmatrix} + g_y \begin{pmatrix} 0 \\ 1 \\ 0 \end{pmatrix} + g_z \begin{pmatrix} 0 \\ 0 \\ 1 \end{pmatrix} \tag{2.32}$$

Combining Equation (2.29) and Equation (2.32), we obtain the following analogy:

$$g(t) = \int_{-\infty}^{+\infty} g(t_0)\delta(t - t_0)dt_0 \leftrightarrow \begin{pmatrix} g_x \\ g_y \\ g_z \end{pmatrix} = g_x \begin{pmatrix} 1 \\ 0 \\ 0 \end{pmatrix} + g_y \begin{pmatrix} 0 \\ 1 \\ 0 \end{pmatrix} + g_z \begin{pmatrix} 0 \\ 0 \\ 1 \end{pmatrix} \tag{2.33}$$

In Section 1.3 (Vector Representation in Different Bases), we emphasized that the representation in column vector coordinates is a bit dangerous because we may forget that different columns can represent the same vector, and that we need to know the basis to know which vector the column is representing. Still in that same section, we distinguished between the column vector coordinates representation and the symbol denoting the vector itself, like $\vec{u}$, or, even better, $|u\rangle$. In the same spirit, we need to distinguish between the coordinate representation of a signal, like $g(t)$, and the signal itself. To represent the signal itself, we will use the Dirac notation; so, $|g\rangle$ denotes the signal itself, while $g(t)$ is the coordinate representation of $|g\rangle$ in the basis of impulse functions (in other words, $g(t)$ is the list of coordinates of $|g\rangle$). Often, instead of saying 'basis of impulse functions', we will just say 'the time domain'. Thus, we can say that $g(t)$ is the time domain representation of $|g\rangle$. But the time domain representation $g(t)$ of the signal $|g\rangle$ is not unique. Indeed, as in the Euclidean space, we can have other representations of the same signal $|g\rangle$, that is, we can have representations in other bases. In Section 2.8, we will see an alternative representation of paramount importance: the representation in the frequency domain.

We need a Dirac notation for the basis vectors (just like we have $|x\rangle$, $|y\rangle$, and $|z\rangle$ in the Euclidean space). Thus, a function $\delta(t - t_0)$ is a time domain coordinate representation of the ket $|\delta(t_0)\rangle$. Remember that each t_0 specifies a different impulse function, 'located' at a specific time $t = t_0$; so $|\delta(t_0)\rangle$ is a specific ket: by definition, it is the ket whose time domain representation is $\delta(t - t_0)$.

Furthermore, recall that the equation:

$$\begin{pmatrix} g_x \\ g_y \\ g_z \end{pmatrix} = g_x \begin{pmatrix} 1 \\ 0 \\ 0 \end{pmatrix} + g_y \begin{pmatrix} 0 \\ 1 \\ 0 \end{pmatrix} + g_z \begin{pmatrix} 0 \\ 0 \\ 1 \end{pmatrix} \tag{2.34}$$

is the column vector coordinate representation in the canonical basis of the equation:

$$|g\rangle = g_x|x\rangle + g_y|y\rangle + g_z|z\rangle \tag{2.35}$$

In the same spirit, the equation:

$$g(t) = \int_{-\infty}^{+\infty} g(t_0)\delta(t - t_0)dt_0 \tag{2.36}$$

is the time domain coordinate representation of the equation:

$$|g\rangle = \int_{-\infty}^{+\infty} g(t_0)|\delta(t_0)\rangle dt_0 \tag{2.37}$$

Thus, whereas Equation (2.36) is analogous to Equation (2.34), Equation (2.37) is analogous to Equation (2.35). These analogies are collected below:

$$g(t) = \int_{-\infty}^{+\infty} g(t_0)\delta(t - t_0)dt_0 \quad \longleftrightarrow \quad \begin{pmatrix} g_x \\ g_y \\ g_z \end{pmatrix} = g_x \begin{pmatrix} 1 \\ 0 \\ 0 \end{pmatrix} + g_y \begin{pmatrix} 0 \\ 1 \\ 0 \end{pmatrix} + g_z \begin{pmatrix} 0 \\ 0 \\ 1 \end{pmatrix}$$

$$|g\rangle = \int_{-\infty}^{+\infty} g(t_0)|\delta(t_0)\rangle dt_0 \quad \longleftrightarrow \quad |g\rangle = g_x|x\rangle + g_y|y\rangle + g_z|z\rangle \tag{2.38}$$

A few lines earlier we saw that changing t is analogous to changing the position of the slot in the column of coordinates (see Equation (2.31)). What about changing t_0? Compare Equation (2.37) with Equation (2.35). When we change t_0 in Equation (2.37), we are changing the basis ket $|\delta(t_0)\rangle$ (and, of course, its coordinate $g(t_0)$), so changing t_0 is akin to changing from $|x\rangle$ to $|y\rangle$, or from $|y\rangle$ to $|z\rangle$, and so on. Likewise, in the coordinates representation (Equation (2.36) and Equation (2.34)), changing t_0 is akin to changing the columns representing the basis vectors, that is, changing t_0 in $\delta(t - t_0)$ is like changing from $\begin{pmatrix} 1 \\ 0 \\ 0 \end{pmatrix}$ to $\begin{pmatrix} 0 \\ 1 \\ 0 \end{pmatrix}$ and so on.

To summarize: changing t is akin to changing the slot inside the same column (different coordinates of the same vector), whereas changing t_0 is akin to changing from one column to another column (different basis vectors).

2.6 The Scalar Product

Our next step in building the algebraic notions of signals is to obtain an expression for the scalar product between signals. We can do it formally by invoking the closure relation, but before we do that, let us ponder about how the scalar product between two signals $|u\rangle$ and $|v\rangle$ should look like. We already know that the scalar product in the two-dimensional Euclidean space is given by Equation (1.78), which I repeat below for convenience:

Equation (1.78)

$$\langle v|u \rangle = v_x^* u_x + v_y^* u_y$$

So, what is the recipe taught by Equation (1.78)? It is saying: take the coordinates that are in the first slot of the column vector representation (that is, u_x and v_x), conjugate the one associated with the bra (that is, v_x^*), and take the product between them; then do the same for the second slot; and then add them up.

Thus, a scalar product involves products between coordinates belonging to the same slot, but there is no product between coordinates of different slots (there are no x-coordinates multiplying y-coordinates). In the function space, the time t indexes the slot. Thus, if $u(t)$ and $v(t)$ are, respectively, the coordinates of the time domain representation of $|u\rangle$ and $|v\rangle$, then we should expect to have products between coordinates indexed by the same time (thus the same slot), that is, a product of the type $v^*(t)u(t)$. Notice that the same time t is involved in the two functions, so there is no mixing of slots (like there would be if we had, for example, $v^*(t)u(t+0.3)$). We also need to add up these products between the coordinates of all slots. Of course, since t is a continuous index, we have a continuum of slots, which means that the sum in Equation (1.78) must be turned into an integral involving all slots. By 'involving all slots' I mean that t must be allowed to vary from $-\infty$ to ∞. Thus, we should expect that:

$$\langle v|u \rangle = \int_{-\infty}^{+\infty} v^*(t)u(t)dt \tag{2.39}$$

Equation (2.39) is the analogue of Equation (1.78).

Now let us prove that this is indeed the case by invoking the closure relation.

In Section 1.8 (The Closure Relation), we saw that the basis vectors satisfy a closure relation. In Section 1.10 (The Dirac Notation), we saw that the closure relation of the canonical basis in a two-dimensional space is:

$$I = |x\rangle\langle x| + |y\rangle\langle y| \tag{2.40}$$

Equation (2.40) is like Equation (1.100) applied to the canonical basis.

We have also seen that the impulse functions form a basis of the function space. Thus, they must also satisfy the closure relation. The closure relation for the continuous space, however, cannot be a sum, like Equation (2.40); instead, it must be an integral ranging over the continuum of basis functions. Thus, the closure relation of the basis of impulse functions reads:

$$I = \int_{-\infty}^{+\infty} dt_0 |\delta(t_0)\rangle\langle\delta(t_0)| \tag{2.41}$$

where I is, of course, the identity operator. Notice that the recipe of Equation (2.41) is the same as in Equation (2.40).

With the help of the closure relation, we can easily obtain an expression for the scalar product. First, let us check what happens if we apply the closure relation to a general ket $|g\rangle$. If I is the identity operator, it follows immediately that:

$$|g\rangle = I|g\rangle \tag{2.42}$$

Next, we use Equation (2.41) into Equation (2.42):

$$|g\rangle = \int_{-\infty}^{+\infty} dt_0 |\delta(t_0)\rangle\langle\delta(t_0)|g\rangle \tag{2.43}$$

But what is $\langle\delta(t_0)|g\rangle$? It is the result of a scalar product, so it must be a scalar. Thus, for a fixed t_0, it must be a number. Which number? Well, compare Equation (2.43) with Equation (2.37), which I repeat below for convenience.
Equation (2.37)

$$|g\rangle = \int_{-\infty}^{+\infty} g(t_0)|\delta(t_0)\rangle dt_0$$

In both equations, $|g\rangle$ is expressed as a linear combination of the basis kets $|\delta(t_0)\rangle$. Thus, if we have the same $|g\rangle$ and the same basis, then we must have the same coordinates, which entails that:

$$\langle\delta(t_0)|g\rangle = g(t_0) \tag{2.44}$$

Equation (2.44) makes a lot of sense: if we take the scalar product between a vector and a basis vector, then the result is the coordinate with respect to that basis vector (like in Equation (1.34) of Euclidean space, or its equivalent in Dirac notation: $\langle x|u\rangle = u_x$).

Thus, Equation (2.43) recovers Equation (2.37). Indeed:

$$|g\rangle = I|g\rangle = \int_{-\infty}^{+\infty} dt_0 |\delta(t_0)\rangle\langle\delta(t_0)|g\rangle = \int_{-\infty}^{+\infty} dt_0 |\delta(t_0)\rangle g(t_0) = \int_{-\infty}^{+\infty} dt_0 g(t_0)|\delta(t_0)\rangle$$

Now let us evaluate $\langle v|u\rangle$ by plugging the identity operator before the ket u:

$$\langle v|u\rangle = \langle v|I|u\rangle = \langle v|\int_{-\infty}^{+\infty} dt_0 |\delta(t_0)\rangle\langle\delta(t_0)|u\rangle = \int_{-\infty}^{+\infty} dt_0 \langle v|\delta(t_0)\rangle\langle\delta(t_0)|u\rangle$$

According to Equation (2.44), $\langle \delta(t_0)|u\rangle = u(t_0)$. What about $\langle v|\delta(t_0)\rangle$? It is almost the same, but we need to use the property of Equation (1.79). Thus, $\langle v|\delta(t_0)\rangle = \langle \delta(t_0)|v\rangle^* = v^*(t_0)$. Therefore, we conclude that:

$$\langle v|u\rangle = \int_{-\infty}^{+\infty} dt_0 \langle v|\delta(t_0)\rangle\langle \delta(t_0)|u\rangle = \int_{-\infty}^{+\infty} v^*(t_0)u(t_0)dt_0 \qquad (2.45)$$

The last integral in Equation (2.45) is identical to the integral in Equation (2.39), only using a different dummy index.

2.7 Orthonormality of the Basis of Impulse Functions

A significant difference between a continuous and a discrete space is the notion of orthonormality. In a discrete space, a basis is said to be orthonormal, as opposed to only orthogonal, if its vectors are unit vectors. In other words, if the scalar product of a basis vector with itself is equal to one.

But, if we try to apply the same criteria of orthonormality to our basis impulse functions, we quickly run into trouble. To begin with, as you will prove in the exercise list, if we take the scalar product of an impulse function with itself, the result is an infinite number. That means the impulse function is not square-integrable, so it does not belong to the space!! This is the funniest bit of the function space: its basis functions may not belong to the space. But there is still a notion of orthonormality in the function space, and it is related to the unit 'area' of the impulse functions.

Fortunately, we can find a notion of orthonormality using our strategy of drawing analogies between the continuous and discrete spaces. The orthonormality analogy is based on the Kronecker delta, which can be understood as the discrete version of the impulse function. The Kronecker delta is a delta symbol with two subscripts. If the two subscripts are the same, then the Kronecker delta is equal to one. If they are different, then the Kronecker delta is equal to zero. Thus, the definition of the Kronecker delta is:

$$\delta_{mn} = 1 \text{ for } m = n$$

$$\delta_{mn} = 0 \text{ for } m \neq n \qquad (2.46)$$

The Kronecker delta provides a convenient way of expressing orthonormality in a discrete space. For example, in a two-dimensional Euclidean space, an expression of the orthonormality of the canonical basis requires three identities: $\langle x|x\rangle = 1$, $\langle y|y\rangle = 1$, and $\langle x|y\rangle = 0$. But we can group these three identities into a single identity using the Kronecker delta. Thus, by allowing m and n to stand for x and/or y, these three identities can be expressed as:

$$\langle m|n\rangle = \delta_{mn} \qquad (2.47)$$

For example, allowing $m = x$ and $n = x$ gives $\langle x|x \rangle = \delta_{xx}$. But δ_{xx} has the same symbols in the subscript so, according to Equation (2.46), $\delta_{xx} = 1$. Likewise, Equation (2.47) and Equation (2.46) imply that $\langle x|y \rangle = \delta_{xy} = 0$.

The orthonormality in the function space can be expressed in a way analogous to Equation (2.47), but instead of the Kronecker delta we use the Dirac delta (that is, an impulse function). Thus, for any basis (not only the basis of impulse functions), the criterion of orthonormality is:

$$\langle m|n \rangle = \delta(m - n) \tag{2.48}$$

In Equation (2.48), m and n are real numbers, as it is appropriate to a basis formed by a continuum of functions.

What is the message of Equation (2.48)? It is saying that, if we get two different functions, that is, if $m \neq n$, then the scalar product between them is zero, since the delta function is zero everywhere, except at the origin, and we only hit the origin if $m = n$. Thus, according to Equation (2.48), if $m \neq n$, then $\langle m|n \rangle = 0$. That is a statement of orthogonality.

But, if $m = n$, then the scalar product is not one, as in a discrete space; instead, the scalar product is the infinite number $\delta(0)$. Notice that it is not any infinite number; it is a special infinite number: it is the infinite number of the Dirac delta, that is, the peak of a rectangular function with unit area and vanishingly small width.

To gain further insight into the meaning of Equation (2.48), let us apply it to the basis of impulse functions. In this case, m and n stand for two possible 'locations' of the impulse functions. Thus, for this particular case, Equation (2.48) reads:

$$\langle \delta(t_1)|\delta(t_2) \rangle = \delta(t_1 - t_2) \tag{2.49}$$

Thus, what is the message of Equation (2.49)? It must be the same of Equation (2.48). First, Equation (2.49) is affirming that, if we take the scalar product between two different delta functions, that is, if $t_1 \neq t_2$, then the result is zero. That makes a bucketload of sense: we have just seen that the scalar product involves product of coordinates (see Equation (2.39)). But, if we take the product of one arrow 'located' at t_1 and another arrow 'located' at t_2, then this product must be zero, because there is no overlap between the arrows: where one is nonzero, the other is zero, and vice versa. But, if we take two arrows 'located' at exactly the same time, that is, if we take $t_1 = t_2$, then the two arrows overlap, and the scalar product is the infinite number $\delta(0)$ (see the worked exercise below for a rigorous proof of Equation (2.49)).

In short, Equation (2.49) is a statement of orthonormality of the basis of impulse functions. It is this expression that guarantees that the scalar product between a signal and a basis function results in the corresponding coordinate (which is the statement of Equation (2.44)), which is the proof of the orthonormality pudding (see Exercise 5 of Chapter 2).

Worked Exercise: Proof of Orthonormality of the Basis of Impulse Functions

Prove Equation (2.49).

Solution:

We need to evaluate $\langle \delta(t_1)|\delta(t_2)\rangle$. Using Equation (2.39), and the fact that the impulse functions are real (which means that $\delta^*(t - t_1) = \delta(t - t_1)$), we get:

$$\langle \delta(t_1)|\delta(t_2)\rangle = \int_{-\infty}^{+\infty} \delta(t - t_1)\delta(t - t_2)dt \tag{2.50}$$

Our goal is to prove that $\langle \delta(t_1)|\delta(t_2)\rangle = \delta(t_1 - t_2)$, which means that we need to prove that $\langle \delta(t_1)|\delta(t_2)\rangle$ does what an impulse function does; in other words, that it has the sifting property described by Equation (2.21), which I repeat below for convenience:
Equation (2.21)

$$g(t_0) = \int_{-\infty}^{+\infty} g(t)\delta(t - t_0)dt$$

To adapt Equation (2.21) to our problem, let us change the symbol t for t_1 and the symbol t_0 for t_2:
Still Equation (2.21)

$$g(t_2) = \int_{-\infty}^{+\infty} g(t_1)\delta(t_1 - t_2)dt_1$$

Thus, to prove that $\langle \delta(t_1)|\delta(t_2)\rangle = \delta(t_1 - t_2)$, we need to prove that:

$$g(t_2) = \int_{-\infty}^{+\infty} g(t_1)\langle \delta(t_1)|\delta(t_2)\rangle dt_1 \tag{2.51}$$

So, let us evaluate the integral in Equation (2.51) and see whether it results in $g(t_2)$. With the help of Equation (2.50):

$$\int_{-\infty}^{+\infty} g(t_1)\langle \delta(t_1)|\delta(t_2)\rangle dt_1 = \int_{-\infty}^{+\infty} g(t_1) \int_{-\infty}^{+\infty} \delta(t - t_1)\delta(t - t_2)dtdt_1$$

The integral over t_1 involves only $g(t_1)$ and the impulse function $\delta(t - t_1)$, so it is easy to evaluate it using the sifting property:

$$\int_{-\infty}^{+\infty} g(t_1)\delta(t - t_1)dt_1 = \int_{-\infty}^{+\infty} g(t_1)\delta(t_1 - t)dt_1 = g(t)$$

Thus, first integrating over t_1, and then over t, we get:

$$\int_{-\infty}^{+\infty} g(t_1)\langle\delta(t_1)|\delta(t_2)\rangle dt_1 = \int_{-\infty}^{+\infty}\int_{-\infty}^{+\infty} [g(t_1)\delta(t-t_1)dt_1]\delta(t-t_2)dt$$

$$= \int_{-\infty}^{+\infty} g(t)\delta(t-t_2)dt = g(t_2)$$

The last integral was also evaluated using the sifting property.

Thus, we have proved that $\int_{-\infty}^{+\infty} g(t_1)\langle\delta(t_1)|\delta(t_2)\rangle dt_1 = g(t_2)$, which entails that:

$$\langle\delta(t_1)|\delta(t_2)\rangle = \delta(t_1 - t_2)$$

Now we are ready to make the acquaintance of another basis of the function space.

2.8 Exponentials as a Basis: The Frequency Domain Representation of Signals

In this section, we will learn that the set of functions of the type $e^{i2\pi ft}$, where i is the unit imaginary number ($i^2 = -1$) and f is a real number, also forms a basis of the function space.

From now on, we will be dealing intensely with these exponentials. If you are not acquainted with exponentials with imaginary numbers, no need to worry, they are just complex numbers, of the form:

$$e^{i\theta} = cos(\theta) + i \cdot sin(\theta) \tag{2.52}$$

Equation (2.52) is known as Euler's formula. Notice that it entails that the magnitude of $e^{i\theta}$ is one:

$$|e^{i\theta}| = \sqrt{(e^{i\theta})(e^{i\theta})^*} = \sqrt{e^{i\theta}e^{-i\theta}} = \sqrt{e^{i(\theta-\theta)}} = \sqrt{e^{i0}} = 1 \tag{2.53}$$

Or, if you prefer:
$$|e^{i\theta}| = \sqrt{cos^2(\theta) + sin^2(\theta)} = 1 \tag{2.54}$$

So, according to Euler's formula, $e^{i2\pi ft} = cos\,(2\pi ft) + i \cdot sin\,(2\pi ft)$, from which we immediately infer that the parameter f is a frequency (and, if the unit of t is seconds, then the unit of f must be Hertz).

In this section, we need to evaluate the following (and quite useful) integral involving exponentials:

$$\int_{-\infty}^{+\infty} e^{i2\pi ft}df =?$$

Notice that the time t is a constant inside the integral. But, of course, the result of the integral depends on which t is chosen. Thus, the result must be a function of time t. Let us call this function $x(t)$:

$$\int_{-\infty}^{+\infty} e^{i2\pi ft} df = x(t) \tag{2.55}$$

Our main job in this section is to figure out which function $x(t)$ is. As soon as we find it, it will be clear that the exponentials form a basis.

Let us begin by gaining an intuition about what to expect. Recall that $e^{i2\pi ft} = cos(2\pi ft) + i \cdot sin(2\pi ft)$. Therefore, Equation (2.55) is a sum over cosines and sines with all possible frequencies (by 'all possible frequencies' I mean that the integral ranges from $f = -\infty$ to $f = \infty$). Thus, using Euler's formula:

$$x(t) = \int_{-\infty}^{+\infty} e^{i2\pi ft} df = \int_{-\infty}^{+\infty} cos(2\pi ft) df + i \int_{-\infty}^{+\infty} sin(2\pi ft) df \tag{2.56}$$

To help visualize what to expect from the integrals of Equation (2.56), a plot of three cosines with three different frequencies and another plot of three sines with three different frequencies are shown in Figure 2.12. The integrals of Equation (2.56) are sums over cosines and sines with all possible frequencies. Now, pick any specific time $t = t_1$ and focus your attention on it (but it cannot be the origin $t = 0$). Now imagine that the cosine of a certain frequency gives you a certain value at $t = t_1$; for example, the value 0.6. Surely, there must be another frequency giving you the value -0.6 at this same time $t = t_1$. To see that this must be true, imagine that the cosines and sines are like springs attached to the vertical axis (that is, attached to the axis crossing the origin at $t = 0$), and that you can stretch or compress these springs. Stretching is tantamount to increasing the period (so reducing the frequency), and compressing is tantamount to reducing the period (so increasing the frequency). Thus, all you need to do to get -0.6 is to stretch or compress a spring until the value -0.6 lands at the time $t = t_1$. Thus, you have found a frequency that gives you a contribution of -0.6 at $t = t_1$. Of course, when you add up the contribution of the frequency giving you 0.6 with the contribution of the frequency giving you -0.6, they cancel each other. And this must be true for any frequency: for any frequency you pick, there will be one cancelling its contribution at $t = t_1$. Consequently, $x(t_1)$ must be zero. And this must be true for all time t, unless there is any time t that is somehow special.

But is there any special time t? Look again at Figure 2.12: at the origin $(t = 0)$ all cosines have their peaks aligned! Thus, when $t = 0$ is chosen, each one of these cosines contributes with a number 1 in the integral. Since there are infinitely many cosines (because there are infinitely many frequencies), and each one of these cosines contributes with a unit amplitude to the integral evaluated at $t = 0$, so the result of the integral must be an infinite number.

Notice that the integral involving the sines evaluated at $t = 0$ is zero, because the sines are all zero at the origin.

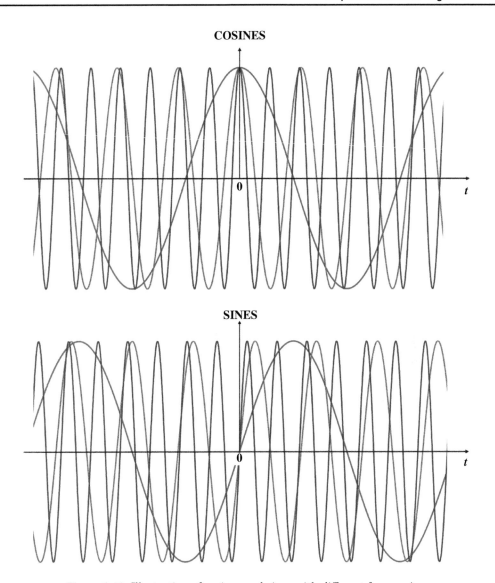

Figure 2.12 Illustration of cosines and sines with different frequencies.

Thus, we have concluded that we should expect $x(t)$ to be zero everywhere, except at the origin, where it should be an infinite number.

Now let us check if that is indeed the case. We begin by evaluating $x(t)$ at the origin. From Equation (2.56):

$$x(0) = \int_{-\infty}^{+\infty} cos(0)df + i \int_{-\infty}^{+\infty} sin(0)df = \int_{-\infty}^{+\infty} 1df + i \int_{-\infty}^{+\infty} 0df = \int_{-\infty}^{+\infty} 1df = \infty$$

(2.57)

As we already expected.

Now we need to evaluate $x(t)$ for $t \neq 0$. That is a bit tricky, because if we just naively solve the integral, like this $\int_{-\infty}^{+\infty} e^{i2\pi ft}df = (e^{i2\pi ft}/i2\pi t)\Big|_{-\infty}^{+\infty}$, we end up with nonsense: what on earth is the value of $e^{i2\pi ft}$ when $f = \pm\infty$? This value is undefined, because $e^{i2\pi ft}$ is a periodic function. So, we don the artful dodger hat and, instead of solving the integral directly, we solve another integral, which we know beforehand that will converge, and then check the limit of this other integral that approaches the one we want. Let me show you: first, we split it up into two parts:

$$x(t) = \int_{-\infty}^{+\infty} e^{i2\pi ft}df = \int_{-\infty}^{0} e^{i2\pi ft}df + \int_{0}^{+\infty} e^{i2\pi ft}df \qquad (2.58)$$

Then we solve each part independently, but to ensure convergence, we replace $e^{i2\pi ft}$ with $e^{\pm\eta f}e^{i2\pi ft}$, where η is a real and positive number, and then check the limit of the integrals when $\eta \to 0$.

As the domain of the first integral extends from $-\infty$, we need to pick $+\eta$ to ensure convergence. In the second, the domain goes to $+\infty$, which requires $-\eta$. Thus:

$$x(t) = \lim_{\eta \to 0} \left[\int_{-\infty}^{0} e^{+\eta f}e^{i2\pi ft}df + \int_{0}^{+\infty} e^{-\eta f}e^{i2\pi ft}df \right] \qquad (2.59)$$

Solving the integrals:

$$x(t) = \lim_{\eta \to 0} \left[\frac{e^{+\eta f}e^{i2\pi ft}}{\eta + i2\pi t}\bigg|_{-\infty}^{0} + \frac{e^{-\eta f}e^{i2\pi ft}}{-\eta + i2\pi t}\bigg|_{0}^{\infty} \right] = \lim_{\eta \to 0} \left[\frac{1}{\eta + i2\pi t} - \frac{1}{-\eta + i2\pi t} \right] \qquad (2.60)$$

Notice that the only difference between the two terms in the last equality of Equation (2.60) is that, in one case, the denominator involves $+\eta$, and in the other it involves $-\eta$. So, when $\eta \to 0$ the two terms approach each other. Consequently, $x(t) \to 0$ when $\eta \to 0$. Thus, we have concluded that $x(t) = 0$ when $t \neq 0$.

Collecting the two pieces of information we obtained so far:

$$x(t) = 0, for\, t \neq 0$$

$$x(0) = \infty \qquad (2.61)$$

This function $x(t)$ smells of impulse function, does it not? It is zero everywhere, except at the origin, where it goes to infinity. But, alas, smells can be deceitful! Have not we all had the surreal experience of walking into an elevator and encountering an organic compound smelling of vanilla, which compound turned out to be a person and not vanilla at all? So, we need to be wary of smells and check whether $x(t)$ is indeed an impulse function.

Look at Equation (2.18), which summarizes the properties of the impulse function, and which I repeat below for convenience:

Equation (2.18)

$$\delta(t) = 0, for\, t \neq 0$$

$$\delta(0) = \infty$$

$$\int_{-\infty}^{\infty} \delta(t)dt = 1$$

We have just seen that $x(t)$ satisfies the first two properties (see Equation (2.61)). Thus, we need to check whether it satisfies the third, that is, if it has unit 'area'. If it does, then $x(t)$ not only smells of impulse function, but it is indeed the impulse function.

So, let us evaluate its area:

$$\int_{-\infty}^{\infty} x(t)dt = ?$$

We already know that $x(t)$ is zero everywhere, except at the origin. So, the integral above can be evaluated over a finite region around the origin. We pick a real and positive number ε, and define it as the 'distance' from the origin within which we will evaluate the integral. Thus:

$$\int_{-\infty}^{\infty} x(t)dt = \int_{-\varepsilon}^{\varepsilon} x(t)dt \tag{2.62}$$

Notice that in Equation (2.62) we are taking advantage of $x(t)$ being zero outside the origin to replace the pesky infinite numbers with finite numbers. Now we plug in the definition of $x(t)$ (Equation (2.55)):

$$\int_{-\infty}^{\infty} x(t)dt = \int_{-\varepsilon}^{\varepsilon} x(t)dt = \int_{-\varepsilon}^{\varepsilon} \int_{-\infty}^{+\infty} e^{i2\pi ft} df dt \tag{2.63}$$

Still in the spirit of avoiding infinites like the plague, we first solve the integral over time:

$$\int_{-\infty}^{\infty} x(t)dt = \int_{-\infty}^{\infty} \left[\int_{-\varepsilon}^{+\varepsilon} e^{i2\pi ft} dt \right] df = \int_{-\infty}^{\infty} \left[\frac{e^{i2\pi ft}}{i2\pi f} \Big|_{-\varepsilon}^{+\varepsilon} \right] df$$

Therefore:

$$\int_{-\infty}^{\infty} x(t)dt = \int_{-\infty}^{\infty} \left[\frac{e^{i2\pi f\varepsilon} - e^{-i2\pi f\varepsilon}}{i2\pi f} \right] df \tag{2.64}$$

Euler's formula (Equation (2.52)) entails that:

$$\frac{e^{i\theta} - e^{-i\theta}}{2i} = sin(\theta) \tag{2.65}$$

Using the identity of Equation (2.65) in Equation (2.64), we get:

$$\int_{-\infty}^{\infty} x(t)dt = \int_{-\infty}^{\infty} \frac{sin(2\pi f\varepsilon)}{\pi f} df \tag{2.66}$$

We are almost there: we just need to make a change of variables to cast Equation (2.66) into a form that will allow us to borrow a well-known mathematical result. Define:

$$\alpha = 2\pi\varepsilon f \tag{2.67}$$

And recast Equation (2.66) in terms of α:

$$\int_{-\infty}^{\infty} x(t)dt = \int_{-\infty}^{\infty} \frac{sin(\alpha)}{\frac{\alpha}{2\varepsilon}} \frac{d\alpha}{2\pi\varepsilon} = \frac{1}{\pi}\int_{-\infty}^{\infty} \frac{sin(\alpha)}{\alpha}d\alpha \tag{2.68}$$

The last integral in Equation (2.68) is known as 'Dirichlet integral'. A discussion on how to evaluate it would take us too far afield, and there is nothing in it particularly pertinent to our purposes (more than one strategy to evaluate it can be found in G. Barton's 'Elements of Green's function and Propagation'). So, I will just quote the result. Interestingly, it is the number π:

$$\int_{-\infty}^{\infty} \frac{sin(\alpha)}{\alpha}d\alpha = \pi \tag{2.69}$$

With the help of Equation (2.69), we have concluded that:

$$\int_{-\infty}^{\infty} x(t)dt = \frac{\pi}{\pi} = 1 \tag{2.70}$$

There you go: this time the smell was not a trickster: $x(t)$ ticks all the three boxes in the list of properties of the impulse function (Equation (2.18)), which means that $x(t) = \delta(t)$. Thus, from Equation (2.55) and the identity $x(t) = \delta(t)$, we have concluded that:

$$\int_{-\infty}^{+\infty} e^{i2\pi ft}df = \delta(t) \tag{2.71}$$

This beautiful result is tremendously important, because it is saying that we can build up an impulse function by adding up all functions of the type $e^{i2\pi ft}$ (each value of f specifies a different function $e^{i2\pi ft}$). And if we can do that for one impulse function, then we can do that for all impulse functions. Indeed, Equation (2.71) immediately entails that:

$$\int_{-\infty}^{+\infty} e^{i2\pi f(t-t_0)}df = \delta(t - t_0) \tag{2.72}$$

According to Equation (2.72), we can construct any impulse $\delta(t - t_0)$ from a linear combination of all functions of the type $e^{i2\pi ft}$, each one of these multiplying a coordinate $e^{-i2\pi ft_0}$ (recall that $e^{i2\pi f(t-t_0)} = e^{-i2\pi ft_0}e^{i2\pi ft}$).

If we can build any impulse function using linear combinations of functions $e^{i2\pi ft}$, and if the impulse functions form a basis, then the set of all functions $e^{i2\pi ft}$ must also form a basis (recall that this is a set of functions because each value of f specifies a different function). In Section 2.9, we will find how to convert a signal representation in the basis of impulse function into a representation in this new basis of exponential functions.

The theory we are developing in this textbook is strongly dependent on this new basis functions $e^{i2\pi ft}$. So, let us already ascribe kets for them. Say you pick a certain frequency f_0, where the subscript 0 is a reminder that f_0 is just a number. So, $e^{i2\pi f_0 t}$ is a function of time, and as such it is the time domain coordinate representation of a ket, which we denote by $|e(f_0)\rangle$. Since, by definition, $e^{i2\pi f_0 t}$ is the list of coordinates of $|e(f_0)\rangle$ with respect to the basis of impulse functions, it follows that:

$$\langle \delta(t_0)|e(f_0)\rangle = e^{i2\pi f_0 t_0} \tag{2.73}$$

Equation (2.73) is a particular application of Equation (2.44).

We have just seen that Equation (2.71) entails that the functions $e^{i2\pi ft}$ form a basis. But the same equation also entails that this basis is orthonormal. To see that this must be true, we swap the roles of time t and frequency f in Equation (2.71), thus resulting in:

$$\int_{-\infty}^{+\infty} e^{i2\pi ft} dt = \delta(f) \tag{2.74}$$

Notice that Equation (2.74) entails that:

$$\int_{-\infty}^{+\infty} e^{i2\pi (f_1 - f_2)t} dt = \delta(f_1 - f_2) \tag{2.75}$$

Recall that, to prove that our new basis is orthonormal, we need to prove that it satisfies Equation (2.48), which I repeat below for convenience:

Equation (2.48)

$$\langle m|n\rangle = \delta(m - n)$$

In our new basis, Equation (2.48) reads:

Equation (2.48) applied to the new basis

$$\langle e(f_2)|e(f_1)\rangle = \delta(f_2 - f_1)$$

To prove that our new basis satisfies Equation (2.48), we need to evaluate the scalar product between two basis kets $|e(f_1)\rangle$ and $|e(f_2)\rangle$. Thus, according to Equation (2.39):

$$\langle e(f_2)|e(f_1)\rangle = \int_{-\infty}^{+\infty} (e^{i2\pi f_2 t})^* e^{i2\pi f_1 t} dt \tag{2.76}$$

But $(e^{i2\pi f_2 t})^* = e^{-i2\pi f_2 t}$. Therefore:

$$\langle e(f_2)|e(f_1)\rangle = \int_{-\infty}^{+\infty} e^{i2\pi (f_1 - f_2)t} dt = \delta(f_1 - f_2) \tag{2.77}$$

where Equation (2.75) was used.

Since $\delta(f_1 - f_2) = \delta(f_2 - f_1)$, we have proven that:

$$\langle e(f_2)|e(f_1)\rangle = \delta(f_2 - f_1) \tag{2.78}$$

which concludes the proof that our new basis is indeed orthonormal.

So, we have a new orthonormal basis: the basis of exponentials. The representation of signals in this new basis is called 'the frequency domain' representation. In Section 2.9, we will learn how to change from the time domain representation (basis of impulse functions) to the frequency domain representation of signals (basis of exponentials), and vice versa. Such a change is carried out by the famous Fourier transform.

2.9   The Fourier Transform

The Fourier transforms are a pair of equations mapping the representation of a signal in the basis of impulse functions, into its representation in the new basis of exponential functions, and vice versa. Fourier transforms are among the most useful mathematical tools in engineering and physics, so we will dedicate the remainder of this chapter to make sure we understand them well.

We begin with the expression of the time domain representation of a signal, that is, the representation in terms of a linear combination of impulse functions (Equation (2.24), which I repeat below for convenience):

Equation (2.24)

$$g(t) = \int_{-\infty}^{+\infty} g(t_0)\delta(t - t_0)dt_0$$

Now, we use Equation (2.72) in Equation (2.24):

$$g(t) = \int_{-\infty}^{+\infty} g(t_0)\delta(t - t_0)dt_0 = \int_{-\infty}^{+\infty} g(t_0) \left[\int_{-\infty}^{+\infty} e^{i2\pi f(t-t_0)}df \right] dt_0$$

Rearranging the order of the integrals:

$$g(t) = \int_{-\infty}^{+\infty} \int_{-\infty}^{+\infty} g(t_0)e^{i2\pi f(t-t_0)}dt_0 df$$

We want to evaluate the integral over t_0, so let us gather the terms that depend on it:

$$g(t) = \int_{-\infty}^{+\infty} e^{i2\pi ft} \left[\int_{-\infty}^{+\infty} g(t_0)e^{-i2\pi ft_0}dt_0 \right] df \qquad (2.79)$$

Notice that f is constant within the integral over t_0. But the result of the integral must depend on the value of f, so the result of the integral is a function of f. Let us call this function $G(f)$. So, by definition:

$$G(f) = \int_{-\infty}^{+\infty} g(t_0)e^{-i2\pi ft_0}dt_0 \qquad (2.80)$$

Notice that t_0 is a dummy index, so we can use any symbol we want in its place. With the cheekiness of one who already knows the result, I choose the dummy index

t to write Equation (2.80):

$$G(f) = \int_{-\infty}^{+\infty} g(t)e^{-i2\pi ft}dt \tag{2.81}$$

I emphasize, once again, that changing dummy indexes does not change anything, so Equation (2.81) and Equation (2.80) are identical.

With this definition of $G(f)$, we can recast Equation (2.79) in a cleaner form. Thus, substituting Equation (2.80) into Equation (2.79), we find:

$$g(t) = \int_{-\infty}^{+\infty} G(f)e^{i2\pi ft}df \tag{2.82}$$

Equation (2.82) and Equation (2.81) form a Fourier transform pair. Equation (2.81) is the direct Fourier transform, while Equation (2.82) is the inverse Fourier transform. We say that $G(f)$ is the Fourier transform of $g(t)$. Notice that the roles played by time and frequency in the Fourier transforms are symmetrical. Indeed, by swapping t for f, we can go back and forth between the direct and inverse transforms, with the only difference that the inverse transform has a negative sign in the argument of the exponential. By the way, it does not matter which transform has the negative sign in the argument of the exponential: we could have put it in the direct transform instead of putting it in the inverse. What matters is that the arguments in the two transforms have opposite signs: if one has a positive sign, then the other necessarily must be negative.

According to the inverse Fourier transform (Equation (2.82)), $g(t)$ can be expressed as a linear combination of exponentials, each multiplied by $G(f)$. In other words, $G(f)$ are the coordinates with respect to the basis of exponentials. Thus, $g(t)$ and $G(f)$ are coordinates with respect to two different bases, and the Fourier pair teaches how to go from one representation to the other: the direct Fourier transform (Equation (2.81)) teaches how to obtain $G(f)$ from $g(t)$ and the inverse Fourier transform (Equation (2.82)) teaches how to obtain $g(t)$ from $G(f)$.

Fourier transforms are so important in engineering and physics that they deserve a frame:

FOURIER TRANSFORMS
DIRECT FOURIER TRANSFORM

$$G(f) = \int_{-\infty}^{+\infty} g(t)e^{-i2\pi ft}dt$$

INVERSE FOURIER TRANSFORM

$$g(t) = \int_{-\infty}^{+\infty} G(f)e^{i2\pi ft}df \tag{2.83}$$

We will spend the rest of this chapter looking at the meaning of Fourier transforms from different points of view. We already know that it is just a change of representation from the basis of impulse functions to the new basis of exponentials, but in Section 2.10 we will further explore its algebraic meaning. Then we will discuss its 'physical' meaning in Section 2.11, and finally we will see a bunch of its properties, which are extremely useful, in Section 2.12.

Worked Exercise: The Fourier Transform of the Rectangular Function

Find the Fourier transform of the rectangular function.

Solution:

We defined the rectangular function in Equation (2.7), which I repeat below for convenience:
Equation (2.7)

$$rect(t) = 1 \quad for - c \le t \le c$$

$$rect(t) = 0 \; oustide \; the \; interval - c \le t \le c$$

Denoting its Fourier transform as $S(f)$, we get:

$$S(f) = \int_{-\infty}^{+\infty} rect(t) e^{-i2\pi ft} dt = \int_{-c}^{+c} e^{-i2\pi ft} dt = -\left. \frac{e^{-i2\pi ft}}{i2\pi f} \right|_{-c}^{+c}$$

Thus:

$$S(f) = \frac{e^{i2\pi fc} - e^{-i2\pi fc}}{i2\pi f} = \frac{sin(2\pi fc)}{\pi f} \tag{2.84}$$

The last equality above uses Equation (2.65).
We want to express Equation (2.84) in terms of a *sinc* function, which is defined as:

$$sinc(x) = \frac{sin(x)}{x} \tag{2.85}$$

By the way, we have already encountered the *sinc* function in Equation (2.68). Rewriting Equation (2.84) in terms of the *sinc* function:

$$S(f) = \frac{2c}{2c} \frac{sin(2\pi fc)}{\pi f} = 2c \cdot sinc(2\pi fc) \tag{2.86}$$

We will come back to this Fourier transform in Section 2.11 (The Physical Meaning of Fourier Transform), where we will also see some plots of the *sinc* function.

2.10   The Algebraic Meaning of Fourier Transforms

The inverse Fourier transform (Equation (2.82)) is an expression of $g(t)$ as a linear combination of exponential functions. Each one of these functions is multiplied by a different number $G(f)$, so $G(f)$ must be the coordinates of the signal with respect to this new basis of exponentials. Thus, $G(f)$ is the representation of the signal $|g\rangle$ in the new basis of exponentials: it is the frequency domain representation.

To gain insight into the algebraic meaning of Fourier transforms, we use our strategy of finding analogues in the Euclidean space. We begin with the analogue of the inverse Fourier transform (Equation (2.82)).

We have seen that the basis of impulse functions is analogous to the canonical basis; likewise, the 'new' basis of exponentials is analogous to the 'new' basis that we had denoted by $\hat{a}_1$ and $\hat{a}_2$ in Section 1.3 (Vector Representation in Different Bases). I will denote them using the Dirac notation, and I will again assume a space of three dimensions. Thus, we denote the kets of the 'new' basis in Euclidean space by $|a_1\rangle$, $|a_2\rangle$, and $|a_3\rangle$ and we express these kets in terms of linear combinations of the canonical basis vectors:

$$|a_1\rangle = a_{1x}|x\rangle + a_{1y}|y\rangle + a_{1z}|z\rangle$$

$$|a_2\rangle = a_{2x}|x\rangle + a_{2y}|y\rangle + a_{2z}|z\rangle$$

$$|a_3\rangle = a_{3x}|x\rangle + a_{3y}|y\rangle + a_{3z}|z\rangle \qquad (2.87)$$

The inverse Fourier transform (Equation (2.82)) is an equation involving the time domain coordinates $g(t)$ on the left-hand side, and a linear combination of time domain coordinates $e^{i2\pi ft}$ on the right-hand side. Thus, on the one hand, the inverse Fourier transform is an equation involving time domain coordinates. On the other hand, as we have seen in Section 2.5 (Impulse Functions as a Basis: The Time Domain Representation of Signals), the time domain basis, that is, the basis of impulse functions, is analogous to the canonical basis in the Euclidean space. Consequently, an equation involving time domain coordinates in the function space is analogous to an equation involving canonical coordinates in the Euclidean space. Thus, the inverse Fourier transform (Equation (2.82)) must be analogous to a column vector equation in the canonical basis.

The analogue of the left-hand side of Equation (2.82), $g(t)$, was given in Equation (2.29), which I repeat below for convenience:

Equation (2.29)

$$g(t) \leftrightarrow \begin{pmatrix} g_x \\ g_y \\ g_z \end{pmatrix}$$

We still need to find the analogue of the right-hand side of Equation (2.82). Recall that $e^{i2\pi f_0 t}$ are the time domain coordinates of $|e(f_0)\rangle$, and that $|e(f_0)\rangle$ is a ket of the 'new' basis. In Euclidean space, we are denoting the 'new' basis kets as $|a_1\rangle$,

$|a_2\rangle$, and $|a_3\rangle$. So, $|e(f_0)\rangle$ must be analogous to either $|a_1\rangle$, $|a_2\rangle$, or $|a_3\rangle$. Moreover, since the functions $e^{i2\pi f_0 t}$ are the coordinates of a 'new' basis ket with respect to the basis of impulse functions, they must be analogous to the coordinates of either $|a_1\rangle$, $|a_2\rangle$, or $|a_3\rangle$ with respect to the canonical basis. Therefore, the coordinates $e^{i2\pi f t}$, without specifying the frequency, are analogous to the column representation of a 'new' basis vector, without specifying the basis vector. Thus:

$$e^{i2\pi f t} \leftrightarrow \begin{pmatrix} a_{nx} \\ a_{ny} \\ a_{nz} \end{pmatrix} \tag{2.88}$$

where n stands for 1, 2, or 3, and the column vector coordinates of the 'new basis' functions were defined in Equation (2.87). Notice that, as each f indexes a different basis function, changing f is akin to changing the index n form 1 to 2 or 3 in Equation (2.88). Likewise, changing f_0 in $|e(f_0)\rangle$ changes the 'new' basis ket, so it is analogous to changing from $|a_1\rangle$ to $|a_2\rangle$ and so on.

The integral in Equation (2.82) is over f, which means that the integral is a linear combination of all 'new' basis functions (each f indexing a different exponential, that is, a different basis function). Thus, its analogue must be a linear combination of the column vector representation of the new basis vectors in the canonical basis:

$$\int_{-\infty}^{+\infty} G(f)e^{i2\pi f t}df \leftrightarrow G_1 \begin{pmatrix} a_{1x} \\ a_{1y} \\ a_{1z} \end{pmatrix} + G_2 \begin{pmatrix} a_{2x} \\ a_{2y} \\ a_{2z} \end{pmatrix} + G_3 \begin{pmatrix} a_{3x} \\ a_{3y} \\ a_{3z} \end{pmatrix} \tag{2.89}$$

Combining Equation (2.29) with Equation (2.89), we obtain the column vector analogue of the inverse Fourier transform:

$$g(t) = \int_{-\infty}^{+\infty} G(f)e^{i2\pi f t}df \quad \leftrightarrow \quad \begin{pmatrix} g_x \\ g_y \\ g_z \end{pmatrix} = G_1 \begin{pmatrix} a_{1x} \\ a_{1y} \\ a_{1z} \end{pmatrix} + G_2 \begin{pmatrix} a_{2x} \\ a_{2y} \\ a_{2z} \end{pmatrix} + G_3 \begin{pmatrix} a_{3x} \\ a_{3y} \\ a_{3z} \end{pmatrix} \tag{2.90}$$

I emphasize, one more time, that as the inverse Fourier transform is a time domain equation (it is a linear combination of the functions of time $e^{i2\pi f t}$), so its analogue is a linear combination of column vectors in the canonical basis.

Recall that we have already seen a column vector equation of the same form of Equation (2.90) in Equation (1.70). Furthermore, notice that fixing the time in the integral in Equation (2.90) is akin to fixing the vector slot in the sum in Equation (2.90). Thus, for example, computing the integral for a specific time is akin to computing the sum for a specific column slot, for example, $g_x = G_1 a_{1x} + G_2 a_{2x} + G_3 a_{3x}$. Likewise, computing the integral for another time is akin to computing the sum for another column slot, like $g_y = G_1 a_{1y} + G_2 a_{2y} + G_3 a_{3y}$.

In the spirit of distinguishing between a signal and its representation in a particular basis, we can express the inverse Fourier transform using the Dirac notation.

We follow the same approach we followed for the time domain. First, recall that the column vector equation:

$$\begin{pmatrix} g_x \\ g_y \\ g_z \end{pmatrix} = G_1 \begin{pmatrix} a_{1x} \\ a_{1y} \\ a_{1z} \end{pmatrix} + G_2 \begin{pmatrix} a_{2x} \\ a_{2y} \\ a_{2z} \end{pmatrix} + G_3 \begin{pmatrix} a_{3x} \\ a_{3y} \\ a_{3z} \end{pmatrix} \tag{2.91}$$

Is the column vector coordinate representation in the canonical basis of the equation:

$$|g\rangle = G_1 |a_1\rangle + G_2 |a_2\rangle + G_3 |a_3\rangle \tag{2.92}$$

Likewise, the inverse Fourier transform:
Equation (2.82)

$$g(t) = \int_{-\infty}^{+\infty} G(f) e^{i2\pi ft} df$$

is the time domain representation of the equation:

$$|g\rangle = \int_{-\infty}^{+\infty} G(f_0) |e(f_0)\rangle df_0 \tag{2.93}$$

Combining these analogies together:

$$g(t) = \int_{-\infty}^{+\infty} G(f_0) e^{i2\pi f_0 t} df_0 \quad \leftrightarrow \quad \begin{pmatrix} g_x \\ g_y \\ g_z \end{pmatrix} = G_1 \begin{pmatrix} a_{1x} \\ a_{1y} \\ a_{1z} \end{pmatrix} + G_2 \begin{pmatrix} a_{2x} \\ a_{2y} \\ a_{2z} \end{pmatrix} + G_3 \begin{pmatrix} a_{3x} \\ a_{3y} \\ a_{3z} \end{pmatrix}$$

$$|g\rangle = \int_{-\infty}^{+\infty} G(f_0) |e(f_0)\rangle df_0 \quad \leftrightarrow \quad |g\rangle = G_1 |a_1\rangle + G_2 |a_2\rangle + G_3 |a_3\rangle \tag{2.94}$$

Notice that, to make it easier to connect the coordinate representation with the equation in the Dirac notation, I used the dummy index f_0 in the integrals of Equation (2.93) and Equation (2.94).

It is instructive to compare Equation (2.94) with Equation (2.38), which I repeat below for convenience. Compare them closely, keeping in mind that they are different representations of the same signal $|g\rangle$.

Equation (2.38)

$$g(t) = \int_{-\infty}^{+\infty} g(t_0) \delta(t - t_0) dt_0 \quad \leftrightarrow \quad \begin{pmatrix} g_x \\ g_y \\ g_z \end{pmatrix} = g_x \begin{pmatrix} 1 \\ 0 \\ 0 \end{pmatrix} + g_y \begin{pmatrix} 0 \\ 1 \\ 0 \end{pmatrix} + g_z \begin{pmatrix} 0 \\ 0 \\ 1 \end{pmatrix}$$

$$|g\rangle = \int_{-\infty}^{+\infty} g(t_0) |\delta(t_0)\rangle dt_0 \quad \leftrightarrow \quad |g\rangle = g_x |x\rangle + g_y |y\rangle + g_z |z\rangle$$

It is important to always keep in mind that a Fourier transform is a change of representation of the same signal. It is NOT a change of signals. A Fourier transform does NOT transform one signal into another signal. So $g(t)$ and $G(f)$ are two different representations of the very same signal $|g\rangle$. That means one can specify what the signal $|g\rangle$ is by specifying either $g(t)$ or $G(f)$. But, once the signal $|g\rangle$ has been specified in a basis, its coordinates in the other basis have been automatically specified as well. In other words, if I give you $G(f)$, then $g(t)$ is automatically given: the latter is the function obtained with the inverse Fourier transform (Equation (2.82)). Likewise, if I give you $g(t)$, then $G(f)$ is automatically given: the latter is the function obtained with the direct Fourier transform (Equation (2.81)).

To summarize: $g(t)$ and $G(f)$ are, respectively, the time domain and the frequency domain representations of the signal $|g\rangle$.

We have just seen that the inverse Fourier transform is a linear combination of 'new' basis vectors expressed in the time domain coordinates (Equation (2.90)). What about the direct Fourier transform? What is its algebraic meaning?

This question is quite easy to answer: it is a scalar product. To see that it is a scalar product, recall that $G(f)$ are the coordinates of the ket $|g\rangle$ in the basis of kets $|e(f)\rangle$ (see Equation (2.93)). Since we have already proved that this is an orthonormal basis (Equation (2.78)), it follows that a particular coordinate $G(f_1)$ can be 'selected' by the scalar product $\langle e(f_1)|g\rangle$:

$$\langle e(f_1)|g\rangle = G(f_1) \tag{2.95}$$

Notice that Equation (2.95) is analogous to the Euclidean space equation $G_1 = \langle a_1|g\rangle$, where the coordinates and kets were defined in Equation (2.92).

Worked Exercise: Projection on the Basis of Exponentials

Prove Equation (2.95) by projecting Equation (2.93) on the bra $\langle e(f_1)|$.

Solution:

Projecting Equation (2.93) on the bra $\langle e(f_1)|$:

$$\langle e(f_1)|g\rangle = \int_{-\infty}^{+\infty} G(f_0)\langle e(f_1)|e(f_0)\rangle df_0$$

According to Equation (2.78):

$$\langle e(f_1)|e(f_0)\rangle = \delta(f_1 - f_0)$$

Recalling that $\delta(f_1 - f_0) = \delta(f_0 - f_1)$, we obtain:

$$\langle e(f_1)|g\rangle = \int_{-\infty}^{+\infty} G(f_0)\delta(f_0 - f_1)df_0 = G(f_1)$$

We used the sifting property of impulse functions in the last equality.

Now, let us evaluate this scalar product in the basis of impulse functions by inserting the closure relation of impulse functions (Equation (2.41)):

$$\langle e(f_1)|g\rangle = \langle e(f_1)|I|g\rangle = \langle e(f_1)| \int_{-\infty}^{+\infty} dt_0 |\delta(t_0)\rangle\langle\delta(t_0)|g\rangle$$

$$= \int_{-\infty}^{+\infty} dt_0 \langle e(f_1)|\delta(t_0)\rangle\langle\delta(t_0)|g\rangle$$

But $\langle\delta(t_0)|e(f_1)\rangle = e^{i2\pi f_1 t_0}$ (see Equation (2.73)), so $\langle e(f_1)|\delta(t_0)\rangle = e^{-i2\pi f_1 t_0}$. Furthermore, $\langle\delta(t_0)|g\rangle = g(t_0)$. Thus:

$$\langle e(f_1)|g\rangle = \int_{-\infty}^{+\infty} g(t_0)e^{-i2\pi f_1 t_0} dt_0 \tag{2.96}$$

The integral in Equation (2.96) is the direct Fourier transform using the dummy index t_0 (as in Equation (2.80)), evaluated at the specific frequency $f = f_1$. So, we have confirmed that the direct Fourier transform is a scalar product between the signal and a function of the 'new' basis. That makes sense: the result of the direct Fourier transform is precisely the coordinate of this signal with respect to the 'new' basis. Summarizing:

$$G(f_1) = \langle e(f_1)|g\rangle = \int_{-\infty}^{+\infty} g(t)e^{-i2\pi f_1 t} dt \tag{2.97}$$

The only difference between Equation (2.97) and Equation (2.81) is that Equation (2.97) specifies that the scalar product is being taken with a specific bra indexed by the specific frequency f_1, instead of a general bra indexed by a general frequency f. Thus, a general form like Equation (2.81) can be interpreted as denoting infinitely many scalar products: each different value of f indexing a different scalar product. Likewise, a specific form like Equation (2.97) is interpreted as denoting a single scalar product, involving only the frequency $f = f_1$. The only reason I insisted on using f_1 instead of a general f in Equation (2.97) is to bring out this difference between an equation denoting infinitely many integrals (or infinitely many scalar products) and an equation denoting a single integral. But, of course, we can leave it as a general f (as in Equation (2.81)), as long as we do not forget that each value of f specifies a different scalar product.

To conclude this section, we notice that, since the kets $|e(f_0)\rangle$ form an orthonormal basis, then they also satisfy the closure relation:

$$I = \int_{-\infty}^{+\infty} df_0 |e(f_0)\rangle\langle e(f_0)| \tag{2.98}$$

The closure relations of our two bases afford a convenient way of expressing the scalar product between signals. For example, suppose we want to evaluate $\langle v|u\rangle$. If we want to do that in the time domain, then we insert the closure relation with the impulse functions, as was done in Section 2.6 (The Scalar Product) (see

Equation (2.45)). But if we want to evaluate the very same $\langle v|u \rangle$ in the 'new' basis, then we insert the closure relation of the 'new' basis:

$$\langle v|u \rangle = \langle v|I|u \rangle = \langle v| \int_{-\infty}^{+\infty} df_0 |e(f_0)\rangle\langle e(f_0)|u \rangle = \int_{-\infty}^{+\infty} df_0 \langle v|e(f_0)\rangle\langle e(f_0)|u \rangle$$

Denoting $\langle e(f_0)|u \rangle = U(f_0)$ and $\langle v|e(f_0)\rangle = \langle e(f_0)|v \rangle^* = V^*(f_0)$, we get:

$$\langle v|u \rangle = \int_{-\infty}^{+\infty} V^*(f_0)U(f_0)df_0 \tag{2.99}$$

Since this is the very same scalar product of Equation (2.45), it follows that:

$$\langle v|u \rangle = \int_{-\infty}^{+\infty} V^*(f_0)U(f_0)df_0 = \int_{-\infty}^{+\infty} v^*(t_0)u(t_0)dt_0 \tag{2.100}$$

Of course, f_0 and t_0 in Equation (2.100) are dummy indexes, so we can recast them as:

$$\int_{-\infty}^{+\infty} V^*(f)U(f)df = \int_{-\infty}^{+\infty} v^*(t)u(t)dt \tag{2.101}$$

Notice that, since $u(t)$ is the time domain representation of $|u\rangle$ and $U(f)$ is the frequency domain representation of the same $|u\rangle$, it follows that $U(f)$ is the Fourier transform of $u(t)$ (see Equation (2.97)). Likewise, $V(f)$ is the Fourier transform of $v(t)$.

As we will see later in this chapter, Equation (2.101) is a general form of Parseval's theorem.

2.11   The Physical Meaning of Fourier Transforms

In Section 2.10, we discussed the algebraic meaning of Fourier transforms. Now we look at their physical meaning.

Suppose that $g(t)$ is a description of your favourite song: $g(t)$ may represent the amplitude of the speaker's membrane as a function of time, or the amplitude of air vibration. In any case, unless you are fond of chemical substances of debatable moral standing, your song cannot have imaginary parts: $g(t)$ must be a real function. Thus, to discuss the physical meaning of Fourier transforms, we will assume that $g(t)$ is real, which allows the inverse Fourier transform to be recast in a more intuitive form.

To begin with, notice that, even if $g(t)$ is real, that does not mean that its transform $G(f)$ is also real. Later in Section 2.12 (Properties of Fourier Transforms), we will investigate how the reality of $g(t)$ affects $G(f)$, but for now it suffices to keep in mind that, in general, $G(f)$ is complex, even when $g(t)$ is real.

There are two ways one can write a complex number: the 'Cartesian' way and the 'polar' way. The Cartesian way is the more familiar form $c = a + ib$. But one can also write the same complex number c in the polar form $c = |c|e^{i\theta}$. The relationship between the Cartesian and polar forms can be found using Euler's formula: $c = |c|e^{i\theta} = |c|cos(\theta) + i \cdot |c|sin(\theta)$, from which we readily infer that $a = |c|cos(\theta)$ and $b = |c|sin(\theta)$. Since $G(f)$ is a complex function (or, if you prefer, a collection of complex numbers, each one indexed by a different f), we can also write it in polar form:

$$G(f) = |G(f)|e^{i\theta(f)} \tag{2.102}$$

Now we write the inverse Fourier transform using the polar form:

$$g(t) = \int_{-\infty}^{+\infty} G(f)e^{i2\pi ft}df = \int_{-\infty}^{+\infty} |G(f)|e^{i[2\pi ft+\theta(f)]}df \tag{2.103}$$

With the help of Euler's formula, we obtain:

$$g(t) = \int_{-\infty}^{+\infty} |G(f)|cos[2\pi ft + \theta(f)]df + i\int_{-\infty}^{+\infty} |G(f)|sin[2\pi ft + \theta(f)]df \tag{2.104}$$

By assumption, $g(t)$ is real, so the integral involving sines must vanish (in Section 2.12 [Properties of Fourier Transforms], we will find what conditions $G(f)$ must satisfy for this integral to vanish). Therefore, the inverse Fourier transform of a real signal reduces to:

$$g(t) = \int_{-\infty}^{+\infty} |G(f)|cos[2\pi ft + \theta(f)]df \tag{2.105}$$

The form of Equation (2.105) is more convenient to investigate the physical meaning of $G(f)$.

First, recall that, since $G(f)$ is a complex number, there are two degrees of freedom for each f; thus, for each f, there are two pieces of information stored in $G(f)$: its magnitude $|G(f)|$ and its phase $\theta(f)$.

So, what is the message of Equation (2.105)? Its integral is a sum over cosines with different frequencies, amplitudes, and phases. Thus, the message of Equation (2.105) is that, to generate your favourite song $g(t)$, you can generate a whole bunch of cosine waves, each one with a different frequency (actually, you need a continuum of frequencies, as Equation (2.105) is an integral), and combine them together. So, imagine you have a super generator of cosine waves. Each one of these waves, when played on its own, sounds horrible, a bit like the *tu-tu-tu-tu* of calling phones. But, if you properly adjust the amplitude and phase of each one of these horribly sounding cosines and play them all together, out comes your beautiful song. The information on how you need to adjust the amplitude and phase of each horribly sounding cosine to create your beautiful song is stored in $G(f)$: the amplitude of the cosine with frequency f must be $|G(f)|$ and its phase must be $\theta(f)$. This is the message of Equation (2.105). So, in a sense, the Fourier transform is a recipe to make art from trash. Piero Manzoni would be proud.

The recipe of Equation (2.105) is illustrated in Figure 2.13. A sketch of the time domain representation of a signal is shown in Figure 2.13a, and schematic

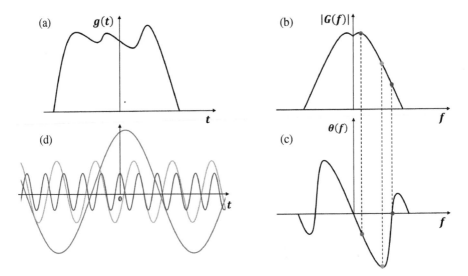

Figure 2.13 Illustration of the recipe of Equation (2.105). The illustration is only schematic and does not show an actual Fourier transform.

representations of the magnitude and phase of its Fourier transform are shown in Figure 2.13b and Figure 2.13c, respectively (this is not an actual Fourier transform: it is just a schematics to convey the meaning of Equation (2.105)). Three cosines contributing to the integral in Equation (2.105) are shown in Figure 2.13d, and the amplitude and phase of each of these cosines are shown by dots with the corresponding colour in Figure 2.13b and Figure 2.13c, respectively. In the example of Figure 2.13b, the red cosine has the highest amplitude, followed by the green and blue cosines. Still in Figure 2.13b, the farther away from the origin the dot is, the higher the frequency is. So, the blue cosine has the highest frequency, and the red has the lowest. Consequently, in Figure 2.13d, the blue cosine has the shortest period, and the red cosine has the longest period. When you put together the continuum of cosines following the recipe set by $G(f)$, and add them up (so the red, blue, and green cosines, and all the others that were not drawn, are added together), the result is $g(t)$. Notice that the phase of each cosine is specified by the phase of $G(f)$, as illustrated in Figure 2.13c (for example, the blue dot coincides with a point of zero phase; consequently, the blue cosine of Figure 2.13d has its peak aligned with the origin).

You may be wondering why, in this example, I put colourful dots only in the positive frequency part of the plots of Figure 2.13b and Figure 2.13c. The reason I did not consider the negative frequency part is that it carries the same information of the positive frequency part. Indeed, $|G(f)|$ is an even function, and $\theta(f)$ is an odd function. As we will prove in Section 2.12 (Properties of Fourier Transforms), this is always true when $g(t)$ is real.

Often one hears statements to the effect that the Fourier transform of a signal carries the information of its frequency content. But what exactly does this term

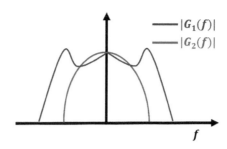

Figure 2.14 Schematic illustration of the difference between a high-pitched song (blue curve) and a low-pitched song (red curve).

'information of frequency content' mean? Well, it is information about the magnitude and phase of the cosines forming the signal. In other words, the frequency content of $g(t)$ is $G(f)$. So, the frequency content is the Fourier transform itself.

Let us go back to our example where $g(t)$ is a song. The way we perceive high and low frequencies is through the pitch: high frequencies are perceived as high-pitched sounds, and low frequencies are perceived as low-pitched sounds. So, let us say you take the Fourier transforms of two songs: song $g_1(t)$ and song $g_2(t)$. Say that $g_1(t)$ is a heavy metal song, with guitars in fury and screechy vocals. Guitars produce high-pitched sounds, so we should expect $g_1(t)$ to feature high frequencies. What does that mean? It means that $|G_1(f)|$ is predominantly high for high f. In other words, to generate $g_1(t)$, we need high amplitudes in high frequencies. Now, say that $g_2(t)$ is an only bass song. Now the pitch is quite low, so we should expect that $|G_2(f)|$ is predominantly high for low frequencies. Thus, in this example, the low frequency regions feature high magnitude of the Fourier transform. These differences between a high-pitched song and a low-pitched song are illustrated schematically in Figure 2.14.

These discussions about high and low pitches carry the notion of how 'fast' or 'slow' a signal is. Roughly speaking, a signal is 'fast' when its amplitude changes significantly in a short period of time, and a signal is 'slow' when it takes a long time for its amplitude to change significantly (we need the quotation marks in 'fast' and 'slow' because speed proper is a rate of change of space with time, but here the signal amplitude is playing the role of space). Thus, the ultra-short pulses travelling through optical fibres and connecting our computers through the Internet are extremely fast signals. But puffs of smoke taking five minutes to disappear are slow signals.

A 'fast' signal is a signal whose amplitude changes a lot in a short period of time. As such, to generate a fast signal, one needs cosines with high frequencies, because a high-frequency cosine changes 'quickly' – its period is short. Thus, 'fast' signals have Fourier transforms featuring high amplitudes in high frequencies. Likewise, 'slow' signals have Fourier transforms featuring high amplitudes in low frequencies.

This notion connecting the 'speed' of a signal with its Fourier transform can be nicely illustrated using a rectangular function. In the worked exercise of Section 2.9

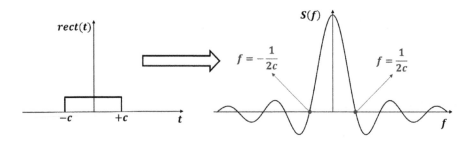

Figure 2.15 Illustration of a rectangular function and its Fourier transform.

(The Fourier Transform), we found that the Fourier transform of a rectangular function is a *sinc* function (see Equation (2.86)). These two domains are illustrated in Figure 2.15. Notice that the *sinc* function has a high amplitude around the origin, and then decays in an oscillatory manner. Thus, we can use the point where the *sinc* first crosses the horizontal axis as a measure of its width. As indicated in Figure 2.15, these points are inversely proportional to the width of the rectangular function (you can find these points by requiring the argument of the sine function in the numerator of the *sinc* function to be π).

Thus, the wider the rectangular function in the time domain is (that is, the larger c is), the thinner its Fourier transform is. But a wide rectangular function is a 'slow' function: it goes up, and then stays up for a long time, and then comes back down. Moreover, a thin Fourier transform entails that the magnitude of the Fourier transform is significant only for low frequencies. If, on the other hand, the rectangular function is thin (that is, if c is a small number), then it is a 'fast' signal: it goes up and quickly comes back down. Thus, it requires higher amplitudes in higher frequencies to be able to go up and down quickly. Consequently, its Fourier transform is wider, entailing that the region within the first zeros now involve higher frequencies (in other words, the first zeros are pushed away from the origin as c is made smaller).

This connection between 'speed' in the time domain and 'width' in the frequency domain is illustrated in Figure 2.16: the red signal is slower, so it has a narrower Fourier transform, while the blue signal is faster, so it has a wider Fourier transform.

By the way, you may be wondering what happens if we allow the rectangular function to keep broadening, all the way to infinity. If we do that, then we get a

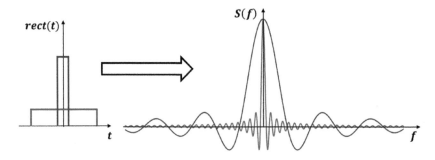

Figure 2.16 Comparison between a slow signal (red) and a fast signal (blue).

constant function in the time domain ($g(t) = 1$). What about the frequency domain? Well, the *sinc* gets thinner and thinner, taller and taller, until it converges to an impulse function. Thus, the Fourier transform of $g(t) = 1$ is the impulse function $G(f) = \delta(f)$. By the same token, if we let the rectangular function get thinner and thinner, taller and taller, then it gets closer and closer to an impulse function. In the frequency domain, the *sinc* gets wider and wider until it becomes a constant function. Thus, the Fourier transform of $g(t) = \delta(t)$ is the constant function $G(f) = 1$. You will be asked to prove these relations in the exercise list.

If the Fourier transform of a constant signal is $\delta(f)$, then there is no cosine involved in a constant signal, since $\delta(f) = 0$ for $f \neq 0$. That makes sense: there is no oscillation in a constant signal. This feature is encapsulated by the first property of Fourier transforms, to be treated in Section 2.12.

2.12   Properties of Fourier Transforms

In this section, we will deduce a bunch of properties of Fourier transforms. These properties are quite useful, and often save a lot of work. One property in particular, the property of convolution, is of paramount importance in the analysis of systems, as we will see in Chapter 3.

2.12.1   Fourier Transform and the DC level

We begin with a quite straightforward property concerning the Direct Current (DC) level of a signal.

The DC level is the zero-frequency component of a signal. Denoting the DC level of $g(t)$ as g_{DC}, we can define it as:

$$g_{DC} = \int_{-\infty}^{+\infty} g(t)dt \tag{2.106}$$

Thus, the DC level is intimately related to the 'average' of a signal: if its average is zero, so is the DC level.

The property of this section concerns the identification of the DC level in the Fourier transform of $g(t)$. Recall that its Fourier transform is:
Equation (2.81)

$$G(f) = \int_{-\infty}^{+\infty} g(t)e^{-i2\pi ft}dt$$

For $f = 0$, we get:

$$G(0) = \int_{-\infty}^{+\infty} g(t)e^{-i2\pi 0t}dt = \int_{-\infty}^{+\infty} g(t)dt \tag{2.107}$$

Comparing Equation (2.107) with Equation (2.106), we conclude that:

$$g_{DC} = G(0) \tag{2.108}$$

Thus, according to Equation (2.108), the origin of the Fourier spectrum (a Fourier spectrum is just a plot of the Fourier transform), that is, $G(0)$, is the DC level of the signal. In other words, the DC level is the zero-frequency component of the signal.

2.12.2   Property of Reality

If $g(t)$ is a real function, then $g(t) = g^*(t)$. This latter equality is the 'signature' of a real function. The property of reality connects the 'signature' of a real function in the time domain with its 'signature' in the frequency domain. In other words, it answers the question of how $g(t) = g^*(t)$ is manifested in $G(f)$.

To answer this question, consider the inverse Fourier transform:
Equation (2.82)

$$g(t) = \int_{-\infty}^{+\infty} G(f)e^{i2\pi ft}df$$

The complex conjugate of $g(t)$ is:

$$g^*(t) = \int_{-\infty}^{+\infty} G^*(f)e^{-i2\pi ft}df \tag{2.109}$$

Now, define $f' = -f$ and express the integral above in terms of f':

$$g^*(t) = \int_{-\infty}^{+\infty} G^*(f)e^{-i2\pi ft}df = -\int_{+\infty}^{-\infty} G^*(-f')e^{i2\pi f't}df' = \int_{-\infty}^{+\infty} G^*(-f')e^{i2\pi f't}df' \tag{2.110}$$

Since f' is a dummy index, we can replace this index with f in the last integral in Equation (2.110) (this is just a dummy index replacement, not a change of variables, so we are no longer considering $f' = -f$). Thus:

$$g^*(t) = \int_{-\infty}^{+\infty} G^*(-f)e^{i2\pi ft}df \tag{2.111}$$

By assumption, $g(t)$ is real, that is, $g(t) = g^*(t)$. But, on the one hand, according to Equation (2.82), the coordinates of $g(t)$ in the basis of exponentials are $G(f)$; on the other hand, according to Equation (2.111), the coordinates of $g^*(t)$ in the basis of exponentials are $G^*(-f)$. But, if $g(t) = g^*(t)$, then their coordinates must be the same. In other words, $g(t) = g^*(t)$ entails that:

if:

$$g(t) = g^*(t)$$

then:

$$G(f) = G^*(-f) \tag{2.112}$$

Equation (2.112) expresses how the condition of reality in the time domain is manifested in the frequency domain. It is instructive to recast this condition in polar form. In general, if:

$$G(f) = |G(f)|e^{i\theta(f)}$$

then:

$$G^*(f) = |G(f)|e^{-i\theta(f)}$$

So, the condition $G(f) = G^*(-f)$ entails that:

$$|G(f)|e^{i\theta(f)} = |G(-f)|e^{-i\theta(-f)} \tag{2.113}$$

There are two equalities embedded in Equation (2.113).

One equality asserts that the magnitude of $G(f)$ is an even function (see the example of Figure 2.13b), that is:

$$|G(f)| = |G(-f)| \tag{2.114}$$

The other equality asserts that the phase of $G(f)$ is an odd function (see the example of Figure 2.13c), that is:

$$\theta(f) = -\theta(-f) \tag{2.115}$$

Thus, as had already been commented in Section 2.11 (The Physical Meaning of Fourier Transforms), one region of the Fourier spectrum (recall that a Fourier spectrum is just a plot of the Fourier transform) carries redundant information: if we know $G(f)$ for positive f, then we already know $G(f)$ for negative f: we can infer one from the other through Equation (2.113).

Why is that? Why half of the frequency domain coordinates are redundant when $g(t)$ is real? This is an interesting question, and it has to do with the degrees of freedom of a signal. We have seen that, to specify a signal, we need to specify its coordinates. In the time domain, each t indexes a different coordinate. If $g(t)$ was complex, then each coordinate would be a complex number. But a complex number has two degrees of freedom (in other words, we need two numbers to specify a complex number: the real and imaginary part or, equivalently, the magnitude and phase). Thus, if $g(t)$ was complex, each coordinate would carry two degrees of freedom. What about the frequency domain? Well, it is the same thing: each f indexes a different coordinate, and if each coordinate is a complex number, then each coordinate carries two degrees of freedom. That makes sense: since we are not changing the signal, only the representation, we must have the same number of degrees of freedom in both representations.

Now, what happens if $g(t)$ is real? Now the coordinates are real, so we lost one degree of freedom in each coordinate, which means that we lost half of the degrees of freedom. But we have seen that $G(f)$ is still complex; so, does that mean that $G(f)$ has twice as many degrees of freedom as $g(t)$? That cannot be true: they are equivalent representations of the same signal. And it is not true: they have the same number of degrees of freedom. Indeed, whereas $g(t)$ lost half of its degrees of freedom by becoming real, $G(f)$ lost half of its degrees of freedom by making half of its coordinates dependent on the other half. This is what is entailed by Equation (2.113): only half of the coordinates (say, the coordinates indexed by positive f) constitute degrees of freedom: the other half (the coordinates indexed by negative f) are automatically specified through Equation (2.113). Consequently, $g(t)$ and $G(f)$ still have the same number of degrees of freedom: $g(t)$ lost half of them by being real, and $G(f)$ lost half of them by being symmetric (that is, by obeying Equation (2.114) and Equation (2.115)).

In the list of exercises, you will be asked to prove that Equation (2.113) ensures that the sine integral in Equation (2.104) vanishes.

2.12.3 Symmetry Between Time and Frequency

As mentioned in Section 2.9 (The Fourier Transform), time and frequency play symmetric roles in the Fourier transform. This symmetry entails that, if $G_1(f)$ is the Fourier transform of $g_1(t)$, then a function of time with the same form of $G_1(f)$ has a Fourier transform of the same form of $g_1(t)$. In other words, if $g_2(t) = G_1(t)$, then $G_2(f) = g_1(-f)$. This property can be checked by inspection of the Fourier transform pair. As an example, we have seen that the Fourier transform of a rectangular function is the *sinc* function (Equation (2.86), see also Figure 2.15). Thus, the Fourier transform of a *sinc* function must be a rectangular function, as illustrated in Figure 2.17.

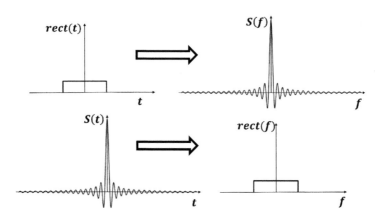

Figure 2.17 Example of the symmetry between time and frequency. The Fourier transform of a rectangular function is a *sinc* function. Thus, the Fourier transform of a *sinc* function must be a rectangular function.

2.12.4 Time Shifting

Our goal in this section is to find a relationship between the Fourier transform of $g(t)$ and the Fourier transform of its time-shifted version $g_2(t) = g(t - t_0)$.

The Fourier transform of $g_2(t)$ is:

$$G_2(f) = \int_{-\infty}^{+\infty} g_2(t)e^{-i2\pi ft}dt = \int_{-\infty}^{+\infty} g(t - t_0)e^{-i2\pi ft}dt \qquad (2.116)$$

We define $t' = t - t_0$ and recast Equation (2.116) in terms of t':

$$G_2(f) = \int_{-\infty}^{+\infty} g(t')e^{-i2\pi f(t'+t_0)}dt' \qquad (2.117)$$

Extracting the term that does not depend on t':

$$G_2(f) = e^{-i2\pi ft_0}\int_{-\infty}^{+\infty} g(t')e^{-i2\pi ft'}dt' \qquad (2.118)$$

Notice that the integral in Equation (2.118) is the direct Fourier transform of $g(t)$, which we denote by $G(f)$. Thus, we have concluded that:

if:
$$g_2(t) = g(t - t_0)$$

then:
$$G_2(f) = e^{-i2\pi ft_0}G(f) \qquad (2.119)$$

According to Equation (2.119), a delay in $g(t)$ by t_0 induces a linear phase modulation in its Fourier transform. In other words:

$$\theta_2(f) = \theta(f) - 2\pi t_0 f \qquad (2.120)$$

where $\theta_2(f)$ and $\theta(f)$ are the phases of $G_2(f)$ and $G(f)$, respectively. Moreover, Equation (2.119) entails that:

$$|G_2(f)| = |G(f)| \qquad (2.121)$$

Does that make sense? Suppose, again, that $g(t)$ is your favourite song. Then, $g_2(t)$ is still your song, but now played at a different time. For example, say that we set our time scale so that $t = 0$ corresponds to the $0:00$ h of your birthday. And suppose that $g(t)$ is only nonzero for $0 < t < 120$ s. In this way, $g(t)$ represents your song, which is 120 s long, and starts playing at $0:00$ h of your birthday. Now, suppose that $t_0 = 600$ s. In this case, $g_2(t) = g(t - 600)$ is still your song, but now starting a bit later, at $t = 600$ s. Since they are the same song, just played at different times, the magnitudes of their Fourier transforms must be the same (otherwise one of them could have more bass than guitar, for example). So, Equation (2.121) makes a lot of sense. Moreover, Equation (2.120) teaches that their phase relationship is well defined: although their phases are not the same (because they start playing at different times), their phases must obey Equation (2.120); otherwise, they would not be the same song.

2.12.5 Spectral Shifting

If $g_2(t) = e^{i2\pi f_0 t} g(t)$, what is the relationship between $G_2(f)$ and $G(f)$?
The Fourier transform of $g_2(t)$ is:

$$G_2(f) = \int_{-\infty}^{+\infty} g_2(t) e^{-i2\pi ft} dt = \int_{-\infty}^{+\infty} e^{i2\pi f_0 t} g(t) e^{-i2\pi ft} dt$$

Rearranging:

$$G_2(f) = \int_{-\infty}^{+\infty} g(t) e^{-i2\pi(f-f_0)t} dt \tag{2.122}$$

The integral in Equation (2.122) is the Fourier transform of $g(t)$ evaluated at the frequency $f - f_0$. Indeed, if:

$$G(f) = \int_{-\infty}^{+\infty} g(t) e^{-i2\pi ft} dt$$

then:

$$G(f - f_0) = \int_{-\infty}^{+\infty} g(t) e^{-i2\pi(f-f_0)t} dt \tag{2.123}$$

Thus, from Equation (2.122) and Equation (2.123), we conclude that:

if:

$$g_2(t) = e^{i2\pi f_0 t} g(t)$$

then:

$$G_2(f) = G(f - f_0) \tag{2.124}$$

Compare Equation (2.124) with Equation (2.119): they look like mirror images of each other. That is not accidental: it is a consequence of the symmetry between time and frequency.

Worked Exercise: The Property of Spectral Shifting and AM Modulation

The property of spectral shifting is of paramount importance in communication theory because it underlies the process of modulation. For example, suppose that you and your worst enemy work at rival radio stations. The two of you want each to send a different special song to a third person, equally unrequitedly beloved by both competitors. So, you and your worst enemy convert the songs into electro-magnetic waves, which propagate through open space, until they reach the antenna of the alluring third person's car. But, if they reached the antenna at the same time, what will the third person listen to? If there was no modulation of the audio signal, the third person would listen to both songs played at the same time; not ideal. But that is not how it works: before sending the songs, each of you choose a different frequency to 'carry' the signal. This carrier frequency is the radio station frequency, so the third person can choose which song will be listened to by choosing the radio station frequency. The notion of a carrier frequency is related to the idea of modulation.

The notion of modulation is quite simple, and there are many types of modu-lation. Here, we will discuss the Amplitude Modulation (AM), because it is more directly related to spectral shifting.

So, suppose that the Fourier transforms of the audios before modulation are $G_1(f)$ and $G_2(f)$. Before modulation, these functions overlap around the origin, as illustrated in Figure 2.18a. So, to transmit them at the radio frequency, we need to shift them to the frequency carriers f_1 and f_2, as illustrated in Figure 2.18b. Now, the modulated functions $G_{1m}(f)$ and $G_{2m}(f)$ do not overlap, which means that the audios can be demodulated without mixing them up (if you are wondering why the spectral shift must go to both positive and negative frequencies, recall that the magnitude of the Fourier transform of a real signal is an even function).

From Figure 2.18, we infer that the relationships between modulated and unmodulated signals are:

$$G_{1m} = G_1(f - f_1) + G_1(f + f_1)$$

$$G_{2m} = G_2(f - f_2) + G_2(f + f_2) \tag{2.125}$$

In this exercise, we want to find the relationships between the modulated and unmodulated signals in the time domain.

Solution:

Using Equation (2.124) in Equation (2.125), we get:

$$g_{1m}(t) = g_1(t)\, e^{i2\pi f_1 t} + g_1(t)\, e^{-i2\pi f_1 t}$$

$$g_{2m}(t) = g_2(t)\, e^{i2\pi f_2 t} + g_2(t)\, e^{-i2\pi f_2 t} \tag{2.126}$$

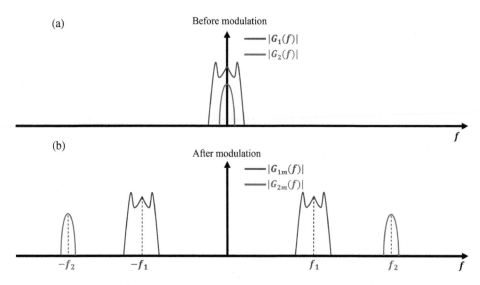

Figure 2.18 Illustration of AM modulation. Spectrum before modulation (a) and spectrum after modulation (b).

Euler's formula (Equation (2.52)) entails that:

$$cos(\theta) = \frac{e^{i\theta} + e^{-i\theta}}{2} \qquad (2.127)$$

Using Equation (2.127) in Equation (2.126), we obtain:

$$g_{1m}(t) = g_1(t) \cdot [2 \cdot cos(2\pi f_1 t)]$$

$$g_{2m}(t) = g_2(t) \cdot [2 \cdot cos(2\pi f_2 t)] \qquad (2.128)$$

Thus, according to Equation (2.128), to modulate a signal we need to multiply it by a cosine function with the carrier frequency.

2.12.6 Differentiation

The property of differentiation is one of the simplest, and yet most useful, of our list. It relates the Fourier transform of a signal $g(t)$ with its time domain derivative $g_2(t) = dg/dt$. We start with the inverse Fourier transform:

$$g(t) = \int_{-\infty}^{+\infty} G(f)e^{i2\pi ft}df$$

And take the derivative on both sides:

$$\frac{dg(t)}{dt} = \frac{d\int_{-\infty}^{+\infty} G(f)e^{i2\pi ft}df}{dt} = \int_{-\infty}^{+\infty} G(f)\frac{de^{i2\pi ft}}{dt}df = \int_{-\infty}^{+\infty} i2\pi fG(f)e^{i2\pi ft}df$$

$$(2.129)$$

By assumption, $g_2(t) = dg/dt$, so:

$$g_2(t) = \int_{-\infty}^{+\infty} i2\pi f G(f) e^{i2\pi ft} df \tag{2.130}$$

But the inverse Fourier transform of $G_2(f)$ is:

$$g_2(t) = \int_{-\infty}^{+\infty} G_2(f) e^{i2\pi ft} df \tag{2.131}$$

According to Equation (2.130), the coordinates of $g_2(t)$ in the basis of exponentials are $i2\pi f G(f)$. But, according to Equation (2.131), these coordinates are the Fourier transform of $g_2(t)$, which we are denoting as $G_2(f)$. Therefore, these must be the same coordinates. In other words:

if:

$$g_2(t) = \frac{dg(t)}{dt}$$

then:

$$G_2(f) = i2\pi f G(f) \tag{2.132}$$

Notice that Equation (2.132) entails that $G_2(0) = i2\pi 0 \cdot G(f) = 0$. Recall that $G_2(0)$ is the DC level of $g_2(t)$, so $G_2(0) = 0$ implies that $g_2(t)$ has no DC level. That makes sense: differentiation 'kills' the constant part of a function.

2.12.7 Integration

We define the integral function as:

$$g_2(t) = \int_{-\infty}^{t} g(\tau) d\tau \tag{2.133}$$

As usual, our job is to find the relationship between $G_2(f)$ and $G(f)$. Of course, Equation (2.133) entails that:

$$\frac{dg_2(t)}{dt} = g(t) \tag{2.134}$$

which entails that (see Equation (2.132)):

$$i2\pi f G_2(f) = G(f)$$

And, consequently:

$$G_2(f) = \frac{G(f)}{i2\pi f}$$

This equation is almost correct, but it is incomplete because we lost information about the DC level when we went from Equation (2.133) to Equation (2.134). So, we need to include this lost bit in the expression above. This time, we will not prove it rigorously, only motivate it. So, I will first tell you the result and then motivate it. The correct expression is:

if:

$$g_2(t) = \int_{-\infty}^{t} g(\tau)d\tau$$

then:

$$G_2(f) = \frac{G(f)}{i2\pi f} + \frac{G(0)}{2}\delta(f) \qquad (2.135)$$

Notice that, unless $G(0) = 0$, the DC level of $g_2(t)$ is an infinite number. To make sense of why this must be true, consider a general function $g(t)$, as illustrated in Figure 2.19a. If $g(t)$ is a representation of a physical signal, then it must begin at some time and end at some time. We denote the time when $g(t)$ begins by t_1, and the time when it ends by t_2. How does its integral $g_2(t)$ look like? Well, before t_1 it must be zero $(g_2(t_1) = \int_{-\infty}^{t_1} g(\tau)d\tau = \int_{-\infty}^{t_1} 0 d\tau = 0)$, then it increases monotonically all the way to t_2, and from this point on it must be a constant, as represented in Figure 2.19b (it is a constant because, for $t > t_2$, we have $g_2(t) = \int_{-\infty}^{t} g(\tau)d\tau = \int_{-\infty}^{t_2} g(\tau)d\tau + \int_{t_2}^{t} g(\tau)d\tau = \int_{-\infty}^{t_2} g(\tau)d\tau + \int_{t_2}^{t} 0 d\tau = \int_{-\infty}^{t_2} g(\tau)d\tau = g_2(t_2))$.

The value at which $g_2(t)$ becomes a constant is denoted by B in Figure 2.19b. This is the DC value of $g(t)$. Indeed:

$$B = g_2(t_2) = \int_{-\infty}^{t_2} g(\tau)d\tau = \int_{-\infty}^{+\infty} g(\tau)d\tau = g_{DC} \qquad (2.136)$$

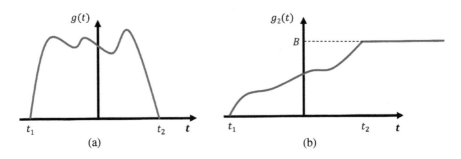

Figure 2.19 Illustration of relationship between a function (a) and its integral (b).

Thus, unless $g_{DC} = 0$, the function $g_2(t)$ will stay 'on' at the value B forever and ever. Consequently, the DC value of $g_2(t)$ is an infinite number.

But no physical signal can be on forever and ever. For example, if $g_2(t)$ is the output of an integrator, at some time the circuit will break, the power will go off, the engineer in charge will retire, people will die, life on Earth will cease, the sun will die, and the universe will turn into a dismal sea of photons, the unavoidable culmination of the dastardly second law of thermodynamics. In short, at some time, $g_2(t)$ will go back to zero. Thus, a relation like Equation (2.133) is a mathematical idealization of a signal that stays 'on' for a long, long time, and consequently has a huge DC level, which is idealized by the impulse function in Equation (2.135).

Now I want to give you an idea of why the impulse function is multiplied by $G(0)/2$. We have already established that $B = g_{DC}$. But $g_{DC} = G(0)$ – see Equation (2.108) – so $B = G(0)$.

Now, suppose that we want to separate the function $g_2(t)$ into one part wherein the signal stays on for negative infinite time at a value that compensates the positive infinite time, and another constant part. We denote the compensated part by $g_{2\_comp}(t)$ and the constant part by $g_{2\_DC}(t)$. These two parts are illustrated in Figure 2.20a and Figure 2.20b, respectively. With this separation, we have:

$$g_2(t) = g_{2\_comp}(t) + g_{2\_DC}(t) \tag{2.137}$$

Notice that, to get $g_{2\_comp}(t)$, we need to 'lower' $g_2(t)$ until we get a function that stays on for infinite negative time at a value that compensates its value at infinite positive time. In other words, as illustrated in Figure 2.20a, we need to lower $g_2(t)$ by $B/2$ (that is, $g_{2\_comp}(t) = g_2(t) - B/2$). But, if $g_{2\_comp}(t) = g_2(t) - B/2$, then $g_{2\_DC}(t) = B/2$, as illustrated in Figure 2.20b. One of the exercises at the end of this chapter asks you to prove that the Fourier transform of the constant function $x(t) = 1$ is $X(f) = \delta(f)$. Thus, the Fourier transform of $g_{2\_DC}(t) = B/2$ must be $(B/2)\delta(f)$. We have already found that $B = G(0)$, so we can identify the term $(G(0)/2)\delta(f)$ in Equation (2.135) as being the Fourier transform of the constant part of $g_2(t)$, that is, of $g_{2\_DC}(t)$.

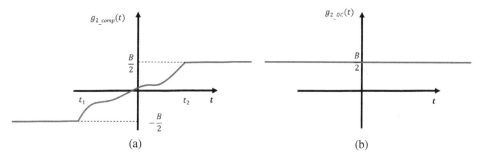

Figure 2.20 Illustration of the separation of $g_2(t)$ into a signal where the value at negative time compensates the value at positive time (a) and a constant signal (b).

2.12.8 Convolution in the Time Domain

The property of convolution in the time domain is particularly relevant to signals and systems, as will be clear in Chapter 3. It asserts that the convolution of two functions in the time domain correspond to their product in the frequency domain.

To prove this relation, we begin by defining:

$$g(t) = a(t) * b(t) = \int_{-\infty}^{\infty} a(t_0)b(t - t_0)dt_0 \qquad (2.138)$$

and now we ask: what is the relationship between $G(f)$, $A(f)$, and $B(f)$, where these functions are the Fourier transforms of $g(t)$, $a(t)$, and $b(t)$, respectively?

To find this relationship, we evaluate the Fourier transform of $g(t)$. Thus:

$$G(f) = \int_{-\infty}^{+\infty} g(t)e^{-i2\pi ft}dt = \int_{-\infty}^{+\infty} \left[\int_{-\infty}^{\infty} a(t_0)b(t - t_0)dt_0 \right] e^{-i2\pi ft}dt \qquad (2.139)$$

It is advantageous to evaluate the integral over t first, since it does not involve $a(t_0)$. So, rearranging:

$$G(f) = \int_{-\infty}^{+\infty} g(t)e^{-i2\pi ft}dt = \int_{-\infty}^{\infty} a(t_0) \left[\int_{-\infty}^{+\infty} b(t - t_0)e^{-i2\pi ft}dt \right] dt_0 \qquad (2.140)$$

The integral inside the brackets (over t) is the Fourier transform of $b(t - t_0)$. From the property of time shifting (Equation (2.119)), the Fourier transform of $b(t - t_0)$ is $e^{-i2\pi ft_0}B(f)$. Thus:

$$G(f) = \int_{-\infty}^{\infty} a(t_0) \left[\int_{-\infty}^{+\infty} b(t - t_0)e^{-i2\pi ft}dt \right] dt_0 = \int_{-\infty}^{\infty} a(t_0)e^{-i2\pi ft_0}B(f)dt_0 \qquad (2.141)$$

Isolating $B(f)$:

$$G(f) = B(f) \int_{-\infty}^{\infty} a(t_0)e^{-i2\pi ft_0}dt_0 \qquad (2.142)$$

Notice that the integral in Equation (2.142) is the Fourier transform of $a(t)$. Therefore:

$$G(f) = B(f)A(f) \qquad (2.143)$$

Thus, we have concluded that convolution in the time domain results in product in the frequency domain. In other words:

if:

$$g(t) = a(t) * b(t)$$

then:

$$G(f) = A(f)B(f) \qquad (2.144)$$

Incidentally, since $A(f)B(f) = B(f)A(f)$, it follows that $a(t) * b(t) = b(t) * a(t)$, which proves Equation (2.6).

2.12.9 Product in the Time Domain

If convolution in the time domain results in product in the frequency domain, then, from the property of symmetry between time and frequency, we can expect that product in the time domain results in convolution in the frequency domain. Let us prove it.
If:

$$g(t) = a(t)b(t) \tag{2.145}$$

then:

$$G(f) = \int_{-\infty}^{+\infty} g(t)e^{-i2\pi ft}dt = \int_{-\infty}^{+\infty} a(t)b(t)e^{-i2\pi ft}dt \tag{2.146}$$

Now we express $a(t)$ in terms of its inverse Fourier transform:

$$G(f) = \int_{-\infty}^{+\infty} \left[\int_{-\infty}^{+\infty} A(f')e^{i2\pi f't}df'\right] b(t)e^{-i2\pi ft}dt \tag{2.147}$$

Rearranging:

$$G(f) = \int_{-\infty}^{+\infty} \int_{-\infty}^{+\infty} A(f')b(t)e^{-i2\pi(f-f')t}df'dt \tag{2.148}$$

Swapping the order of integrals and collecting the terms depending on time:

$$G(f) = \int_{-\infty}^{+\infty} A(f')\left[\int_{-\infty}^{+\infty} b(t)e^{-i2\pi(f-f')t}dt\right] df' \tag{2.149}$$

Notice that the integral over time (inside the brackets) is the Fourier transform of $b(t)$ evaluated at the frequency $f - f'$. Indeed, if $B(f)$ is the Fourier transform of $b(t)$, then it follows that:

$$B(f - f') = \int_{-\infty}^{+\infty} b(t)e^{-i2\pi(f-f')t}dt \tag{2.150}$$

Substituting Equation (2.150) into Equation (2.149), we get:

$$G(f) = \int_{-\infty}^{+\infty} A(f')B(f - f')df' \tag{2.151}$$

Notice that the integral above is the convolution in the frequency domain, that is:

$$G(f) = \int_{-\infty}^{+\infty} A(f')B(f - f')df' = A(f) * B(f)$$

Thus, we have concluded that:

if:
$$g(t) = a(t)b(t)$$

then:
$$G(f) = A(f) * B(f) \tag{2.152}$$

Worked Exercise: The Fourier Transform of a Physical Sinusoidal Wave

An 'ideal' sinusoidal wave cannot represent a physical signal, because an 'ideal' sinusoidal wave never begins and never ends: it extends towards infinite negative time and towards infinite positive time. But no physical signal can be eternal: it must begin at some time and end at some time.

In this exercise, we will use the property of product in the time domain (Equation (2.152)) to find the Fourier transform of a 'physical' or 'chopped' sinusoidal wave, that is, a sinusoidal wave that begins at some time and ends at some time.

Solution:

As illustrated in Figure 2.21a and Figure 2.21b, a physical, or chopped sinusoidal wave can be described mathematically as the product between an ideal sinusoidal wave and a rectangular function. Thus, denoting the chopped sinusoidal wave as $g(t)$:

$$g(t) = sin(2\pi f_0 t)rect(t) \tag{2.153}$$

where $rect(t)$ was defined in Equation (2.7).

To find the Fourier transform of $g(t)$, we will use Equation (2.152). So, we need the Fourier transform of the sine function and of the rectangular function. We have already found the Fourier transform of the rectangular function: it is the $sinc$ function, which we denoted by $S(f)$ – see Equation (2.86). The magnitude of the $sinc$ function is shown in Figure 2.21c (black line). We also need the Fourier transform of the sine function. Defining:

$$a(t) = sin(2\pi f_0 t) \tag{2.154}$$

Then:
$$A(f) = \int_{-\infty}^{+\infty} a(t)e^{-i2\pi ft}dt = \int_{-\infty}^{+\infty} sin(2\pi f_0 t)e^{-i2\pi ft}dt \tag{2.155}$$

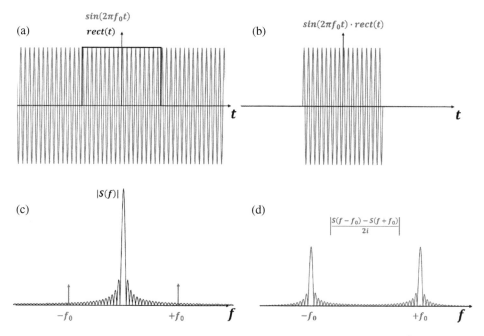

Figure 2.21 Illustration of the effect of chopping a sinusoidal wave. (a) Ideal sinusoidal wave (red) superimposed on a rectangular function (black); (b) chopped sinusoidal wave; (c) magnitude of Fourier transform of the rectangular function (black) and ideal sinusoidal wave (red); and (d) Fourier transform of chopped sinusoidal wave.

With the help of Equation (2.65):

$$A(f) = \int_{-\infty}^{+\infty} \left[\frac{e^{i2\pi f_0 t} - e^{-i2\pi f_0 t}}{2i} \right] e^{-i2\pi f t} dt$$

$$= \frac{1}{2i} \int_{-\infty}^{+\infty} e^{i2\pi(f_0 - f)t} dt - \frac{1}{2i} \int_{-\infty}^{+\infty} e^{-i2\pi(f_0 + f)t} dt \qquad (2.156)$$

According to Equation (2.75):

$$\int_{-\infty}^{+\infty} e^{i2\pi(f_0 - f)t} dt = \delta(f_0 - f)$$

The impulse function is even (Equation (2.22)), so $\delta(f_0 - f) = \delta(f - f_0)$. Therefore:

$$\int_{-\infty}^{+\infty} e^{i2\pi(f_0 - f)t} dt = \delta(f - f_0) \qquad (2.157)$$

Likewise:

$$\int_{-\infty}^{+\infty} e^{-i2\pi(f + f_0)t} dt = \delta(-f_0 - f) = \delta(f + f_0) \qquad (2.158)$$

Substituting Equation (2.158) and Equation (2.157) into Equation (2.156), we obtain:

$$A(f) = \frac{\delta(f - f_0) - \delta(f + f_0)}{2i} \qquad (2.159)$$

Thus, the Fourier transform of a sine function with frequency f_0 consists of two impulse functions located at f_0 and $-f_0$, as illustrated by the red arrows in Figure 2.21c. This is an important result, and I will comment on it soon.

Now we are ready to evaluate the Fourier transform of $g(t)$ – as defined in Equation (2.153). Using the convolution in the frequency domain (Equation (2.152)), we obtain:

$$G(f) = A(f) * S(f) = \frac{\delta(f - f_0) * S(f) - \delta(f + f_0) * S(f)}{2i} \qquad (2.160)$$

We already know that the convolution of a function with the impulse function results in the function itself (see Equation (2.25)). Furthermore, according to Equation (2.26), $g(t) * \delta(t - a) = g(t - a)$, that is, the convolution with an impulse function 'located' at $t = a$ results in a time-shifted function. Using this result in the frequency domain, we conclude that:

$$G(f) = \frac{\delta(f - f_0) * S(f) - \delta(f + f_0) * S(f)}{2i} = \frac{S(f - f_0) - S(f + f_0)}{2i} \qquad (2.161)$$

Thus, the Fourier transform of the chopped sinusoidal function consists of two *sinc* functions 'located' at f_0 and $-f_0$, as illustrated in Figure 2.21d.

Let us interpret this result.

An ideal sinusoidal wave is a 'pure frequency' wave (it can also be a cosine or a combination of both, as long as they have the same frequency). That means that it is a function with only one frequency f_0 (if we wanted to be overly strict, we would say that $e^{i2\pi f_0 t}$ is the actual 'pure frequency' function, because it has only the f_0 component, whereas the sine also has the $-f_0$: but $e^{i2\pi f_0 t}$ is a complex function, and we always need the $-f_0$ component to get a real function). The fact that it is a 'pure frequency' function is manifested in the Fourier domain: the Fourier transform of a sine wave consists of two impulse functions, 'located' at f_0 and $-f_0$ (again, we need the $-f_0$ component: the magnitude of the Fourier transform of a real function is an even function, so if there is an impulse at f_0, there must be one at $-f_0$ as well).

But a 'pure frequency' is only a mathematical idealization because for a function to be a 'pure frequency' wave it must be a 'perfect' sinusoidal oscillation: it must never begin, and never end.

Consequently, any physical sinusoidal wave is a composite of frequencies. This is what Equation (2.161) and Figure 2.21d are teaching: the Fourier transform of a 'chopped' sinusoidal wave, that is, one with a beginning and an end, is no longer an impulse function (so, no longer a 'pure frequency'), but a *sinc* function centred around the frequency of oscillation. The longer the sinusoidal wave is let to oscillate, that is, the wider the rectangular function doing the chopping is, then the thinner the *sinc* function is (see discussion about fast and slow functions in Section 2.11 [The Physical Meaning of Fourier Transforms]). Thus, the more oscillations, the 'purer' the frequency content of the sinusoidal wave is (because the thinner the *sinc* function is). If the rectangular function is allowed to broaden to infinity, then the *sinc* function converges to the impulse function, and a 'pure frequency' is obtained. On the other hand, the less oscillations, that is, the thinner the rectangular function is, then the wider the *sinc* function is. But a wide *sinc* function implies that the chopped sinusoidal wave contains a wide range of frequencies, and as such it is not pure.

2.12.10   The Energy of a Signal and Parseval's Theorem

This last property relates the energy of a signal in the time and frequency domains.

By definition, the energy of a signal $|g\rangle$ is the scalar product of the signal with itself: $\langle g|g\rangle$. But, since a scalar product does not depend on the basis, we are free to evaluate it either in the time domain or in the frequency domain. Indeed, in Section 2.10 (The Algebraic Meaning of Fourier Transforms), we found that, in general:

Equation (2.100)

$$\langle v|u\rangle \int_{-\infty}^{+\infty} V^*(f)U(f)df = \int_{-\infty}^{+\infty} v^*(t)u(t)dt$$

Parseval's theorem is Equation (2.100) applied to the particular case of $|u\rangle = |v\rangle$. Indeed, according to Equation (2.100), the scalar product of $|g\rangle$ with itself reduces to:

$$\langle g|g\rangle = \int_{-\infty}^{+\infty} G^*(f)G(f)df = \int_{-\infty}^{+\infty} g^*(t)g(t)dt$$

Or, equivalently:

$$\langle g|g\rangle = \int_{-\infty}^{+\infty} |G(f)|^2 df = \int_{-\infty}^{+\infty} |g(t)|^2 dt \qquad (2.162)$$

And that is it. From an algebraic point of view, Parseval's theorem is nothing to write home about: it is just a particular statement of the basis independence of the scalar product. But that is not to say that Parseval's theorem is not useful: it comes in handy when we need to evaluate difficult integrals of complicated functions, but whose Fourier transforms have a simpler form. Some examples are given in the exercise list.

2.13   The Fourier Series

The Fourier series can be interpreted as the Fourier transform of a periodic signal. To highlight the connection between the two types of transforms, we will adopt a strategy where we start from a nonperiodic function, and then repeat it indefinitely to create a periodic function. In this way, it will be easy to see the relationship between the Fourier transform of the 'original', nonperiodic function, and its periodic version.

The connection between a nonperiodic function and its periodic version can be obtained by means of an impulse train. As illustrated in Figure 2.22a, an impulse train is a periodic series of impulse functions, each located at a multiple of the period T. Denoting the impulse train as $imt(t)$, its mathematical definition reads:

$$imt(t) = \sum_{n=-\infty}^{\infty} \delta(t - nT) \tag{2.163}$$

In Equation (2.163), n is an integer number. Thus, each value of n defines an impulse function 'located' at $t = nT$.

We also need its Fourier transform $IMT(f)$, which is given by:

$$IMT(f) = \frac{1}{T} \sum_{k=-\infty}^{\infty} \delta\left(f - k\frac{1}{T}\right) \tag{2.164}$$

In Equation (2.164), k is an integer number.

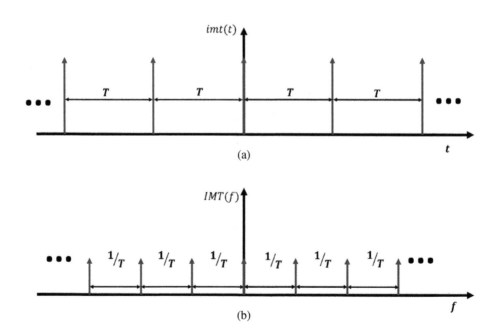

Figure 2.22 Illustration of an impulse train in the time domain (a) and in the frequency domain (b).

A rigorous proof of $IMT(f)$ would take us too far away from our path, so I will only motivate it. Thus, taking the Fourier transform of $imt(t)$, we get:

$$IMT(f) = \int_{-\infty}^{\infty} imt(t)e^{-i2\pi ft}dt = \int_{-\infty}^{\infty} \left[\sum_{n=-\infty}^{\infty} \delta(t-nT) \right] e^{-i2\pi ft}dt$$

Inverting the order of the integral and sum:

$$IMT(f) = \sum_{n=-\infty}^{\infty} \left[\int_{-\infty}^{\infty} e^{-i2\pi ft}\delta(t-nT)dt \right]$$

Using the sifting property of the impulse function:

$$IMT(f) = \sum_{n=-\infty}^{\infty} e^{-i2\pi fnT} \tag{2.165}$$

It can be proved that the sum in Equation (2.165) reduces to the sum in Equation (2.164) (in other words: $\sum_{n=-\infty}^{\infty} e^{-i2\pi fnT} = \frac{1}{T}\sum_{k=-\infty}^{\infty} \delta(f-k\frac{1}{T})$). Though we are not going to evaluate the sum in Equation (2.165) explicitly, we can obtain a reasonable intuition about why it reduces to the sum in Equation (2.164) by noticing that, if $f \neq k/T$, then the exponentials add up to zero (similarly to why Equation (2.74) adds up to zero if $f \neq 0$: the exponentials cancel each other). However, if f is a multiple of $1/T$, that is, if $f = k/T$, then the arguments of all exponentials become a multiple of 2π, which means that all exponentials take on the value one. So, at these points, we are adding one infinite times, which, of course, results in an infinite number. This is the origin of the impulse functions located at multiples of $1/T$ (see Figure 2.22b).

According to Equation (2.26), $g(t) * \delta(t-a) = g(t-a)$. In other words, convolving a function with an impulse function 'located' at $t = a$ shifts the function to $t = a$. So, what happens if we convolve a function with the impulse train $(g(t) * imt(t))$? Now we have a series of impulse functions located at $t = nT$, so, as illustrated in Figure 2.23, the convolution with each of these impulse functions creates a copy of $g(t)$ centred at $t = nT$. Indeed:

$$g(t) * imt(t) = g(t) * \sum_{n=-\infty}^{n=\infty} \delta(t-nT) = \sum_{n=-\infty}^{n=\infty} g(t) * \delta(t-nT) = \sum_{n=-\infty}^{n=\infty} g(t-nT)$$
$$\tag{2.166}$$

Thus, we can create a periodic version of $g(t)$ by convolving it with the impulse train. Let us denote this periodic version of $g(t)$ by $g_p(t)$. Thus:

$$g_p(t) = g(t) * imt(t) \tag{2.167}$$

Our goal is to find the Fourier transform of the periodic function $g_p(t)$ in terms of the Fourier transform of $g(t)$ and $imt(t)$. Using the property of convolution in

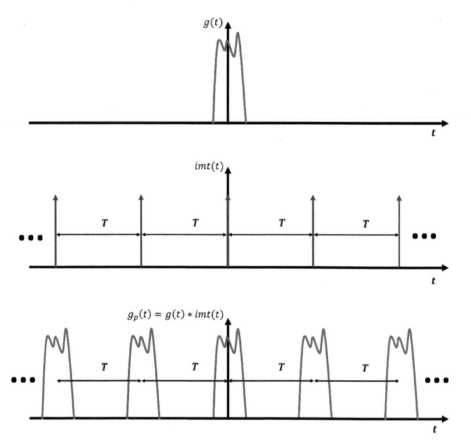

Figure 2.23 Representation of the creation of a periodic function through the convolution with an impulse train.

the time domain (Equation (2.144)), we get:

$$G_p(f) = G(f)IMT(f) = \frac{G(f)}{T} \sum_{k=-\infty}^{\infty} \delta\left(f - k\frac{1}{T}\right) \tag{2.168}$$

Notice that we used Equation (2.164) in Equation (2.168). According to Equation (2.168), the Fourier transform of a periodic function $g_p(t)$ consists of an impulse train modulated by the Fourier transform $G(f)$. Thus, $G_p(f)$ is zero everywhere, expect at frequencies which are multiples of $1/T$. This motivates the definition of the 'fundamental frequency' f_T:

$$f_T = \frac{1}{T} \tag{2.169}$$

Recasting the argument of the delta functions in Equation (2.168) in terms of the fundamental frequency, we obtain:

$$G_p(f) = G(f)IMT(f) = \frac{G(f)}{T} \sum_{k=-\infty}^{\infty} \delta(f - kf_T) \qquad (2.170)$$

Now we evaluate the inverse Fourier transform of $G_p(f)$.

Due to the presence of the impulse train in the frequency domain, the inverse Fourier transform of the periodic function is reduced to a sum. Indeed:

$$g_p(t) = \int_{-\infty}^{+\infty} G_p(f)e^{i2\pi ft}df = \int_{-\infty}^{+\infty} \left[\frac{G(f)}{T} \sum_{k=-\infty}^{\infty} \delta(f - kf_T) \right] e^{i2\pi ft}df$$

Inverting the order of the integral and sum:

$$g_p(t) = \int_{-\infty}^{+\infty} G_p(f)e^{i2\pi ft}df = \frac{1}{T} \sum_{k=-\infty}^{\infty} \int_{-\infty}^{+\infty} G(f)e^{i2\pi ft}\delta(f - kf_T)df$$

Using the sifting property of the impulse functions, we get:

$$g_p(t) = \frac{1}{T} \sum_{k=-\infty}^{\infty} G(kf_T)e^{i2\pi kf_T t} \qquad (2.171)$$

According to Equation (2.171), the inverse Fourier transform of a periodic function is a sum involving 'samples' of $G(f)$ taken at multiples of the fundamental frequency $(f = kf_T)$. The sum over k motivates the definition of a discrete function $G[k]$, where the brackets denote that the function is discrete, that is, its argument must be an integer number. We define $G[k]$ as being the discrete function whose values are samples of $G(f)$ at $f = kf_T$:

$$G[k] = \frac{1}{T}G(kf_T) \qquad (2.172)$$

Notice that we have incorporated the term $1/T$ in Equation (2.172).

With this definition, Equation (2.171) can be expressed as:

$$g_p(t) = \sum_{k=-\infty}^{\infty} G[k]e^{i2\pi kf_T t} \qquad (2.173)$$

Furthermore, notice that Equation (2.172) entails that:

$$G[k] = \frac{1}{T}G(kf_T) = \frac{1}{T} \int_{-\infty}^{\infty} g(t)e^{-i2\pi kf_T t}dt \qquad (2.174)$$

But, by construction, $g(t)$ is limited to the central period of $g_p(t)$ (see Figure 2.23), so it must be zero outside the interval $-T/2 < t < T/2$. Therefore, the integral in Equation (2.174) can be expressed as:

$$G[k] = \frac{1}{T}G(kf_T) = \frac{1}{T}\int_{-\frac{T}{2}}^{\frac{T}{2}} g(t)e^{-i2\pi kf_T t}dt \qquad (2.175)$$

But $g_p(t) = g(t)$ inside the interval $-T/2 < t < T/2$; so, we can replace $g(t)$ with $g_p(t)$ inside the integral in Equation (2.175), which now reads:

$$G[k] = \frac{1}{T}\int_{-\frac{T}{2}}^{\frac{T}{2}} g_p(t)e^{-i2\pi kf_T t}dt \qquad (2.176)$$

Equation (2.176) and Equation (2.173) constitute a Fourier series pair:

FOURIER SERIES

DIRECT FOURIER SERIES

$$G[k] = \frac{1}{T}\int_{-\frac{T}{2}}^{\frac{T}{2}} g_p(t)e^{-i2\pi kf_T t}dt$$

INVERSE FOURIER SERIES

$$g_p(t) = \sum_{k=-\infty}^{\infty} G[k]e^{i2\pi kf_T t}$$

As is apparent from the mathematical derivation we followed, the Fourier series is a particular case of the Fourier transform: it is the Fourier transform applied to a periodic function (one can also interpret the Fourier transform as the Fourier series in the limit of $T \to \infty$; you will be asked to demonstrate this in the exercise list at the end of this chapter).

The most significant difference between the Fourier transforms and the Fourier series is that the latter involves only frequencies that are multiples of the fundamental frequency. We have seen that this is due to the product with the impulse train in the frequency domain, but we can get a more intuitive understanding with the help of Figure 2.24. It shows three components with different frequencies, highlighting their relationship with the fundamental period T, which is marked by the green boxes. Both Figure 2.24a and Figure 2.24b show three fundamental periods (three green boxes). In Figure 2.24a, all components have frequencies which are multiples of the fundamental frequency (the frequency of the black component is the fundamental frequency, the frequency of the red component is twice the fundamental frequency, and the frequency of the blue component is thrice the fundamental

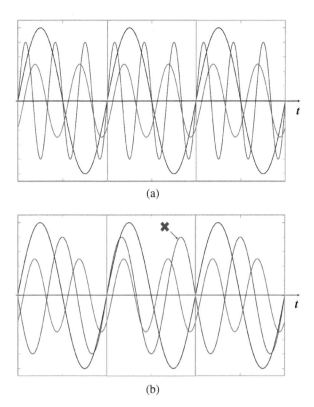

(a)

(b)

Figure 2.24 Illustration of the reason why the Fourier series involve only multiples of the fundamental frequency. The green boxes mark the fundamental period. (a) All components have frequencies which are multiples of the fundamental frequency; consequently, their patterns are perfectly repeated in each period; (b) the blue component's frequency is not a multiple of the fundamental frequency; consequently, its form is not repeated in each period.

frequency). Consequently, all three components 'fit' inside the periods; in other words, the pattern of each component is perfectly replicated within each green box. Thus, when we add these three components together, we get a pattern that is also perfectly replicated inside each green box. In other words, if we add these three components together, we get a periodic function with period T (recall that T is the 'length' of each green box).

Now look at Figure 2.24b, where the black and red components still have frequencies which are multiples of the fundamental frequency, but not the blue component: in this example the frequency of the blue component is NOT a multiple of the fundamental frequency. Consequently, the blue component features different patterns in different green boxes (for instance, there is only one positive peak in the first green box, but two positive peaks in the second). Thus, if these three components are added together, the result is not a periodic function with period T because the contribution of the blue component spoils the periodicity.

Thus, a periodic function can only be built from multiples of the fundamental frequency because only multiples of the fundamental frequency 'fit' inside the fundamental period.

Another significant feature of the Fourier series is that, according to Equation (2.172) (repeated below), the Fourier series components $G[k]$ are samples of the Fourier transform $G(f)$:

Equation (2.172)

$$G[k] = \frac{1}{T}G(kf_T)$$

Recall that $G(f)$ is the Fourier transform of $g(t)$ and, by construction, $g(t)$ is the 'central period' of $g_p(t)$ – that is, it is the period around the origin, corresponding to $n = 0$ in Equation (2.166). Thus, if we already know $G(f)$, then we do not need to calculate $G[k]$ from scratch: we can use Equation (2.172).

The example below illustrates this idea.

Worked Exercise: The Fourier Series of a Square Wave

Find the Fourier series components of the square wave in Figure 2.25, assuming that the amplitude of the square wave is one.

Solution:

The square wave is a periodic wave, whose central period is the rectangular function. We have already found the Fourier transform of the rectangular function in the Worked Exercise: The Fourier Transform of the Rectangular Function, of this chapter. The result is a *sinc* function, which I repeat below for convenience:

Equation (2.86)

$$S(f) = 2c \cdot sinc(2\pi fc) = 2c \cdot \frac{sin(2\pi fc)}{2\pi fc}$$

The square wave in Figure 2.25 has a period of:

$$T = 4c \tag{2.177}$$

So, the fundamental frequency is (see Equation (2.169)):

$$f_T = \frac{1}{4c} \tag{2.178}$$

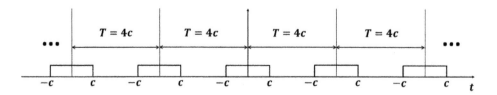

Figure 2.25 A square wave.

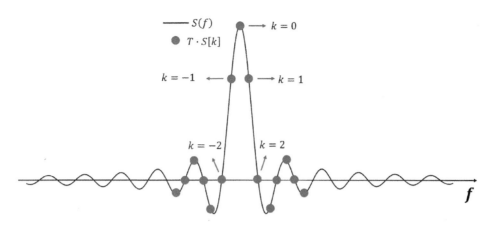

Figure 2.26 Illustration of the relationship between the Fourier transform of a rectangular function $S(f)$ and the Fourier series of the square wave $S[k]$. The illustration shows $S[k]$ for $k = 0, \pm 1, \pm 2, \pm 3, \pm 4, \pm 5, \pm 6$ and ± 7.

From Equation (2.172), we get:

$$S[k] = \frac{1}{T}S(kf_T) = \frac{1}{4c}S(kf_T)$$ (2.179)

Using Equation (2.86) in Equation (2.179), we find:

$$S[k] = \frac{1}{2}\frac{sin(2\pi k f_T c)}{2\pi k f_T c}$$ (2.180)

Finally, substituting Equation (2.178) into Equation (2.180), we obtain:

$$S[k] = \frac{1}{2}\frac{sin\left(2\pi k\frac{1}{4c}c\right)}{2\pi k\frac{1}{4c}c} = \frac{1}{2}\frac{sin\left(\frac{\pi k}{2}\right)}{\frac{\pi k}{2}} = \frac{1}{2}sinc\left(\frac{\pi k}{2}\right)$$ (2.181)

The relationship between $S(f)$ and a few values of $S[k]$ is illustrated in Figure 2.26. Notice that $S[k] = 0$ for $k = \pm 2, \pm 4, \pm 6 \ldots$. That makes sense: the argument of the sine function in the numerator of $S[k]$ is a multiple of π for these values of k, and the sine of a multiple of π is zero.

2.14 Exercises

Exercise 1
Evaluate the convolution between $a(t) = rect(t)$ – see Equation (2.7) – and $b(t) = e^{-\sigma t}u(t)$, where σ is a real and positive number and $u(t)$ is the Heaviside step function ($u(t) = 1$ for $t \geq 0$ and $u(t) = 0$ for $t < 0$).

Exercise 2

Suppose that two signals $a(t)$ and $b(t)$ begin at $t = -a$ and end at $t = +a$. Both signals are zero outside this interval. Find the interval within which their convolution is different from zero.

Exercise 3

Consider the following equation:

$$b(t) = a(t) * q(t)$$

where $q(t)$ is defined in Equation (2.14).

(a) Sketch a signal $a(t)$ and, based on it, sketch the signal $b(t)$.

(b) Analyze qualitatively the relationship between $a(t)$ and $b(t)$ for different values of T, as defined in Equation (2.14).

Exercise 4

Prove that $\int_{-\infty}^{+\infty} \delta^2(t)dt = \infty$.

  Hint: evaluate $\int_{-\infty}^{+\infty} q^2(t)dt$ and then take the limit $T \to 0$.

Exercise 5

Projecting Equation (2.37) on $\langle \delta(t_1)|$, prove that $\langle \delta(t_1)|g \rangle = g(t_1)$.

  Hint: use the orthonormality of the basis of impulse functions (Equation (2.49)).

Exercise 6

Are the exponential functions $e^{i2\pi ft}$ square integrable (see Equation (2.1))? Do they belong to the function space?

Exercise 7

We have seen that the direct Fourier transform is a scalar product. But the inverse Fourier transform (Equation (2.82)) can also be interpreted as a scalar product. Prove that this is true by evaluating $\langle \delta(t_0)|g \rangle$ in the frequency domain.

  Hint: insert the closure relation of the exponential basis (Equation (2.98)).

Exercise 8

Find the analogue in Euclidean space of the direct Fourier transform (Equation (2.81)).

Exercise 9

Find the analogue in Euclidean space of Equation (2.75).

  Hint: recall that this equation expresses the orthonormality of the basis of exponentials.

Exercise 10

Find the Fourier transform of $\delta(t)$. Interpret your result.

Exercise 11

Find the Fourier transform of $\delta(t - t_0)$. Interpret your result.

Exercise 12

Find the Fourier transform of the constant function $g(t) = 1$. Sketch a plot of the result and compare it with the Fourier transform of $\delta(t)$ from the point of view of the symmetry between time and frequency.

Hint: to evaluate the Fourier transform of $g(t) = 1$, treat it as $g(t) = e^{i2\pi f_0 t}$ with $f_0 = 0$, and use the orthonormality of the basis of exponentials (Equation (2.77)).

Exercise 13

Find the Fourier transform of the exponential $e^{i2\pi f_0 t}$. Interpret your result algebraically: what is its analogue in the Euclidean space?

Exercise 14

In Equation (2.105), we expressed the inverse Fourier transform resulting in a real function $g(t)$ in terms of the magnitude and phase:

Equation (2.105)

$$g(t) = \int_{-\infty}^{+\infty} |G(f)|\cos[2\pi ft + \theta(f)]df$$

Sometimes this relationship is expressed in the following equivalent form:

$$g(t) = \int_{-\infty}^{+\infty} A(f)\cos(2\pi ft)df + \int_{-\infty}^{+\infty} B(f)\sin(2\pi ft)df$$

Express $A(f)$ and $B(f)$ in terms of $|G(f)|$ and $\theta(f)$.

Exercise 15

Prove that the sine integral in Equation (2.104) vanishes if $G(f) = G^*(-f)$ – see Equation (2.112).

Exercise 16

We have seen that the Fourier transform of a real function $g(t)$ obeys the symmetry relation $G(f) = G^*(-f)$. What is the symmetry relation of a purely imaginary function $g(t)$?

Exercise 17

Find the condition that a real function $g(t)$ must satisfy so that its Fourier transform $G(f)$ is also a real function.

Exercise 18

Suppose that $g_g(t)$ is a Gaussian function:

$$g_g(t) = A \cdot e^{-\left(\frac{t}{\sigma}\right)^2} \tag{2.182}$$

Interestingly, the Fourier transform of a Gaussian function is another Gaussian function:

$$G_g(f) = A \cdot \sigma\sqrt{\pi}e^{-(\pi\sigma f)^2} \tag{2.183}$$

Now suppose that $g_{gm}(t)$ is a modulated Gaussian function:

$$g_{gm}(t) = g_g(t) \cdot \cos(2\pi f_0 t)$$

Assuming that $f_0 \gg 1/\sigma$, sketch the Fourier transform of $g_{gm}(t)$.
If the energy of $g_g(t)$ is E, what is the energy of $g_{gm}(t)$?
Hint: evaluate the energy in the frequency domain (using Parseval's theorem) and take advantage of the fact that there is no spectral overlap when $f_0 \gg 1/\sigma$.

Exercise 19
Prove that the Fourier transform can be obtained from the Fourier series in the limit of $T \to \infty$.
Hint: notice that $1/T = \Delta f$, and that Equation (2.171) is the Riemann sum that converges to the inverse Fourier transform in the limit of $\Delta f \to 0$.

Exercise 20
Without solving any integral, prove that the result of the convolution of the function $S(f) = 2c \cdot sinc(2\pi f c)$ with itself is itself.
Hint: use the results of Worked Exercise: The Fourier Transform of a Rectangular Function.

Exercise 21
Using Parseval's theorem, prove that:

$$\int_{-\infty}^{+\infty} [2c \cdot sinc(2\pi f c)]^2 df = 2c$$

3

Representation of Systems

Learning Objectives

Chapter 2 dealt with the representation of signals, which are vectors in the function space. In this chapter, we deal with the representation of systems, which play the role of operators in the function space. We are interested in a particular class of systems, the so-called linear and time invariant (LTI) systems. Such systems are ubiquitous in engineering and physics, and the main subject of this chapter is the theory of their representation and properties.

We will learn that exponentials are eigenvectors of LTI operators; consequently, their representation in the frequency domain is diagonal and, as such, it is greatly simplified. This feature leads to the concept of 'frequency response' and to the key property of frequency conservation in LTI systems. Frequency conservation is one of the most useful and powerful concepts in engineering and physics. At the end of the chapter, we will also catch a glimpse of how the notion of frequency conservation appears in other fields.

3.1 Introduction and Properties

As discussed in the Interlude: Signals and Systems: What is it About?, a system is anything that modifies a signal. We define the system's input as the signal that will be acted upon by the system, and the system's output as the signal resulting from the action of the system on the input.

Recall that, in linear algebra, an operator is a mathematical object that transforms a vector into another vector. If a signal is a vector, and a system is something that transforms a signal into another signal, then the mathematical object describing a system must be an operator. Following the notation of Chapter 1 (Review of

Essentials of Signals and Systems, First Edition. Emiliano R. Martins.
© 2023 John Wiley & Sons Ltd. Published 2023 by John Wiley & Sons Ltd.
Companion website: www.wiley.com/go/martins/essentialsofsignalsandsystems

Figure 3.1 Representation of the action of a system.

Linear Algebra), we denote operators acting on signals by a capital letter followed by curly brackets, like $T\{\}$. Thus, if $x(t)$ is the time domain representation of an input signal, then the time domain representation of the output signal $y(t)$ is denoted by:

$$y(t) = T\{x(t)\} \tag{3.1}$$

Equation (3.1) can be schematically represented as shown in Figure 3.1.

Since a system can be anything that modifies a signal, there is an enormous range of systems in the world. We are mainly interested in systems that obey two properties: linearity and time invariance. We begin by spelling out the meanings of these two properties (and of a third one: causality). Then, in Sections 3.2–3.9, we focus attention on systems obeying the properties of linearity and time invariance.

3.1.1 Linearity

The property of linearity is quite straightforward and is formally identical to the definition of linearity of an operator in linear algebra (see Equation (1.40)). If you skipped Chapter 1 (Review of Linear Algebra), I suggest you take a look at the discussion about the physical meaning of linearity, which follows Equation (1.40) in Section 1.4 (Linear Operators).

Thus, just like in Equation (1.40), an operator $T\{\}$ is linear if and only if:

$$T\{ax_1(t) + bx_2(t)\} = aT\{x_1(t)\} + bT\{x_2(t)\} \tag{3.2}$$

where a and b are constants.

Notice that I am not saying that every system in the universe obeys Equation (3.2). What I am saying is that the ones that do obey it are classified as linear.

3.1.2 Time Invariance

The concept of time invariance is also quite straightforward. Let me first define time invariance mathematically, and then we spell out its meaning.

If:

$$y(t) = T\{x(t)\}$$

then $T\{\}$ is time invariant if and only if:

$$T\{x(t - t_0)\} = y(t - t_0) \tag{3.3}$$

According to Equation (3.3), a system is time invariant if the system itself does not change in time. Thus, if the system itself does not change in time, and if you put $x(t)$ in its input and obtain $y(t)$ in the output, then you should get the same result if you repeat the same experiment some time t_0 later. In other words, if $x(t - t_0)$ is the input (same signal as before, only plugged in some time t_0 later), you should get $y(t - t_0)$ (same output as before, but coming out some time t_0 later).

Let us exemplify this property using songs again. Suppose that you have a certain song $x(t)$ which has an annoying high-frequency part that you want to get rid of. This part is depicted as the fast oscillation in Figure 3.2a. If you want to remove the high-frequency part of a signal, you must pass it through a filter. Later, in Chapter 5 (Filters), we will look at filters in more detail. For now, just suppose that it worked fine, and the pesky bit was removed, as represented in Figure 3.2b. So, Figure 3.2a is the input of the system, and Figure 3.2b is the output.

If the system is time invariant, it does not matter when you put the signal in. So, suppose you put the same signal as before, but some time t_0 later, as represented in Figure 3.2c. If the system is time invariant, and only if it is time invariant,

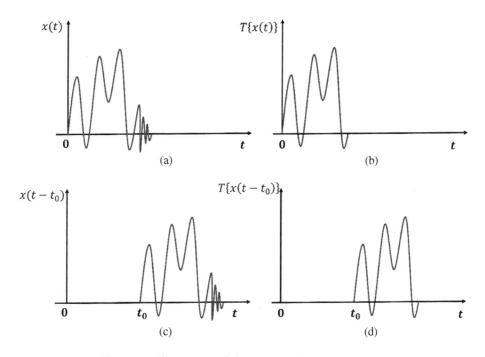

Figure 3.2 Illustration of the concept of time invariance.

then the output is a time-shifted version of the previous output, as represented in Figure 3.2d. If the system was not time invariant, then you would compare $T\{x(t)\}$ with $T\{x(t - t_0)\}$ and find that they have nothing to do with each other: they would be completely different signals.

Worked Exercise: Example of a Time Invariant System

$T\{\}$ is defined as:

$$T\{x(t)\} = \frac{dx}{dt}$$

Determine whether $T\{\}$ is time invariant or not.

Solution:

To test if it is time invariant, we have to compare two cases. In the first case, we put a signal in the input, obtain the output, and then time shift the output. In the second case, we put a time shifted signal in the input and obtain the output. Then we compare both outputs: if they are identical, then the system is time invariant; if not, it is not.

So, let us test. First, we put $x(t)$ in the input, and call the output $y(t)$:

$$y(t) = T\{x(t)\} = \frac{dx}{dt} \tag{3.4}$$

Now we time shift the output:

$$y(t - t_0) = \frac{dx(t - t_0)}{dt} \tag{3.5}$$

This is the first case. Now we move on to the second case, in which we time shift the input. To avoid getting lost in the notation, let us define $g(t) = x(t - t_0)$ and let us denote the output of the time shifted input by $z(t)$. Thus:

$$z(t) = T\{g(t)\} = \frac{dg}{dt}$$

But $g(t) = x(t - t_0)$, so:

$$z(t) = \frac{dg}{dt} = \frac{dx(t - t_0)}{dt} \tag{3.6}$$

That was the second case. Now we compare them: is $z(t)$ identical to $y(t - t_0)$? Comparing Equation (3.6) with Equation (3.5), we conclude that they are indeed identical, which means that the system is time invariant. That makes sense: the system is just taking the derivative of the signal, and a derivative does not depend on absolute time, just on how the signal **varies** in time.

Worked Exercise: An Example of a Time Variant System

$T\{\}$ is defined as:

$$T\{x(t)\} = x(t) + t \tag{3.7}$$

Determine whether $T\{\}$ is time invariant or not.

Solution:

We follow the same recipe as the previous example. In the first case, we put the signal $x(t)$ in the input, and obtain the output $y(t)$:

$$y(t) = T\{x(t)\} = x(t) + t$$

And now we time shift the output $y(t)$:

$$y(t - t_0) = x(t - t_0) + t - t_0 \tag{3.8}$$

That was the first case. Now we put the time shifted signal $g(t) = x(t - t_0)$ in the input:

$$z(t) = T\{g(t)\} = g(t) + t \tag{3.9}$$

Notice that we obeyed the 'command' of the definition of $T\{\}$ (Equation (3.7)): the output is whatever goes in the input plus the time t. By construction, $g(t) = x(t - t_0)$, so:

$$z(t) = g(t) + t = x(t - t_0) + t \tag{3.10}$$

Now we compare the two cases: is $z(t)$ identical to $y(t - t_0)$? Comparing Equation (3.10) with Equation (3.8), we conclude that they are not identical. Thus, the system is NOT time invariant. That makes sense: the system depends explicitly on the time t, so the output depends on what time the input signal is plugged in.

3.1.3 Causality

If I push you, you fall. But, unless you are a soccer player, you only fall after I push you, not before. That is because you and I are causal systems. A causal system is a system that responds to the input signal only after the signal is applied, but not earlier (we allow for a causal system to be 'instantaneous', that is, for the output to depend on the input at the same instant). Of course, all physical systems we are aware of at the time of writing of this textbook are causal (with the shameful exception of soccer players).

The concept of causality is this simple, but it has quite interesting consequences in engineering and physics. For example, the way materials reflect light is tied to the way they absorb light, and this is a consequence of causality (this kind of relationship is more generally known as Kramers–Kronig relations). It will also have some interesting consequences in our course, as we will see later.

Notice that, if $y(t)$ is the output of a causal system, then its value at a specific time t_0, that is, $y(t_0)$, can only depend on the input $x(t)$ for $t \leq t_0$. Otherwise, it would be depending on a future value of $x(t)$.

3.2 Operators Representing Linear and Time Invariant Systems

As mentioned in the Interlude: Signals and Systems: What is it About?, many important systems are linear and time invariant (LTI). As a matter of fact, even nonlinear systems often behave as approximately linear when the input signal is not too strong. For example, according to Hooke's law, the restoring force of a spring depends linearly on the distance from the resting point. But this is not really true: if a spring is compressed or stretched too much, then the force is no longer linear. However, most of the time the force is not sufficiently strong for the nonlinear effects to appear, and so the system can be treated as linear. Thus, under certain conditions, the theory of linear systems is applicable even to nonlinear systems. As to time invariance, the laws of physics themselves are time invariant (by the way, that is why energy is conserved, as beautifully captured by Noether's theorem). Since physical systems obey the laws of physics, they must also be time invariant. Notice, however, that often we are interested in only a part of the system (technically, that means we are interested in an open system), and in this case the system may not be time invariant (that is why energy is conserved only in closed systems, but not in open systems). Nevertheless, it is still the case that many systems of interest are time invariant. So, from now on, we assume that the systems we are dealing with are LTI.

The most general mathematical representation of an LTI system is a linear differential equation operator of the form:

$$A\{\} = \sum_{n=1}^{\infty} a_n \frac{d^n}{dt^n} + a_0.$$ (3.11)

where a_0 and a_n are constants.

The 'command' of Equation (3.11) is: what goes inside the curly brackets, goes in the 'slots' on the right-hand side. Thus:

$$A\{x(t)\} = \sum_{n=1}^{\infty} a_n \frac{d^n x(t)}{dt^n} + a_0 \cdot x(t)$$ (3.12)

Notice that, even if an LTI system involves integrals, it can be put in the form of Equation (3.12), with the caveat that $A\{x(t)\}$ may be identified with derivatives of the output (an example is shown in the exercise list). Thus, we can assume a form as in Equation (3.12) without loss of generality.

For pedagogical reasons, we will develop the theory assuming operators involving derivatives up to second order. So, we assume that the operator has the form:

$$A\{\} = a_2\frac{d^2}{dt^2} + a_1\frac{d}{dt} + a_0. \tag{3.13}$$

The reason why we will assume the more restricted form of Equation (3.13) instead of Equation (3.11) is that the notation of Equation (3.13) is less cluttered, which helps to understand the concepts. At the same time, we do not lose anything conceptually in assuming the form of Equation (3.13): what we learn assuming this form can be easily extended to the form of Equation (3.11).

Now we begin to construct the theory of representation of LTI systems, still following our strategy of comparing the mathematical objects in the function space with their analogues in the Euclidean space. Thus, our first task is to identify the analogue of an LTI operator in the Euclidean space.

3.3   Linear Systems as Matrices

In Chapter 1 (Review of Linear Algebra), we learned that a linear operator is represented by a matrix. Furthermore, we have just seen that the operator representing an LTI system is a linear differential equation. So, the Euclidean analogue of a linear differential equation must be a matrix. Furthermore, a differential equation is an operator acting on a function of time. But, in Chapter 2 (Representation of Signals), we learned that a function of time is a particular representation of a signal: it is the time domain representation. In other words, it is the representation in the basis of impulse functions. Thus, since a differential equation is an object that acts on the coordinates with respect to the basis of impulse functions, it must be a particular representation of an operator: it is the representation in the basis of impulse functions. Still in Chapter 2 (Representation of Signals), we saw that the basis of impulse functions is analogous to the canonical basis in the Euclidean space. Thus, we conclude that a differential equation must be analogous to a matrix representation in the canonical basis.

If you have already taken a course in numerical methods, the assertion that a differential equation is akin to a matrix will come as no surprise: after all, that is precisely the business of numerical methods: to solve differential equations in a computer by treating signals as vectors and derivatives as matrices. If, however, you have never dealt with numerical methods, no worries: it is not difficult to see why a differential operator must be akin to a matrix. I will show you explicitly the simplest form: the first-order differential. The others follow a similar idea.

Suppose that we have a first-order operator of the form:

$$T\{\} = \frac{d}{dt} \tag{3.14}$$

Thus, the output–input relationship is:

$$y(t) = T\{x(t)\} = \frac{dx}{dt} \tag{3.15}$$

Now, let us represent the input and output of Equation (3.15) as a column vector equation by discretizing the time coordinates (see Equation (2.29)–Equation (2.31)). Thus, the input representation is of the form:

$$x(t) \approx \begin{pmatrix} \vdots \\ x(t_{-2}) \\ x(t_{-1}) \\ x(t_0) \\ x(t_1) \\ x(t_2) \\ \vdots \end{pmatrix} \tag{3.16}$$

where:

$$t_n = n\Delta t, \tag{3.17}$$

and Δt is the discretization time step.

Our goal is to find the column vector representation of $y(t)$ in terms of the column vector representation of $x(t)$. Recall that:

$$\frac{dx}{dt} \approx \frac{x(t) - x(t - \Delta t)}{\Delta t} \tag{3.18}$$

where the approximation sign turns into an equality in the limit of $\Delta t \to 0$.

Now, suppose that we want to work out $y(t_n)$, that is, the coordinate of the output at the 'slot' t_n (see Equation (2.31)): since $y(t)$ is the derivative of $x(t)$, we have:

$$y(t_n) \approx \frac{x(t_n) - x(t_n - \Delta t)}{\Delta t} \tag{3.19}$$

But Equation (3.17) entails that $t_n - \Delta t = n\Delta t - \Delta t = (n - 1)\Delta t = t_{n-1}$. Thus, Equation (3.19) reduces to:

$$y(t_n) \approx \frac{x(t_n) - x(t_{n-1})}{\Delta t} \tag{3.20}$$

Notice that the 'command' of Equation (3.20) is: to find the output at the slot t_n, subtract the input coordinate at the slot t_{n-1} from the input coordinate at the slot t_n, and then divide by Δt. Thus, Equation (3.20) can be represented by a matrix equation of the form:

$$
\begin{pmatrix} \vdots \\ y(t_{-2}) \\ y(t_{-1}) \\ y(t_0) \\ y(t_1) \\ y(t_2) \\ \vdots \end{pmatrix} = \frac{1}{\Delta t} \begin{pmatrix} \vdots & \vdots & \vdots & \vdots & \vdots \\ \cdots & 1 & 0 & 0 & 0 & 0 & \cdots \\ \cdots & -1 & 1 & 0 & 0 & 0 & \cdots \\ \cdots & 0 & -1 & 1 & 0 & 0 & \cdots \\ \cdots & 0 & 0 & -1 & 1 & 0 & \cdots \\ \cdots & 0 & 0 & 0 & -1 & 1 & \cdots \\ & \vdots & \vdots & \vdots & \vdots & \vdots \end{pmatrix} \begin{pmatrix} \vdots \\ x(t_{-2}) \\ x(t_{-1}) \\ x(t_0) \\ x(t_1) \\ x(t_2) \\ \vdots \end{pmatrix}
\tag{3.21}
$$

From which we readily infer that:

$$
T\{\} = \frac{d}{dt} \approx \frac{1}{\Delta t} \begin{pmatrix} \vdots & \vdots & \vdots & \vdots & \vdots \\ \cdots & 1 & 0 & 0 & 0 & 0 & \cdots \\ \cdots & -1 & 1 & 0 & 0 & 0 & \cdots \\ \cdots & 0 & -1 & 1 & 0 & 0 & \cdots \\ \cdots & 0 & 0 & -1 & 1 & 0 & \cdots \\ \cdots & 0 & 0 & 0 & -1 & 1 & \cdots \\ & \vdots & \vdots & \vdots & \vdots & \vdots \end{pmatrix}
\tag{3.22}
$$

Thus, we have confirmed that a differential equation is akin to a matrix. More precisely, a differential equation is a particular representation of a linear operator: it is the representation in the basis of impulse functions (that is, it is the time domain representation). Of course, this is not the only representation of a linear operator. Indeed, the main goal of this chapter is to show that the representation of the **same** operator in a different basis can greatly simplify the relationship between input and output. But, before we do that, let us introduce the Dirac notation for operators, to highlight the difference between the operator itself and its representation in a certain basis.

3.4   Operators in Dirac Notation

In Dirac notation, the representation of an operator in the function space is identical to its representation in the Euclidean space (for the latter, see Section 1.0 [The Dirac Notation]). Thus, if $|v\rangle$ is the vector acted upon by the operator and $|u\rangle$ is the vector resulting from the action, then the action of the operator A is denoted by:

$$
|u\rangle = A|v\rangle
\tag{3.23}
$$

We have already encountered an example of this relation in Equation (2.42).

Equation (3.23) is a general relation in the sense that it does not specify a basis. Recall from Section 1.5 (Representation of Linear Operators) that to specify a linear operator in the Euclidean space, we need to specify its matrix coefficients in a given basis. By specifying its matrix coefficients, we are specifying how it transforms the coordinates of one vector into the coordinates of another vector. Thus, a matrix specification is essentially a specification of how the operator acts in a specific basis.

Still in Section 1.5 (Representation of Linear Operators), we learned that to extract the coordinate representation of the general relation $\vec{u} = A\{\vec{v}\}$, we take the scalar product with the basis vectors. For example, if we want the x coordinates, we take the scalar product with $\hat{x}$. In other words, we evaluate $\hat{x}\blacksquare\vec{u} = \hat{x}\blacksquare A\{\vec{v}\}$. In Dirac notation, this relation is expressed as $\langle x|u\rangle = \langle x|A|v\rangle$. We say that such a relation was obtained by projecting the relation $|u\rangle = A|v\rangle$ on the bra x.

In the same spirit, to specify an operator in the function space, we need to specify how it acts in a specific basis. So, suppose that we already know how it acts in the basis of impulse functions. That means we know the differential equation that represents the system; in other words, we know how the operator maps the coordinates of $|v\rangle$ with respect to the basis of impulse functions into the coordinates of $|u\rangle$ with respect to the basis of impulse functions. To get this relationship explicitly, we take the scalar product of both sides of Equation (3.23) with a ket belonging to the basis of impulse functions. In other words, we project Equation (3.23) on a bra belonging to the basis of impulse functions. Thus:

$$\langle \delta(t_0)|u\rangle = \langle \delta(t_0)|A|v\rangle \tag{3.24}$$

Recall that $\langle \delta(t_0)|u\rangle = u(t_0)$ – see Equation (2.44). Thus:

$$u(t_0) = \langle \delta(t_0)|A|v\rangle \tag{3.25}$$

We are assuming that the operator has been specified in the basis of impulse functions. In other words, we are assuming that we know the differential equation; for example, that we know the coefficients a_2, a_1, and a_0 in Equation (3.13) (or a_n in Equation (3.11)). Thus, we know that:

$$u(t) = A\{v(t)\} = a_2\frac{d^2v(t)}{dt^2} + a_1\frac{dv(t)}{dt} + a_0 \cdot v(t) \tag{3.26}$$

Comparing Equation (3.26) with Equation (3.25), we conclude that:

$$u(t_0) = \langle \delta(t_0)|A|v\rangle = a_2\frac{d^2v(t_0)}{dt_0{}^2} + a_1\frac{dv(t_0)}{dt_0} + a_0 \cdot v(t_0) \tag{3.27}$$

The algebraic interpretation of Equation (3.27) is that it is specifying the operator A by specifying how it relates the coordinate $v(t_0)$ with the coordinate $u(t_0)$, just like a matrix equation in the Euclidean space. Thus, Equation (3.27) is analogous to the specification of the matrix coefficients in the canonical basis of the Euclidean space. Of course, since this specification is true for any t_0, we can replace it with a general t, which means that Equation (3.27) is mathematically identical to Equation (3.26). There is only a subtle difference of interpretation: whereas

Equation (3.26) is interpreted as connecting any pair of coordinates $v(t)$, $u(t)$, Equation (3.27) is interpreted as connecting a specific pair $v(t_0)$, $u(t_0)$.

But I hasten to remind you that this is not the only way to specify the operator A: we could have specified it in any basis of the space. In particular, we could have specified it in the basis of eigenvectors, which is the topic of Sections 3.5–3.7.

3.5   Statement of the Problem

Suppose that we have an LTI system described by the operator of Equation (3.26), and our job is to find the output in terms of the input. As we will see in some examples later in this chapter, often we work with the inverse operator. In other words, often, in a relationship of the form $u(t) = A\{v(t)\}$, it turns out that the physical input is $u(t)$ and the physical output is $v(t)$, which entails that we want to express $v(t)$ in terms of $u(t)$ – this is the case of electric circuits, for instance, and we will soon see some explicit examples. But, for now, just consider the task that lies ahead of us: we need an expression for $v(t)$ in terms of $u(t)$. In other words, we need to solve a differential equation.

Why is it difficult to solve a differential equation? It is easy to know why it is difficult: it is because it mixes coordinates. That is, the coordinates $u(t)$ depend on coordinates of $v(t)$ at different time slots. For example, in Equation (3.20), the coordinate $y(t_n)$ depends not only on $x(t_n)$, but also on $x(t_{n-1})$. In short, a differential equation is difficult to solve because the operator is not diagonal, and as such it mixes the coordinates.

But this difficulty is identical to the one we encountered with matrix equations in Chapter 1 (Review of Linear Algebra). There, we reviewed the general method for solving matrix equations: first, change the basis to the basis of eigenvectors, thus diagonalizing the matrix; then, find the coordinates of one vector in terms of the other in the basis of eigenvectors, which is easy-peasy because there is no mixing of coordinates; finally, go back to the original – or canonical – basis.

So, we shall not reinvent the wheel here: we are going to approach the same problem with the same strategy. Thus, instead of going about torturing ourselves by trying to solve differential equations with all that faff about homogeneous and particular solutions and whatnot, we follow the good old strategy of changing the basis to the basis of eigenvectors.

3.6   Eigenvectors and Eigenvalues of LTI Operators

But what are the eigenvectors of LTI operators? Maybe you are having a panic attack because you remember from linear algebra courses that finding eigenvectors in a two-dimensional space is already a bit annoying, in a three-dimensional space is physically painful, and in a four-dimensional space is a violation of the human rights. So, finding eigenvectors in a space of infinite dimension, like the function space, must be as easy as finding the meaning of life.

Surprisingly, it is quite easy to find the eigenvectors of LTI operators. Recall that, by definition, an eigenvector is a vector that, when acted upon by the operator, results in a multiple of itself. Thus, if $\varphi(t)$ is an eigenvector of the operator $A\{\}$, it follows that:

$$A\{\varphi(t)\} = \lambda\varphi(t) \tag{3.28}$$

But an LTI operator involves only time derivatives (see Equation (3.11) or Equation (3.13)). So, if we are dealing with LTI operators, then the question 'what function is $\varphi(t)$?' reduces to the question 'what function returns a multiple of itself when it is differentiated?' That is easy: it is the exponential function. Indeed, any function of the form $e^{\alpha t}$ returns a multiple of itself when differentiated. Thus, any function of the form $e^{\alpha t}$ is an eigenvector of any LTI operator. Is not that funny? We do not even need to know what specific operator we are dealing with: if it is LTI, then we know beforehand that $e^{\alpha t}$ is an eigenvector.

But hang on! Knowing that $e^{\alpha t}$ is an eigenvector is not sufficient for our strategy: we need not only an eigenvector, but a basis of eigenvectors. Do we know any set of functions of the type $e^{\alpha t}$ that forms a basis for the function space? Fortunately, we do, as we dedicated some effort in Chapter 2 (Representation of Signals) to prove that functions of the type $e^{i2\pi ft}$ form a basis. Thus, since the set of functions $e^{i2\pi ft}$ form a basis of eigenvectors of LTI operators, the representation in the frequency domain is much simplified, because the operators become diagonal in this basis, and as such there is no mixing of coordinates.

So, we have all the ingredients to find a simple relationship between input and output of an LTI system: go to the frequency domain, solve the problem there, and then come back to the time domain. In Section 3.7, we will spell out each of these steps, making the connections with the analogous steps in the Euclidean space.

3.7   General Method of Solution

Note: this section connects the method of solving differential equations using Fourier transforms with the theory of signals and systems. The main goal of this section is to bring out the notion that the frequency domain representation of an LTI operator is diagonal. Instructors and students who are already comfortable with this idea may opt to skip Section 3.7, and go directly to the representation in terms of convolution in the time domain, and product in the frequency domain, which is the subject of Section 3.8.

3.7.1   Step 1: Defining the Problem

Following the notation used in Section 1.7 (General Method of Solution of a Matrix Equation), we assume an operator A and begin with a general relation of the form:

$$|u\rangle = A|v\rangle \tag{3.29}$$

Now we assume that we are given the representation in the basis of impulse functions (in the function space) and in the canonical basis (in the two-dimensional Euclidean space). Thus:

$$u(t) = a_2 \frac{d^2v(t)}{dt^2} + a_1 \frac{dv(t)}{dt} + a_0 \cdot v(t) \leftrightarrow \begin{pmatrix} u_x \\ u_y \end{pmatrix} = \begin{pmatrix} a_{xx} & a_{xy} \\ a_{yx} & a_{yy} \end{pmatrix} \begin{pmatrix} v_x \\ v_y \end{pmatrix} \qquad (3.30)$$

For bookkeeping, we write each vector explicitly as a linear combination of basis vectors:

$$|v\rangle = \int_{-\infty}^{+\infty} v(t_0)|\delta(t_0)\rangle dt_0 \quad \leftrightarrow \quad |v\rangle = v_x|x\rangle + v_y|y\rangle$$

$$|u\rangle = \int_{-\infty}^{+\infty} u(t_0)|\delta(t_0)\rangle dt_0 \quad \leftrightarrow \quad |u\rangle = u_x|x\rangle + u_y|y\rangle \qquad (3.31)$$

Our goal is to find $v(t)$ in terms of $u(t)$ in the function space, and v_x, v_y in terms of u_x, u_y in the Euclidean space.

3.7.2   Step 2: Finding the Eigenvalues

In the Euclidean space, this step should be called 'finding the eigenvectors and eigenvalues'. But in the function space we already know the eigenvectors, so we only need to find the eigenvalues. You can do that straight away by plugging in the eigenvector $e^{i2\pi ft}$ in the differential equation (in other words, by evaluating $A\{e^{i2\pi ft}\} = \lambda e^{i2\pi ft}$). However, to highlight the algebraic meaning, I will also show you how to do it starting from the general relation.

Following the notation of Section 2.8 (Exponentials as a Basis: The Frequency Domain Representation of Signals), we denote each ket of the basis of eigenvectors by $|e(f_0)\rangle$. By definition, the eigenvalue of the operator A with respect to the eigenvector $|e(f_0)\rangle$ satisfies the relation:

$$A|e(f_0)\rangle = \lambda(f_0)|e(f_0)\rangle \qquad (3.32)$$

By assumption, we know the representation of A in the basis of impulse functions (in other words, we know the differential equation). So, we need to extract the time domain representation of Equation (3.32). To do so, we project Equation (3.32) on a bra of the impulse basis. Thus:

$$\langle\delta(t_0)|A|e(f_0)\rangle = \langle\delta(t_0)|\lambda(f_0)|e(f_0)\rangle = \lambda(f_0)\langle\delta(t_0)|e(f_0)\rangle \qquad (3.33)$$

Recall that $\langle\delta(t_0)|e(f_0)\rangle = e^{i2\pi f_0 t_0}$ – see Equation (2.73). With the help of the last equality in Equation (3.27), Equation (3.33) reduces to:

$$a_2 \frac{d^2 e^{i2\pi f_0 t_0}}{dt_0^2} + a_1 \frac{d e^{i2\pi f_0 t_0}}{dt_0} + a_0 \cdot e^{i2\pi f_0 t_0} = \lambda(f_0) e^{i2\pi f_0 t_0}$$

Evaluating the derivatives, we obtain:

$$[-a_2(2\pi f_0)^2 + ia_1 2\pi f_0 + a_0] \cdot e^{i2\pi f_0 t_0} = \lambda(f_0)e^{i2\pi f_0 t_0}$$

From which we readily infer that:

$$\lambda(f_0) = -a_2(2\pi f_0)^2 + ia_1 2\pi f_0 + a_0 \qquad (3.34)$$

Thus, we have found the eigenvalue $\lambda(f_0)$ with repsect to the ket $|e(f_0)\rangle$. The subscript 0 in f_0 is just a reminder that, in principle, we found the specific eigenvalue $\lambda(f_0)$, with respect to the specific ket $|e(f_0)\rangle$. But, of course, we can leave it in terms of a general eigenvalue, like this:

$$\lambda(f) = -a_2(2\pi f)^2 + ia_1 2\pi f + a_0 \qquad (3.35)$$

Equation (3.35) defines the 'spectrum of eigenvalues' of the operator. Recall from Section 1.9 (Representation of Linear Operators in Terms of Eigenvectors and Eigenvalues) that an operator is fully specified by its eigenvectors and eigenvalues. That is also true for LTI operators, with the simplification that the eigenvectors are the same for all LTI operators. Consequently, $\lambda(f)$ is a complete specification of the operator: if we know $\lambda(f)$, we know everything about the operator. That makes sense: the eigenvalues fully specify the diagonal matrix, that is, the operator in the basis of eigenvectors. In the same way, $\lambda(f)$ is the specification of the operator A in the basis of eigenvectors, that is, in the frequency domain. **Thus, we have two representations of the <u>same</u> operator: the time domain representation (that is, the differential equation) and the frequency domain representation (that is, $\lambda(f)$).**

3.7.3   Step 3: The Representation in the Basis of Eigenvectors

In step 3, we obtain the representation of the relation $|u\rangle = A|v\rangle$ in the basis of eigenvectors. Again, it is helpful to write the kets explicitly in this new basis. Thus:

$$|v\rangle = \int_{-\infty}^{+\infty} V(f_0)|e(f_0)\rangle df_0 \quad \longleftrightarrow \quad |v\rangle = V_1|a_1\rangle + V_2|a_2\rangle$$

$$|u\rangle = \int_{-\infty}^{+\infty} U(f_0)|e(f_0)\rangle df_0 \quad \longleftrightarrow \quad |u\rangle = U_1|a_1\rangle + U_2|a_2\rangle \qquad (3.36)$$

where $|a_1\rangle$ and $|a_2\rangle$ are the eigenvectors in the Euclidean space. I emphasize, once again, that we have not changed the kets: they are the same kets as in Equation (3.31). We only changed the basis. Let us make this explicitly one more time:

$$|v\rangle = \int_{-\infty}^{+\infty} v(t_0)|\delta(t_0)\rangle dt_0 = \int_{-\infty}^{+\infty} V(f_0)|e(f_0)\rangle df_0$$

$$|v\rangle = v_x|x\rangle + v_y|y\rangle = V_1|a_1\rangle + V_2|a_2\rangle$$

$$|u\rangle = \int_{-\infty}^{+\infty} u(t_0)|\delta(t_0)\rangle dt_0 = \int_{-\infty}^{+\infty} U(f_0)|e(f_0)\rangle df_0$$

$$|u\rangle = u_x|x\rangle + u_y|y\rangle = U_1|a_1\rangle + U_2|a_2\rangle \tag{3.37}$$

That was just bookkeeping. The real job is to find the representation of $|u\rangle = A|v\rangle$ in the basis of eigenvectors. I will first show you a way that highlights the algebraic meaning of what we are doing, but it is not the most practical way. Once we have understood the algebraic meaning of the steps, I will show you a short-cut based on the property of differentiation of Fourier transforms.

Our job is to find the representation of $|u\rangle = A|v\rangle$ in the basis of eigenvectors. That means we want the relationships between the coordinates of $|u\rangle$ and $|v\rangle$ in the basis of eigenvectors. To get the coordinates, we project both sides of $|u\rangle = A|v\rangle$ on a bra of the basis of eigenvectors:

$$\langle e(f_1)|u\rangle = \langle e(f_1)|A|v\rangle \tag{3.38}$$

But $\langle e(f_1)|u\rangle = U(f_1)$ – see Equation (2.95). Thus:

$$U(f_1) = \langle e(f_1)|A|v\rangle \tag{3.39}$$

To extract the coordinates of $|v\rangle$, we insert the closure relation (Equation (2.98)) before the ket on the right-hand side of Equation (3.39). Thus:

$$\langle e(f_1)|A|v\rangle = \langle e(f_1)|AI|v\rangle = \int_{-\infty}^{+\infty} df_0\langle e(f_1)A|e(f_0)\rangle\langle e(f_0)|v\rangle \tag{3.40}$$

But $A|e(f_0)\rangle = \lambda(f_0)|e(f_0)\rangle$ – see Equation (3.32) – and $\langle e(f_0)|v\rangle = V(f_0)$. Thus:

$$\langle e(f_1)|A|v\rangle = \int_{-\infty}^{+\infty} V(f_0)\lambda(f_0)\langle e(f_1)|e(f_0)\rangle df_0$$

The basis of exponentials is orthonormal, which means that $\langle e(f_1)|e(f_0)\rangle = \delta(f_0 - f_1)$ – see Equation (2.78). Therefore:

$$\langle e(f_1)|A|v\rangle = \int_{-\infty}^{+\infty} V(f_0)\lambda(f_0)\delta(f_0 - f_1)df_0 = V(f_1)\lambda(f_1) \tag{3.41}$$

Substituting Equation (3.41) into Equation (3.39), we conclude that:

$$U(f_1) = \lambda(f_1)V(f_1) \tag{3.42}$$

See, there is no mixing of coordinates: $U(f_1)$ depends only on $V(f_1)$: there is no $V(f_2)$ or any other coordinate in the equality above. Of course, that is true for any coordinate, so we can recast Equation (3.42) in terms of a general frequency f:

$$U(f) = \lambda(f)V(f) \tag{3.43}$$

Equation ((3.43)) is the representation of $|u\rangle = A|v\rangle$ in the basis of eigenvectors: it is the frequency domain representation.

We have seen that, in the time domain representation, t is the index of the column slot (see Equation (2.31)). Likewise, in the frequency domain representation, f is the index of the column slot. Thus, by discretizing f, we can represent Equation (3.43) explicitly in a column vector equation in the basis of eigenvectors, like this:

$$
\begin{pmatrix} \vdots \\ U(f_{-2}) \\ U(f_{-1}) \\ U(f_0) \\ U(f_1) \\ U(f_2) \\ \vdots \end{pmatrix} = \begin{pmatrix} & \vdots & \vdots & \vdots & \vdots & \vdots & \\ \cdots & \lambda(f_{-2}) & 0 & 0 & 0 & 0 & \cdots \\ \cdots & 0 & \lambda(f_{-1}) & 0 & 0 & 0 & \cdots \\ \cdots & 0 & 0 & \lambda(f_0) & 0 & 0 & \cdots \\ \cdots & 0 & 0 & 0 & \lambda(f_1) & 0 & \cdots \\ \cdots & 0 & 0 & 0 & 0 & \lambda(f_2) & \cdots \\ & \vdots & \vdots & \vdots & \vdots & \vdots & \end{pmatrix} \begin{pmatrix} \vdots \\ V(f_{-2}) \\ V(f_{-1}) \\ V(f_0) \\ V(f_1) \\ V(f_2) \\ \vdots \end{pmatrix} \quad (3.44)
$$

As expected, the representation of the operator in the basis of eigenvectors is a diagonal matrix, with the eigenvalues in the diagonal.

Once we have understood the algebraic meaning of Equation (3.43), we do not need to go through all these steps explicitly. Thus, suppose we have the time domain representation:

Equation (3.30)

$$
u(t) = a_2 \frac{d^2 v(t)}{dt^2} + a_1 \frac{dv(t)}{dt} + a_0 \cdot v(t)
$$

The most practical way to find the frequency domain representation is by applying the Fourier transform to both sides of Equation (3.30). With the help of the property of differentiation (Equation (2.132)), we get:

$$
U(f) = -a_2 (2\pi f)^2 V(f) + i a_1 2\pi f V(f) + a_0 V(f)
$$

Regrouping, we find:

$$
U(f) = [-a_2 (2\pi f)^2 + i a_1 2\pi f + a_0] V(f) \quad (3.45)
$$

But, of course, the term inside the brackets is the spectrum of eigenvalues (see Equation (3.35)), so Equation (3.45) is identical to Equation (3.43).

To conclude this sub-section, we collect these results and their analogues in the two-dimensional Euclidean space:

$$
|v\rangle = \int_{-\infty}^{+\infty} V(f_0)|e(f_0)\rangle df_0 \longleftrightarrow |v\rangle = V_1|a_1\rangle + V_2|a_2\rangle
$$

$$
|u\rangle = \int_{-\infty}^{+\infty} U(f_0)|e(f_0)\rangle df_0 \longleftrightarrow |u\rangle = U_1|a_1\rangle + U_2|a_2\rangle
$$

$$
U(f) = \lambda(f)V(f) \longleftrightarrow \begin{pmatrix} U_1 \\ U_2 \end{pmatrix} = \begin{pmatrix} \lambda_1 & 0 \\ 0 & \lambda_2 \end{pmatrix} \begin{pmatrix} V_1 \\ V_2 \end{pmatrix} \quad (3.46)
$$

3.7.4   Step 4: Solving the Equation and Returning to the Original Basis

By assumption, we want $|v\rangle$ in terms of $|u\rangle$, when $|u\rangle$ has been specified in the time domain, that is, when $u(t)$ has been given. In Equation (3.43), we have a trivial relationship between $U(f)$ and $V(f)$. So, we can get $|v\rangle$ by obtaining $V(f)$ through Equation (3.43): all we need to do is to find $U(f)$ and divide it by $\lambda(f)$. And we already know how to find $U(f)$ from $u(t)$: $U(f)$ is the Fourier transform of $u(t)$.

Thus, if our goal is to find $|v\rangle$ in terms of $|u\rangle$, then we have already completed it. Indeed, we have:

$$|v\rangle = \int_{-\infty}^{+\infty} V(f_0)|e(f_0)\rangle df_0 = \int_{-\infty}^{+\infty} \frac{U(f_0)}{\lambda(f_0)}|e(f_0)\rangle df_0 \qquad (3.47)$$

But it may be relevant to find the time domain representation of $|v\rangle$ explicitly. If that is the case, we need to project $|v\rangle$ on the basis of impulse functions. Thus:

$$v(t_0) = \langle \delta(t_0)|v\rangle = \langle \delta(t_0)| \left[\int_{-\infty}^{+\infty} df_0 V(f_0)|e(f_0)\rangle \right] = \int_{-\infty}^{+\infty} df_0 V(f_0)\langle \delta(t_0)|e(f_0)\rangle$$

But $\langle \delta(t_0)|e(f_0)\rangle = e^{i2\pi f_0 t_0}$ – see Equation (2.73), so:

$$v(t_0) = \int_{-\infty}^{+\infty} V(f_0)e^{i2\pi f_0 t_0} df_0$$

Since this relationship is valid for any index t_0, we can re-express it in terms of a general time t. Thus:

$$v(t) = \int_{-\infty}^{+\infty} V(f_0)e^{i2\pi f_0 t} df_0$$

Moreover, f_0 is just a dummy index, so let us replace it with f:

$$v(t) = \int_{-\infty}^{+\infty} V(f)e^{i2\pi f t} df$$

As fully expected, this is just the inverse Fourier transform of $V(f)$. Expressing $V(f)$ in terms of $U(f)$ explicitly, we get:

$$v(t) = \int_{-\infty}^{+\infty} \frac{U(f)}{\lambda(f)}e^{i2\pi f t} df \qquad (3.48)$$

This beautiful equation encapsulates the entire theory we have been developing. If $|u\rangle$ is the input of our system and $|v\rangle$ is the output, then Equation (3.48) is teaching how the output depends on the input and on the system: $U(f)$ specifies the input and $\lambda(f)$ specifies the system. Equation (3.48) deserves a place next to your heart of hearts, so please print it, copy it, draw it on the mirror after a hot shower. We will find it again through another route in Section 3.8.

Recall that the analogue of the inverse Fourier transform is a column vector sum involving the coordinates of the eigenvectors with respect to the canonical basis (see Equation (2.90)). Using the same notation to denote the coordinates of the eigenvectors that was used in Equation (2.90) (see also Equation (2.88)), the analogue of Equation (3.48) in the Euclidean space reads:

$$v(t) = \int_{-\infty}^{+\infty} \frac{U(f)}{\lambda(f)} e^{i2\pi ft} df \longleftrightarrow \begin{pmatrix} v_x \\ v_y \end{pmatrix} = \frac{U_1}{\lambda_1} \begin{pmatrix} a_{1x} \\ a_{1y} \end{pmatrix} + \frac{U_2}{\lambda_2} \begin{pmatrix} a_{2x} \\ a_{2y} \end{pmatrix} \tag{3.49}$$

Recall that we first encountered this expression in the Euclidean space in Equation (1.70).

In Section 3.8, we are going to explore the physical meaning of the eigenvalues. But, before we do so, let us conclude this section with a big summary of all the steps, with analogies with the three-dimensional Euclidean space, and three worked exercises.

$$|u\rangle = A|v\rangle \quad \Longleftrightarrow \quad |u\rangle = A|v\rangle$$

$$|u\rangle = \int_{-\infty}^{+\infty} u(t_0)|\delta(t_0)\rangle dt_0 \longleftrightarrow |u\rangle = u_x|x\rangle + u_y|y\rangle + u_z|z\rangle$$

$$|v\rangle = \int_{-\infty}^{+\infty} v(t_0)|\delta(t_0)\rangle dt_0 \longleftrightarrow |v\rangle = v_x|x\rangle + v_y|y\rangle + v_z|z\rangle$$

$$|u\rangle = \int_{-\infty}^{+\infty} U(f_0)|e(f_0)\rangle df_0 \longleftrightarrow |u\rangle = U_1|a_1\rangle + U_2|a_2\rangle + U_3|a_3\rangle$$

$$|v\rangle = \int_{-\infty}^{+\infty} V(f_0)|e(f_0)\rangle df_0 \longleftrightarrow |v\rangle = V_1|a_1\rangle + V_2|a_2\rangle + V_3|a_3\rangle$$

$$u(t) = a_2 \frac{d^2 v(t)}{dt^2} + a_1 \frac{dv(t)}{dt} + a_0 \cdot v(t) \Longleftrightarrow \begin{pmatrix} u_x \\ u_y \\ u_z \end{pmatrix} = \begin{pmatrix} a_{xx} & a_{xy} & a_{xz} \\ a_{yx} & a_{yy} & a_{yz} \\ a_{zx} & a_{zy} & a_{zz} \end{pmatrix} \begin{pmatrix} v_x \\ v_y \\ v_z \end{pmatrix}$$

$$U(f) = \lambda(f)V(f) \longleftrightarrow \begin{pmatrix} U_1 \\ U_2 \\ U_3 \end{pmatrix} = \begin{pmatrix} \lambda_1 & 0 & 0 \\ 0 & \lambda_2 & 0 \\ 0 & 0 & \lambda_3 \end{pmatrix} \begin{pmatrix} V_1 \\ V_2 \\ V_3 \end{pmatrix}$$

$$|v\rangle = \int_{-\infty}^{+\infty} \frac{U(f_0)}{\lambda(f_0)} |e(f_0)\rangle df_0 \longleftrightarrow |v\rangle = \frac{U_1}{\lambda_1}|a_1\rangle + \frac{U_2}{\lambda_2}|a_2\rangle + \frac{U_3}{\lambda_3}|a_3\rangle$$

$$v(t) = \int_{-\infty}^{+\infty} \frac{U(f)}{\lambda(f)} e^{i2\pi ft} df \longleftrightarrow \begin{pmatrix} v_x \\ v_y \\ v_z \end{pmatrix} = \frac{U_1}{\lambda_1} \begin{pmatrix} a_{1x} \\ a_{1y} \\ a_{1z} \end{pmatrix} + \frac{U_2}{\lambda_2} \begin{pmatrix} a_{2x} \\ a_{2y} \\ a_{2z} \end{pmatrix} + \frac{U_3}{\lambda_3} \begin{pmatrix} a_{3x} \\ a_{3y} \\ a_{3z} \end{pmatrix}$$

$$\tag{3.50}$$

Worked Exercise: Input is an Eigenvector

Suppose that a system is described by the relation $u(t) = A\{v(t)\}$, where $A\{\}$ is given by Equation (3.13). In this example, we assume that $u(t) = 4e^{i2\pi 8.3t}$, and we are asked to find $v(t)$.

Solution:

In this example, $u(t)$ is already an eigenvector, so we do not need to use Fourier transforms explicitly because we do not need to find the coordinates of $u(t)$ with respect to the basis of eigenvectors. We already have them: they are zero for every frequency, except for the frequency $f = 8.3$, for which the coordinate is 4. So, we can get the solution straight away:

$$v(t) = \frac{4e^{i2\pi 8.3t}}{\lambda(8.3)} \tag{3.51}$$

According to Equation (3.35):

$$\lambda(f) = -a_2(2\pi f)^2 + ia_1 2\pi f + a_0$$

So:

$$\lambda(8.3) = -a_2(2\pi 8.3)^2 + ia_1 2\pi 8.3 + a_0$$

And we have a complete solution in Equation (3.51).

We could also have solved this problem rigorously using Equation (3.48). Let us do that. We already know $\lambda(f)$, so we just need to find $U(f)$. Taking the Fourier transform of $u(t)$:

$$U(f) = \int_{-\infty}^{+\infty} u(t)e^{-i2\pi ft}dt = \int_{-\infty}^{+\infty} 4e^{i2\pi 8.3t}e^{-i2\pi ft}dt = 4\int_{-\infty}^{+\infty} e^{-i2\pi(f-8.3)t}dt$$

Using the orthonormality of the basis of exponentials (Equation (2.75)), we find:

$$U(f) = 4\int_{-\infty}^{+\infty} e^{-i2\pi(f-8.3)t}dt = 4\delta(f-8.3) \tag{3.52}$$

That makes sense: the coordinate representation of a basis vector with respect to its own basis is always an impulse function (see Section 2.7 [Orthonormality of the Basis of Impulse Functions] and Section 2.8 [Exponentials as a Basis: The Frequency Domain Representation of Signals]). Using Equation (3.52) in Equation (3.48), we find:

$$v(t) = \int_{-\infty}^{+\infty} \frac{U(f)}{\lambda(f)}e^{i2\pi ft}df = \int_{-\infty}^{+\infty} \frac{4e^{i2\pi ft}}{\lambda(f)}\delta(f-8.3)df = \frac{4e^{i2\pi 8.3t}}{\lambda(8.3)}$$

Of course, this is the same result we found in Equation (3.51).

Worked Exercise: Input is an Explicit Linear Combination of Eigenvectors

Now consider the same system as in the previous example, but with $u(t) = 4e^{i2\pi 8.3t} + 9ie^{i2\pi 63t}$. Find $v(t)$.

Solution:

Now $u(t)$ is no longer an eigenvector (in the exercise list, you are asked to prove that indeed it is not an eigenvector). But it is still an explicit linear combination of eigenvectors. So, again, we already know the coordinates by inspection: they are zero for every frequency, except for $f = 8.3$ and $f = 63$. And, for these two frequencies, the coordinates are 4 and $9i$, respectively. Since $v(t)$ must be a linear combination with the same eigenvectors (but with different coordinates), we get:

$$v(t) = \frac{4e^{i2\pi 8.3t}}{\lambda(8.3)} + \frac{9ie^{i2\pi 63t}}{\lambda(63)} \tag{3.53}$$

Worked Exercise: An Arbitrary Input

Still considering the same system as before, find $v(t)$ when $u(t) = \exp\left(-\dfrac{t^2}{\sigma^2}\right)$

Solution:

Now $u(t)$ is not an explicit linear combination of eigenvectors, and as such we cannot find its coordinates with respect to the basis of eigenvectors by inspection, as we did in the two previous examples. Now we need the good old Fourier transform to find them; $u(t)$ is a Gaussian function, and the Fourier transform of a Gaussian function is another Gaussian function. If we solve the Fourier integral (it is a bit boring, and requires checking a table of integrals), we get:

$$U(f) = \int_{-\infty}^{+\infty} \exp\left(-\frac{t^2}{\sigma^2}\right) e^{-i2\pi ft} dt = \sigma\sqrt{\pi}\exp[-(\pi\sigma f)^2]$$

Using Equation (3.48), we find:

$$v(t) = \int_{-\infty}^{+\infty} \frac{\sigma\sqrt{\pi}\exp[-(\pi\sigma f)^2]}{\lambda(f)} e^{i2\pi ft} df$$

To make it look even more impressive, let us write $\lambda(f)$ explicitly:

$$v(t) = \int_{-\infty}^{+\infty} \frac{\sigma\sqrt{\pi}\exp[-(\pi\sigma f)^2]}{-a_2(2\pi f)^2 + ia_1 2\pi f + a_0} e^{i2\pi ft} df$$

In practice, we solve ugly integrals like this using a computer. Knowing how to evaluate Fourier transforms using computers is such a crucial skill in engineering and physics, that we dedicate two chapters of this book (Chapters 7 and 8) to learn how to do it efficiently.

3.8 The Physical Meaning of Eigenvalues: The Impulse and Frequency Responses

Now we are going to harvest the fruits of the theory we have developed to gain extremely useful insights into the behaviour of LTI systems.

We begin by making sense of the physical meaning of the spectrum of eigenvalues of an LTI system. By physical meaning I mean something like 'how could we measure the spectrum of eigenvalues'? 'What kind of experiment gives information about this spectrum?' It is this sort of question that we want to probe in this section.

We begin with the time domain relation:

$$y(t) = B\{x(t)\} \tag{3.54}$$

where $x(t)$ and $y(t)$ are the time domain representations of the physical input and physical output, respectively, and $B\{\}$ is the time domain representation of the operator.

Express the input $x(t)$ as a linear combination of impulse functions:

$$x(t) = \int_{-\infty}^{+\infty} x(t_0)\delta(t - t_0)dt_0 \tag{3.55}$$

And substitute it back into Equation (3.54):

$$y(t) = B\left\{ \int_{-\infty}^{+\infty} x(t_0)\delta(t - t_0)dt_0 \right\} \tag{3.56}$$

Recall from Section 2.5 (Impulse Functions as a Basis: The Time Domain Representation of Signals), and especially Equation (2.27), that the only functions of time in the integral of convolution are the impulse functions. Indeed, $x(t_0)$ are not functions of t; instead they are the coordinates of each impulse function $\delta(t - t_0)$. Thus, the property of linearity (Equation (3.2)) allows to express the action of the operator on the integral in terms of the integral of the action on each impulse function. So, using the linearity of $B\{\}$, Equation (3.56) can be recast as:

$$y(t) = B\left\{ \int_{-\infty}^{+\infty} x(t_0)\delta(t - t_0)dt_0 \right\} = \int_{-\infty}^{+\infty} x(t_0)B\{\delta(t - t_0)\}dt_0 \tag{3.57}$$

If it is not clear why we can do this passage, break the integral into the Riemann sum (as in Equation (2.27)), and then apply the property of linearity to the Riemann sum.

According to Equation (3.57), the action of a linear operator $B\{\}$ is fully specified when its action on the basis functions $\delta(t - t_0)$ has been specified. In other words, if we know $B\{\delta(t - t_0)\}$, then we know everything about the system. This is not a new conclusion: we have already seen, in Section 1.5 (Representation of Linear Operators), that linear operators are fully specified by their action on basis vectors. The main difference is that now we have infinitely many basis vectors. We need not despair, however, because time invariance reduces the problem of finding the action of $B\{\}$ on infinitely many basis vectors to the problem of finding the action of $B\{\}$ on a single vector. We define $h(t)$ as the output of the system when the input is the impulse function:

$$h(t) = B\{\delta(t)\} \tag{3.58}$$

Then, since the system is time invariant, it follows that $B\{\delta(t - t_0)\} = h(t - t_0)$ – see Equation (3.3). Consequently, we do not need to work out $B\{\delta(t - t_0)\}$ for every single $\delta(t - t_0)$: if we know $B\{\delta(t)\}$, then we automatically know $B\{\delta(t - t_0)\}$.

Thus, due to time invariance, Equation (3.57) reduces to:

$$y(t) = \int_{-\infty}^{+\infty} x(t_0)B\{\delta(t - t_0)\}dt_0 = \int_{-\infty}^{+\infty} x(t_0)h(t - t_0)dt_0 \tag{3.59}$$

Notice that the last integral in Equation (3.59) is the convolution between $x(t)$ and $h(t)$. Thus:

$$y(t) = h(t) * x(t) \tag{3.60}$$

We have reached a key conclusion: **the relationship between input and output in an LTI system is fully described by $h(t)$.** Indeed, since Equation (3.60) is valid for any input/output, the information about the system must be entirely encapsulated in $h(t)$: according to Equation (3.60), if we know $h(t)$, then we know everything about the system, because we can predict the output for any input from knowledge of $h(t)$. After all, Equation (3.60) is teaching how the operator transforms the coordinates of the input $x(t)$ into the coordinates of the output $y(t)$. As such, **Equation ((3.60)) is a time domain representation of the LTI system.** Thus, it must be analogous to a matrix equation in the canonical basis of the Euclidean space. The only difference between the representation of Equation (3.60) and the time domain representation in terms of differential equations is that in Equation (3.60) the information about the operator is stored in the function $h(t)$, instead of being stored in a differential operator. In the exercise list, you will be asked to further explore the connection between Equation (3.60) and a matrix equation in the Euclidean space.

This function $h(t)$ is quite special, and we usually name special things. According to Equation (3.58), $h(t)$ is the output of the system when the input is the impulse function. Let us be creative and call $h(t)$ 'the impulse response' of the system: it is the output signal when the input is an impulse. Thus the impulse response is the time domain representation of a particular, special signal, that fully specifies the LTI system.

But hang on! We have just seen that the spectrum of eigenvalues fully specifies an LTI system. Now we have learned that the impulse response also fully specifies an LTI system. So, surely, they must be intimately connected. And they are. Recall that convolution in the time domain results in product in the frequency domain (Equation (2.144)). So, if we apply the Fourier transform to both sides of Equation (3.60), we get:

$$Y(f) = H(f)X(f) \tag{3.61}$$

where $H(f)$ is the Fourier transform of the impulse function $h(t)$, $Y(f)$ is the Fourier transform of the output signal $y(t)$ and $X(f)$ is the Fourier transform of the input signal $x(t)$. This Fourier transform $H(f)$ is called the 'frequency response' of the system, and is one of the most important functions in the theory of signals and systems. So, keep in mind that **the frequency response is the Fourier transform of the impulse response**.

Now compare Equation (3.61) with Equation (3.43), which I repeat below for convenience:

Equation (3.43)

$$U(f) = \lambda(f)V(f)$$

They are essentially the same relation, which means that, since $\lambda(f)$ is the spectrum of eigenvalues of the operator $A\{\}$ of Section 3.7 (General Method of Solution), it follows that $H(f)$ is the spectrum of eigenvalues of the operator $B\{\}$ in Equation (3.54).

In the development leading to Equation (3.61), we defined from the outset that $y(t)$ (and hence also $Y(f)$) represents the physical output, and $x(t)$ (and hence also $X(f)$), represents the physical input. So, by definition, the frequency response is the ratio of the Fourier transform of the output to the Fourier transform of the input (in other words, $H(f) = Y(f)/X(f)$).

Most of the time, however, when we are working on finding the differential equation that describes the system, we end up with a relationship of the form $input(t) = A\{output(t)\}$, where $input(t)$ and $output(t)$ are the physical input and output, respectively. This is why in Section 3.7 (General Method of Solution) we expressed $v(t)$ in terms of $u(t)$ when they were related by $u(t) = A\{v(t)\}$. In principle, we can invert this relationship to obtain the form assumed in Equation (3.54): $output(t) = B\{input(t)\}$, where $B\{\}$ is the inverse of $A\{\}$ ($B\{\} = A^{-1}\{\}$). However, in practice, if we have already found $A\{\}$, then we need not bother to find the inverse operator $B\{\}$ (often it is quite hard to find it) explicitly, because it is much easier to find its frequency domain representation straight away. Indeed, if $\lambda(f)$ are the eigenvalues of $A\{\}$, then the eigenvalues of $B\{\}$ are $\lambda^{-1}(f)$ – after all, to invert a diagonal matrix we just need to invert each element of the diagonal. But recall that, even though we may not know the differential equation describing $B\{\}$ in Equation (3.54), we still have a time domain representation of $B\{\}$ in terms of the convolution with the impulse response: this is the representation described in Equation (3.60).

Thus, we have a physical meaning for the spectrum of eigenvalues: the spectrum of eigenvalues of the operator $B\{\}$, as defined in Equation (3.54), is the frequency response of the system. In other words, it is the Fourier transform of the signal that the system spits out when the input is an impulse function (another, and more intiuitve, physical meaning is discussed in the Worked Exercise: How Can the Frequency Response be Measured?). Since the spectrum of eigenvalues is the frequency domain representation of a system, it follows that **the frequency response is the frequency domain representation of an LTI system**.

Notice that, since $H(f)$ is the spectrum of eigenvalues, so it must be the proportion that an eigenvector picks up when it is acted on by the system. Let us test if that is true by plugging the eigenvector $x(t) = e^{i2\pi f_0 t}$ in the input and evaluating the output. Thus:

$$y(t) = h(t) * x(t) = h(t) * e^{i2\pi f_0 t} = \int_{-\infty}^{+\infty} e^{i2\pi f_0 t_0} h(t - t_0) dt_0$$

Now define $t' = -t_0$ and recast the convolution integral in terms of t':

$$y(t) = h(t) * e^{i2\pi f_0 t} = \int_{-\infty}^{+\infty} e^{-i2\pi f_0 t'} h(t' + t) dt' \tag{3.62}$$

This integral is the Fourier transform of $h(t' + t)$ evaluated at the frequency f_0. Thus, using the property of time shifting (Equation (2.119): notice that here $-t$ is playing the role of the delay t_0 in Equation (2.119)), we conclude that:

$$y(t) = h(t) * e^{i2\pi f_0 t} = H(f_0) e^{i2\pi f_0 t} \tag{3.63}$$

Thus, we have confirmed that, if the input is the eigenvector $x(t) = e^{i2\pi f_0 t}$, then the output is the same eigenvector, but now multiplied by $H(f_0)$. Thus, $H(f_0)$ is the eigenvalue with respect to the eigenvector $e^{i2\pi f_0 t}$, as expected.

Finally, notice that, from Equation (3.61), it follows that:

$$y(t) = \int_{-\infty}^{+\infty} Y(f) e^{i2\pi f t} df = \int_{-\infty}^{+\infty} X(f) H(f) e^{i2\pi f t} df \tag{3.64}$$

The relation above is the version of Equation (3.48) expressed in terms of $H(f)$. Notice, once again, that the contributions of the input signal and of the system are neatly separated in the integral above: $X(f)$ represents the input, and $H(f)$ represents the system.

Equation (3.64) is the most common form found in engineering textbooks, and it is the one we use in practice. Now let us see some examples.

Worked Exercise: Impulse and Frequency Responses of a Harmonic Oscillator

A harmonic oscillator, or spring–mass system, is one of the most important systems in engineering and physics because it is a model for a plethora of other systems.

Find the impulse and frequency responses of a damped harmonic oscillator.

Solution:

A harmonic oscillator with mass m, spring constant k, and damping coefficient γ is illustrated in Figure 3.3.

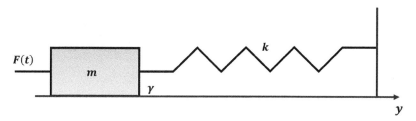

Figure 3.3 Illustration of a harmonic oscillator with mass m, spring constant k, and damping coefficient γ.

The relationship between the external force $F(t)$ and the mass position $y(t)$ can be straightforwardly obtained from Newton's law and Hooke's law:

$$F(t) = m\frac{d^2y(t)}{dt^2} + \gamma\frac{dy(t)}{dt} + ky(t) \tag{3.65}$$

Thus, the operator describing the system is:

$$A\{\} = m\frac{d^2}{dt^2} + \gamma\frac{d}{dt} + k. \tag{3.66}$$

Comparing Equation (3.66) with Equation (3.13), we conclude that $a_2 = m$, $a_1 = \gamma$, and $a_0 = k$.

Usually we are interested in analyzing the way the mass moves when an external force is applied. Thus, $F(t)$ is the physical input and $y(t)$ is the physical output; so, the problem is of the form $input(t) = A\{output(t)\}$; consequently, the frequency response is the spectrum of eigenvalues of the inverse operator $B\{\} = A^{-1}\{\}$. In other words, $H(f) = \lambda^{-1}(f)$, where $\lambda(f)$ is the spectrum of eigenvalues of $A\{\}$.

We can also obtain $H(f)$ straight away from Equation (3.65). Let us do it. By definition, $h(t)$ is the output when $\delta(t)$ is the input. Thus:

$$\delta(t) = m\frac{d^2h(t)}{dt^2} + \gamma\frac{dh(t)}{dt} + kh(t) \tag{3.67}$$

Fourier transforming both sides, we obtain:

$$1 = -m(2\pi f)^2 H(f) + i\gamma 2\pi f H(f) + kH(f) \tag{3.68}$$

where the property of differentiation (Equation (2.132)) and the Fourier transform of $\delta(t)$ ($\int_{-\infty}^{+\infty} \delta(t)e^{-i2\pi ft}dt = e^{-i2\pi f0} = 1$) were used.

Thus, we conclude that:

$$H(f) = \frac{1}{-m(2\pi f)^2 + i\gamma 2\pi f + k} \tag{3.69}$$

Comparing Equation (3.69) with Equation (3.35), and recalling that $a_2 = m$, $a_1 = \gamma$, and $a_0 = k$, we confirm that indeed $H(f) = \lambda^{-1}(f)$.

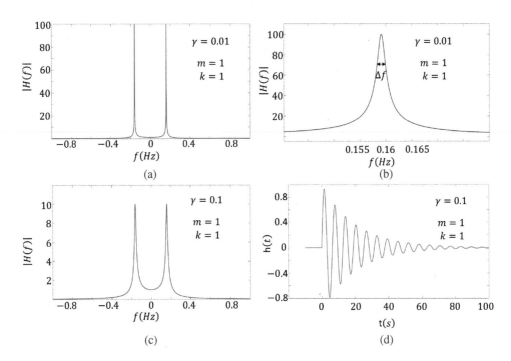

Figure 3.4 (a)–(c) Magnitude of the frequency response of a harmonic oscillator; (b) shows a zoomed in plot of the resonance of (a), and (c) shows the resonance with larger damping coefficient. d) Impulse response of the harmonic oscillator.

As an example, the magnitude of the frequency response of a harmonic oscillator with the parameters $m = k = 1$ and $\gamma = 0.01$ is shown in Figure 3.4a. The shape of the curve is characteristic of a resonant system. The frequency at which the magnitude peaks is the 'resonance frequency'. As you will be asked to prove in the exercise list, for low damping (that is, small γ), the resonance frequency f_0 is:

$$f_0 \approx \frac{1}{2\pi}\sqrt{\frac{k}{m}} \tag{3.70}$$

An exact expression for the resonance frequency will be found later in the Worked Exercise: Impulse Response of a Harmonic Oscillator, found in Section 6.5 (Inverse Laplace Transform by Inspection) – see Equation (6.28).

A zoomed in plot of the positive frequency part of the resonance is shown in Figure 3.4b. The resonance peaks at $f_0 = 0.1591$ Hz, which agrees well with Equation (3.70) with $m = k = 1$.

The resonance width is a measure of its quality-factor (Q). The Q is defined as:

$$Q = \frac{f_0}{\Delta f} \tag{3.71}$$

where Δf is taken between the points at which the peak of the magnitude falls by $1/\sqrt{2}$ of its maximum value. In general, for small γ, the peak is:

$$|H(f_0)| = \frac{1}{\gamma 2\pi f_0} = \frac{1}{\gamma}\sqrt{\frac{m}{k}} \tag{3.72}$$

where Equation (3.70) was used.

Notice that, at the resonance frequency f_0, the real part of the denominator of $H(f)$ vanishes, which explains why $|H(f)|$ peaks at f_0.

In the example of Figure 3.4a and Figure 3.4b, $m = k = 1$ and $\gamma = 0.01$, which results in $|H(f_0)| = 100$. Thus, in this example, Δf is defined by the points where the magnitude falls to $100/\sqrt{2} = 70.71$, as indicated in Figure 3.4b. The width Δf is called the 'full width of half maximum (FWHM)', and the term 'half maximum' is due to the magnitude **square** of the resonance falling by half of its peak value at the points defining Δf.

In general, the Q and the resonance width of a resonant system depend on its losses. In the harmonic oscillator, the losses are due to friction, and are characterized by the damping coefficient γ. The larger γ is, the higher the losses are. Consequently, the larger γ is, the broader the resonances are and the lower the Q is (in the exercise list, you will be asked to prove that $Q = \sqrt{m \cdot k}/\gamma$). Figure 3.4c shows the magnitude of the frequency response assuming a larger γ than it was assumed in Figure 3.4a. Compare Figure 3.4a with Figure 3.4c and notice how the higher losses resulted in a lower peak (see Equation (3.72)) and a broader resonance.

Figure 3.4d shows the impulse response of the harmonic oscillator, $h(t)$, obtained by inverse Fourier transforming the frequency response (that is, by inverse Fourier transforming Equation (3.69)). We will learn a trick to calculate this inverse Fourier transform in Chapter 6 (Introduction to the Laplace Transform, see Equation (6.29)), but for now I just want to show you how it looks like. Notice that the impulse response is a damped oscillation. As is well known from the theory of harmonic oscillators, the frequency of oscillation is the resonance frequency, and the amplitude of $h(t)$ decays exponentially due to damping. Of more immediate interest to our purposes, notice that $h(t) = 0$ for negative time. This is no accident, and you will be asked to explain why in the list of exercises.

Worked Exercise: How can the Frequency Response be Measured?

Consider again the harmonic oscillator of Figure 3.3. How could we find its frequency response experimentally?

Solution:

In the previous worked exercise, we found that the frequency response of a harmonic oscillator is a complex function (Equation (3.69)). A complex function is a function involving imaginary numbers. But all measurable physical quantities are real quantities. So, how could we measure a complex function?

As a matter of fact, there are two ways we could do this.

One way is to first measure the impulse response (Figure 3.4d), which is a real function, and then Fourier transform it to obtain the frequency response. To measure the impulse response, we cannot put an actual impulse in the input, because an impulse function is a mathematical idealization. But we can use an input signal that approximates this mathematical idealization; that is, we can use a signal that is different from zero only for a very short time. In a harmonic oscillator, the input is the force applied to the mass. So, how do we call a force that is applied for a very short time? Well, we call it a flick! To physically approximate the impulse function, we can flick the mass, and then measure its movement after we flick it. If we do a proper flicking job (sorry!), then the mass movement describes the impulse response, and should look like Figure 3.4d. One nice feature is that we do not need to worry too much about our flicking force having 'unit area', like the impulse function, because usually we are interested in the 'shape' of the impulse and frequency responses, not in their absolute amplitudes. That is not to say that the actual amplitude is immaterial. For example, if we flick it too hard, then the restoring force applied by the spring cease to be linear (Hooke's law breaks down), and we enter the domain of nonlinear, or anharmonic, oscillation. Thus, the system is no longer linear, and our theory no longer applies (though we still use parts of it to describe the nonlinear oscillation). But, unless we summon the strongest man in the village to flick the mass, we can use this method.

Now, say we are short of a 'flicker' for the input, but we have a generator of sinusoidal forces. This situation may sound silly when applied to a harmonic oscillator, but it is quite common in other systems. For example, if the system is a circuit, then the 'flicker' is a short pulse generator, and it could well be the case that it cannot generate sufficiently short pulses to be treated as an impulse function (in the list of exercises, you will be asked to discuss further the meaning of 'sufficiently short pulses'). Can we still measure the frequency response in this case? We can, and it is quite instructive to spell out how.

So, suppose we have a 'frequency generator', that is, a source of sinusoidal waves. When I say 'sinusoidal', I mean an oscillation with a well-defined frequency (we will further discuss what 'well-defined frequency' means in the next worked exercise), and it could have any phase we want (the phase is arbitrary, because it depends on what time we define as the origin). Thus, it could well be a cosine wave, or a mixture of both. What matters is that it has a well-defined frequency. For the sake of simplicity, we assume that the input is a cosine with phase zero, unit amplitude, and frequency f_0:

$$F(t) = cos(2\pi f_0 t) \tag{3.73}$$

Now, let us check what the output for this input is. This can be straightforwardly done by using Euler's formula to expand the cosine (as in Equation (2.127)):

$$F(t) = cos(2\pi f_0 t) = \frac{e^{i2\pi f_0 t}}{2} + \frac{e^{-i2\pi f_0 t}}{2} \tag{3.74}$$

Now express the output $y(t)$ in terms of the convolution between the input and the impulse response:

$$y(t) = h(t) * cos(2\pi f_0 t) = \frac{1}{2}h(t) * e^{i2\pi f_0 t} + \frac{1}{2}h(t) * e^{-i2\pi f_0 t} \tag{3.75}$$

With the help of Equation (3.63), Equation (3.75) reduces to:

$$y(t) = \frac{1}{2}H(f_0)e^{i2\pi f_0 t} + \frac{1}{2}H(-f_0)e^{-i2\pi f_0 t} \tag{3.76}$$

We can put Equation (3.76) in a more familiar form by noting that, since $h(t)$ is a real function, then $H(-f_0) = H^*(f_0)$ – see Equation (2.112). Thus:

$$y(t) = \frac{1}{2}H(f_0)e^{i2\pi f_0 t} + \frac{1}{2}H^*(f_0)e^{-i2\pi f_0 t} \tag{3.77}$$

Expressing the frequency response in polar form:

$$H(f_0) = |H(f_0)|e^{i\theta(f_0)} \tag{3.78}$$

Equation (3.77) can be recast as:

$$y(t) = \frac{1}{2}|H(f_0)|e^{i\theta(f_0)}e^{i2\pi f_0 t} + \frac{1}{2}|H(f_0)|e^{-i\theta(f_0)}e^{-i2\pi f_0 t}$$

Rearranging:

$$y(t) = |H(f_0)|\frac{e^{i[2\pi f_0 t + \theta(f_0)]} + e^{-i[2\pi f_0 t + \theta(f_0)]}}{2} \tag{3.79}$$

Finally, using Euler's formula, we conclude that:

$$y(t) = | H(f_0) | cos[2\pi f_0 t + \theta(f_0)]$$

Thus:

If the input is:

$$F(t) \doteq cos(2\pi f_0 t)$$

then the output is:

$$y(t) = | H(f_0) | cos[2\pi f_0 t + \theta(f_0)] \tag{3.80}$$

According to Equation (3.80), when the input is a cosine wave, then the output is also a cosine wave, but with a different amplitude and phase. Importantly, the amplitude and phase of the output carry information about the frequency response evaluated at the input's frequency. In other words, we can measure the complex number $H(f_0) = |H(f_0)|e^{i\theta(f_0)}$ by measuring the amplitude and phase of the output: the amplitude gives $|H(f_0)|$, and the phase difference between output and input gives $\theta(f_0)$. So, we have measured a complex number by measuring two real parameters: amplitude and phase.

Equation (3.80) is a recipe to measure $H(f_0)$ using a frequency generator: choose a frequency f_0, then measure amplitude and phase of the output, and we have $H(f_0)$. Change the frequency and do it again. And again, and again. Thus, we are measuring $H(f)$ by finding its values frequency by frequency.

Now, look again at the plots of $|H(f)|$ in Figure 3.4a and Figure 3.4c. If the system is a resonator, as is the harmonic oscillator, then $|H(f)|$ is high at and around the resonance frequency, but low far away from the resonance frequency. Consequently, the amplitude of the mass oscillation is much higher at the resonance frequency than it is for frequencies away from the resonance frequency. Thus, even if you put a tiny force in the input, but with a frequency coinciding with the resonance frequency, you may get a huge oscillation. Likewise, you may put a huge force in the input, but with the 'wrong frequency', that is, with a frequency far away from the resonance frequency. If you do that, even if the amplitude of the input force is high, the amplitude of the mass oscillation may be tiny.

Resonant systems are ubiquitous in engineering and physics, both for good and for evil. Sensors and filters, for example, often are based on resonance effects. On the evil side, bridges unexpectedly collapsing may also be due to resonances.

Finally, notice that the derivation of Equation (3.80) does not require that the system is a harmonic oscillator: it only requires that the system is LTI, because it requires that the output is the convolutions between the input and the impulse response. Thus, Equation (3.80) is a general result: whenever the input of an LTI system is a sine or cosine wave, the output is also a sine or cosine wave, only with a different amplitude and phase. Since the amplitude and phase carry information about the frequency response, and since the frequency response is a full determination of an LTI system, then one can fully characterize an LTI system by plugging cosines in the input and measuring the output. This is a standard technique used to characterize LTI systems, and it explains why sines and cosines are such ubiquitous functions in engineering and physics.

Worked Exercise: The Transient of a Harmonic Oscillator

In the previous exercise, we learned that the frequency response can be measured using a 'pure frequency' in the input, that is, a sinusoidal input wave. However, we also learned in the Worked Exercise: The Fourier Transform of a Physical Sinusoidal Wave, found in Section 2.12.9 (Product in the Time Domain), that there is no such a thing as a 'pure frequency', because a sinusoidal wave is a mathematical idealization: it has no beginning and no end, whereas all physical signals have a beginning and an end. So, how is it possible to do the measurements described in the previous worked exercise if there is no such a thing as a 'pure frequency'?

Solution:

According to Equation (3.80), if the input is a cosine wave, then the output is also a cosine wave with the same frequency, but with a different amplitude and phase. But what happens when the input is a 'physical' cosine wave, that is, one with a beginning and an end? In this case, there is a transient response: immediately after the input 'begins', the output oscillates in a 'crazy' way for some time, and then it stabilizes into a sinusoidal oscillation. Thus, to get the relationship of Equation (3.80), we need to wait for the transient response to end. After the transient response ends, we measure the amplitude of the output, and the phase difference between output and input, thus obtaining $H(f_0)$.

We can use the theory we have been developing to obtain an expression for the transient response. According to Equation (2.153), a 'physical' sine wave is mathematically described by the product between an ideal sinusoidal function and a rectangular function:

Equation (2.153)

$$g(t) = sin(2\pi f_0 t) rect(t)$$

Using the property of product in the time domain, we concluded that the Fourier transform of a 'physical' sine wave is:

Equation (2.161)

$$G(f) = \frac{S(f - f_0) - S(f + f_0)}{2i}$$

where $S(f)$ is the Fourier transform of the rectangular function. Thus, if $G(f)$, as defined in Equation (2.161), is the frequency domain representation of the input signal, it follows from Equation (3.64) that the output $y(t)$ is given by:

$$y(t) = \int_{-\infty}^{+\infty} Y(f)e^{i2\pi ft}df = \int_{-\infty}^{+\infty} G(f)H(f)e^{i2\pi ft}df$$

$$= \int_{-\infty}^{+\infty} \left[\frac{S(f - f_0) - S(f + f_0)}{2i}\right] H(f)e^{i2\pi ft}df \qquad (3.81)$$

But, according to Equation (2.86):

Equation (2.86)

$$S(f) = 2c \cdot sinc(2\pi fc)$$

From which it follows that:

$$y(t) = \int_{-\infty}^{+\infty} \left[\frac{2c \cdot sinc(2\pi(f - f_0)c) - 2c \cdot sinc(2\pi(f + f_0)c)}{2i}\right] H(f)e^{i2\pi ft}df \quad (3.82)$$

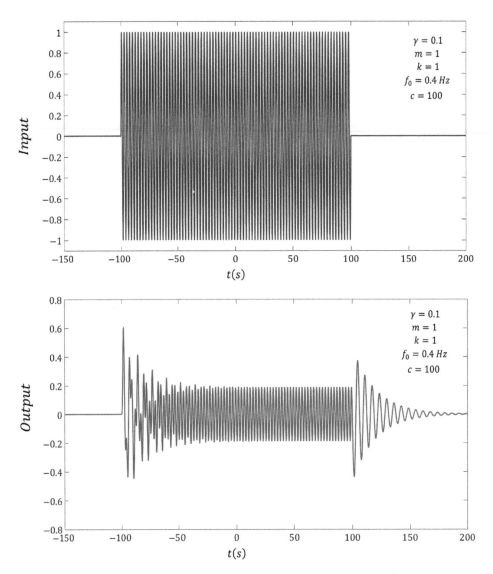

Figure 3.5 Example of transient response of a harmonic oscillator. The input (top plot – blue line) is a sine wave 'beginning' at time $c = -100$ s and ending at time $c = 100$ s. The output (bottom plot – red line) settles into a sinusoidal oscillation after the transient response ends.

With the help of Equation (3.69), we conclude that the output of a harmonic oscillator when the input is a 'physical' sine wave is:

$$y(t) = \int_{-\infty}^{+\infty} \left[\frac{2c \cdot sinc(2\pi(f - f_0)c) - 2c \cdot sinc(2\pi(f + f_0)c)}{2i(-m(2\pi f)^2 + i\gamma 2\pi f + k)} \right] e^{i2\pi ft} df \qquad (3.83)$$

As an example of transient response, Equation (3.83) was numerically integrated (we will learn how to do it in Chapters 7 and 8) assuming the harmonic oscillator parameters of Figure 3.4c and input frequency $f_0 = 0.4$ Hz. The result is shown in Figure 3.5 – bottom figure, red line, together with the input (top figure – blue line).

Notice how the output oscillates in a crazy way when the input is switched on at time $c = -100$ s, and then it settles into a sinusoidal oscillation. When the input is again 'switched off' at $c = 100$ s, the output again oscillates, but this time in a way similar to the impulse response (see Figure 4.3d), until it goes to zero.

Plugging in $m = k = 1$ and $\gamma = 0.1$ into Equation (3.69), we find that the magnitude of the frequency response at $f_0 = 0.4$ Hz is $|H(0.4)| = 0.188$. As expected, this is also the amplitude of the output (red curve) once it settles into a sinusoidal oscillation.

Worked Exercise: Charge and Discharge in an RC Circuit

Consider the circuit of Figure 3.6a. The circuit consists of a resistor in series with a capacitor. The physical input $v_{in}(t)$ and physical output $v_{out}(t)$ are shown in Figure 3.6a. Now consider a particular input signal, which is zero at all times, except between $t = -10$ ms and $t = +10$ ms, when it is one (see Figure 3.6b and Figure 3.6c). Find the output considering $RC = 2$ ms.

Solution:

As illustrated in Figure 3.6b and Figure 3.6c, the input signal is a rectangular function with $c = +10$ ms. Thus, according to Equation (2.86), the Fourier transform of the input signal is:

$$S(f) = 2 \times 10^{-3} \cdot sinc(2\pi f \times 10^{-3}) \qquad (3.84)$$

The frequency response of the RC circuit can be found by applying Kirchhoff's laws. This process, alongside the concepts of impedance and phasors, will be discussed in Chapter 4 (Electric Circuits as LTI Systems). In that chapter, we will use the same circuit of Figure 3.6a as an example and we will find its frequency response using two different techniques. Here, I just quote the result:

$$H(f) = \frac{1}{i2\pi fRC + 1} \qquad (3.85)$$

Combining the input and frequency responses, we find that:

$$v_{out}(t) = \int_{-\infty}^{+\infty} S(f)H(f)e^{i2\pi ft}df = \int_{-\infty}^{+\infty} \frac{2 \times 10^{-3} \cdot sinc(2\pi f \times 10^{-3})}{i2\pi f \times 2 \times 10^{-3} + 1} e^{i2\pi ft}df \qquad (3.86)$$

Solving the integral in Equation (3.86), we find the output shown as the red line in Figure 3.7. Before the unit voltage is plugged in at $t = -10$ ms, the output is zero, as expected. When it is plugged in, the voltage in the capacitor rises exponentially, until it is fully charged, that is, until the output voltage coincides with the input voltage. At time $t = 10$ ms, the resistor and capacitor are short-circuited (see Figure 3.6c). Thus, at $t = 10$ ms the capacitor begins to discharge through the resistor. As is well known from circuit theory, the charge and discharge depend

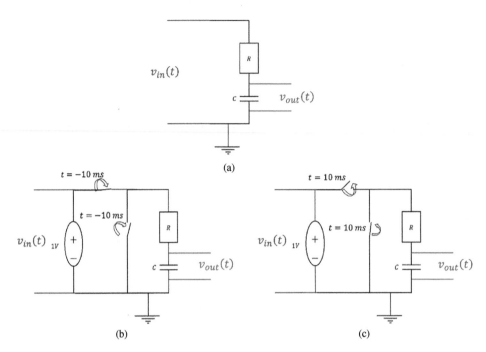

(a)

(b)   (c)

Figure 3.6 (a) RC circuit showing input and output. (b) At time $t = -10\ ms$, a voltage source of 1V is connected to the input. (c) At time $t = 10\ ms$, the voltage is disconnected and the input is short-circuited.

exponentially on time, with a time constant equal to RC. This is the time it takes for the capacitor to discharge a factor of $1/e$ of its initial voltage. In our example, we have $RC = 2$ ms. Thus, after 2 ms, the voltage at the capacitor is $v_{out}(2ms) = 1/e = 0.3679$. These points are highlighted in Figure 3.7.

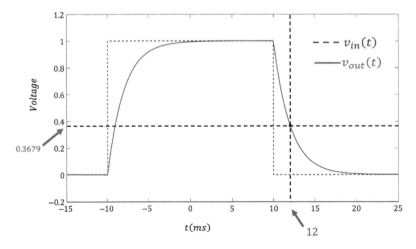

Figure 3.7 Relationship between input and output of an RC circuit. The discharge time $RC = 2\ ms$ is highlighted.

We will derive a mathematical expression for the capacitor discharge in Section 6.7.1 (The differentiation Property of the Unilateral Fourier Transform) (the expression is Equation (6.54)).

3.9   Frequency Conservation in LTI Systems

Conservation of frequency is a key feature of LTI systems. As this is a central concept, we dedicate this section to spell it out.

When we say that an LTI system conserves frequency, we mean that the system cannot create new frequency components that are not already in the input. That this is indeed the case can be immediately seen from Equation (3.61), which I repeat below for convenience:

Equation (3.61)

$$Y(f) = H(f)X(f)$$

Recall that $Y(f)$ and $X(f)$ are the frequency domain representation of the output and input signals, respectively. Now, suppose that a certain frequency f_0 is not in the input. The assertion that f_0 is not in the input is equivalent to the assertion that $X(f_0) = 0$. But, if $X(f_0) = 0$, it follows immediately from Equation (3.61) that $Y(f_0) = 0$. Thus, if a frequency is not in the input, it cannot be in the output. This is what it means to say that an 'LTI system conserves frequency': it cannot create frequencies. Mind you: it can 'kill' frequencies, because $H(f)$ can be zero. But it cannot create frequencies. Therefore, an LTI system can only modify the amplitude and phase of each frequency component of the input: as implied in Equation (3.61), the amplitude of each frequency component is multiplied by the magnitude of $H(f)$, and the phase of $H(f)$ adds to the phase of each input component. Indeed, defining the polar forms of each term in Equation (3.61) as:

$$Y(f) = |Y(f)|e^{i\varphi(f)}$$

$$X(f) = |X(f)|e^{i\alpha(f)}$$

$$H(f) = |H(f)|e^{i\theta(f)} \tag{3.87}$$

It follows immediately from Equation (3.61) that $|Y(f)| = |H(f)||X(f)|$ and $\varphi(f) = \theta(f) + \alpha(f)$. In other words: an LTI system alters the amplitude and phase of each frequency component, but it cannot create new frequency components. This idea is illustrated in Figure 3.8.

Strictly speaking, a pure frequency function is one of the exponential basis functions, that is, it is a function of the from $e^{i2\pi f_0 t}$. However, this is a complex function. So, we also speak of sines and cosines as pure frequency functions, even though they are a composite of $e^{i2\pi f_0 t}$ and $e^{-i2\pi f_0 t}$. In Equation (3.80), we saw that the output of an LTI system when the input is a cosine wave is also a cosine wave, but with a different amplitude and phase (recall that we also had to invoke the property of reality to reach this conclusion). Thus, Equation (3.80) is a 'stronger' statement of conservation of frequency in LTI systems: if the input is a 'pure' frequency, then the output is also the same 'pure' frequency, only with a different amplitude and phase.

Frequency conservation is an extremely useful concept because it gives us prior information about the possible solutions of a system. Thus, in the analysis of LTI systems, one typically assumes that the input is a pure frequency (that is, the input is an exponential of the form $e^{i2\pi f_0 t}$), which guarantees that the output (that is, the solution) has exactly the same time dependence (in other words, the time dependence of the solution is also of the form $e^{i2\pi f_0 t}$). Then, once the output for a single frequency is found, the result can be generalized to an arbitrary input using Fourier transforms. Such a strategy greatly facilitates the analysis of LTI systems, and it is a standard procedure in many fields of engineering and physics, such as electrodynamics, optics, and quantum mechanics, to name only a few. As another example, as we will see in Chapter 4, it leads to the concept of phasors and impedance in the analysis of electric circuits. In Section 3.10, we will catch a glimpse of applications of frequency conservation in other fields.

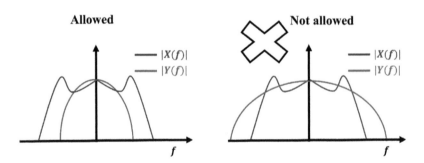

Figure 3.8 Illustration of the concept of frequency conservation in LTI systems. The system can kill frequencies, but it cannot create them, so the output $Y(f)$ cannot have frequency components that are not already in the input $X(f)$.

3.10 Frequency Conservation in Other Fields

Often in engineering and physics, conservation of frequency appears in the guise of spatial frequencies, alongside temporal frequency. The concept is essentially the same, but since there are three dimensions of space as opposed to only one dimension of time, then there are three spatial frequencies components, one for each direction of space. These spatial frequencies are the components of a three-dimensional vector, usually denoted by $\vec{k}$ and called the 'wavevector':

$$\vec{k} = k_x \hat{x} + k_y \hat{y} + k_z \hat{z} \tag{3.88}$$

In linear systems, whereas time invariance leads to conservation of temporal frequency, spatial invariance leads to conservation of spatial frequency. In this section, we will catch a glimpse of two examples of application of conservation of frequency in other fields.

3.10.1 Snell's Law

Our first example is from the field of optics. Maxwell's equations entail that the magnitude of the wavevector depends on the temporal frequency f of the electromagnetic radiation and on the refractive index n as:

$$|\vec{k}| = \frac{2\pi f}{c_0} n \tag{3.89}$$

where c_0 is the speed of light in vacuum.

Now consider the situation depicted in Figure 3.9, where a plane wave is incident from a medium with refractive index n_1, and then suffer reflection and refraction at the boundary with a second medium with refractive index n_2.

The wavevector $\vec{k}$ describes the direction of propagation of the optical ray. From basic trigonometry, we have:

$$sin(\theta_1) = \frac{k_{x1}}{|\vec{k_1}|}$$

$$sin(\theta_2) = \frac{k_{x2}}{|\vec{k_2}|} \tag{3.90}$$

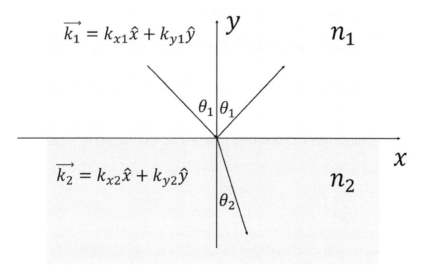

Figure 3.9 Illustration of an optical ray incident on the boundary between two media. Snell's law follows from conservation of temporal and spatial frequencies.

Using Equation (3.89):

$$sin(\theta_1) = \frac{k_{x1}}{\frac{2\pi f_1}{c_0} n_1}$$

$$sin(\theta_2) = \frac{k_{x2}}{\frac{2\pi f_2}{c_0} n_2} \tag{3.91}$$

Since the refractive indexes do not depend on time, the system is time invariant. Therefore, temporal frequency is conserved; in other words: $f_1 = f_2$. Consequently:

$$\frac{k_{x1}}{\frac{2\pi sin(\theta_1)}{c_0} n_1} = \frac{k_{x2}}{\frac{2\pi sin(\theta_2)}{c_0} n_2} \tag{3.92}$$

Rearranging and simplifying, we get:

$$\frac{sin(\theta_1)n_1}{k_{x1}} = \frac{sin(\theta_2)n_2}{k_{x2}} \tag{3.93}$$

We have already used the concept of time invariance to identify that $f_1 = f_2$. Now, we notice that the system is also spatially invariant in the x direction, which means that the refractive indexes do not change as we walk along the x direction (in Figure 3.9, that means the background yellow and green colours representing the refractive indexes do not change when we walk along the horizontal direction). It only changes if we walk along the y direction, but not along the x direction. Consequently, the spatial frequency component in the x direction is conserved; in other words: $k_{x1} = k_{x2}$. Thus, Equation (3.93) simplifies to:

$$sin(\theta_1)n_1 = sin(\theta_2)n_2 \tag{3.94}$$

Equation (3.94) is the famous Snell's law of refraction. As we have just seen, Snell's law of refraction is a consequence of spatial and temporal invariances (by the way, the law of reflection follows the same logic; the only difference is that, in reflection, $n_1 = n_2$, which entails that $\theta_1 = \theta_2$; the latter equality is the mathematical assertion that the angle of reflection is equal to the angle of incidence).

3.10.2 Wavefunctions and Heisenberg's Uncertainty Principle

Our second example is from quantum mechanics. Quantum mechanics is essentially a wave theory: particles are treated as waves, and mathematically represented by wavefunctions, usually denoted by $\psi(\overrightarrow{r}, t)$. In a one-dimensional system, the spatial dependence of the wavefunction reduces to $\psi(x)$. Thus, suppose that $\psi(x)$ represents the spatial dependence of the wavefunction of an electron. According to the Copenhagen interpretation, $|\psi(x)|^2$ is the probability of finding the electron at the position x.

Recall from classical physics that force is the gradient of the potential. A spatially invariant potential is a potential whose gradient is zero, which means that there are no forces. Therefore, a free electron is an electron in an environment where the potential is spatially invariant. But, if the potential is spatially invariant, so is the operator governing the propagation of electrons. Consequently, exponentials of the form $\psi(x) = e^{ikx}$ form a basis of eigenvectors of these operators.

According to the laws of quantum mechanics, the magnitude of the wavevector, which we denote by k, is proportional to the mechanical momentum, which we denote by p. Thus:

$$p = \frac{h}{2\pi}k \qquad (3.95)$$

where h is the Planck constant.

Whereas $|\psi(x)|^2$ is the probability of finding the electron at the position x, the probability of finding the electron with mechanical momentum p is given by the modulus square of the spatial Fourier transform of $\psi(x)$, which we denote by $\Psi(k)$. Thus, $\Psi(k) = \int_{-\infty}^{+\infty} \psi(x)e^{-ikx}dx$, and $|\Psi(k)|^2$ is the probability of finding the electron with mechanical momentum $p = h \cdot k/(2\pi)$.

On the one hand, a wavefunction of the form $\psi(x) = e^{ik_0 x}$, whose Fourier transform – we recall - is $\Psi(k) = \delta(k - k_0)$, describes an electron with completely undefined position, but a well-defined mechanical momentum $p_0 = h \cdot k_0/(2\pi)$. Indeed, in this case, $|\psi(x)|^2 = |e^{ik_0 x}|^2 = 1$, which means that the position probability distribution is a constant function; in other words, there is no preferential position. Its mechanical momentum, however, is well defined, because $\Psi(k) = \delta(k - k_0)$ is zero everywhere, except at the value $k = k_0$, which means that there is zero probability of finding this electron with a mechanical momentum other than $p = p_0$.

On the other hand, a wavefunction of the form $\psi(x) = \delta(x - x_0)$, whose Fourier transform – we recall - is $\Psi(k) = e^{-ikx_0}$, is a solution describing an electron with a well-defined position, but completely undefined mechanical momentum. Indeed, in this case, $\psi(x)$ is zero everywhere, except at the position $x = x_0$. Therefore, there is zero probability of finding this electron anywhere other than at $x = x_0$. Its mechanical momentum, however, is completely undefined as $|\Psi(k)|^2 = |e^{-ikx_0}|^2 = 1$.

These two examples are extreme cases of a general property of quantum particles: as position and momentum are described by functions that are Fourier transforms of each other, a well-defined position entails an ill-defined momentum, and vice versa. This is the same relation between fast and slow signals and their Fourier transforms, as we have seen in Section 2.11 (The Physical Meaning of Fourier Transforms) – see Figure 2.16. Indeed, a wavefunction with a well-defined position is like a fast signal, with space playing the role of time: a wavefunction with well-defined position is different from zero only in a narrow spatial range, just like a fast signal is different from zero only in a narrow temporal range. As a fast signal entails a wide Fourier transform, so a wavefunction with a well-defined position entails a wide Fourier transform, which in turn entails an ill-defined mechanical momentum. A wavefunction with an ill-defined position, however, is like a slow signal, and as such entails a thin Fourier transform, which in turn entails a well-defined mechanical momentum. This is the origin of the famous Heisenberg uncertainty principle, which asserts that certainty in position comes at the expense of certainty in momentum, and vice versa.

3.11 Exercises

Exercise 1
Determine whether the systems below are linear or nonlinear, time variant or time invariant, causal or noncausal.

(a) $y(t) = T\{x(t)\} = x^2(t)$

(b) $y(t) = T\{x(t)\} = x(t) + x(t - t_0),\ \ for\ \ t_0 > 0$

(c) $y(t) = T\{x(t)\} = x(t) + x(t - t_0),\ \ for\ \ t_0 < 0$

(d) $y(t) = T\{x(t)\} = x(t) + t$

(e) $y(t) = T\{x(t)\} = x(t)t$

Exercise 2
Prove that $u(t) = 4e^{i2\pi 8.3t} + 9ie^{i2\pi 63t}$ is not an eigenvector of an LTI system. Is there any condition that would turn $u(t)$ into an eigenvector?

Exercise 3
Using Equation (3.48), solve the Worked Exercise: Input is an Explicit Linear Combination of Eigenvectors.

Exercise 4
In Section 3.2 (Operators Representing Linear and Time Invariant Systems), we noticed that the relationship $y(t) = T\{x(t)\}$ may involve derivatives of the output. We can write this most general form explicitly as:

$$\sum_{n=1}^{\infty} b_n \frac{d^n y(t)}{dt^n} + b_0 \cdot y(t) = \sum_{n=1}^{\infty} a_n \frac{d^n x(t)}{dt^n} + a_0 \cdot x(t)$$

Find the frequency response of this system in terms of the coefficients a_0, b_0, a_n, b_n.

Exercise 5
Suppose that the frequency response of an LTI system is $H(f) = e^{-i\alpha f}$. What is the relationship between the output and input signals in the time domain?

Exercise 6
We have seen that the output of an LTI system when the input is an impulse function is a full description of the system. But what is so special about the impulse function that makes its output a full description of the system?

Exercise 7

Consider a system $T\{\}$ specified as $T\{x(t)\} = x^2(t)$. Find the relationship between output and input in the frequency domain. Is this relationship of the form $Y(f) = H(f)X(f)$? Is frequency conserved in this system?

Exercise 8

We have seen that $h(t)$ in Figure 3.4d is zero for negative times. Is this feature peculiar to the system analyzed in Figure 3.4, or is it a general condition of any LTI system? Justify your answer.

Hint: recall that we are dealing with causal systems.

Exercise 9

We have seen that the magnitude of the frequency response of the harmonic oscillator is an even function (see Figure 3.4a and Figure 3.4c). Is this a necessary condition of LTI systems, or is it peculiar to the harmonic oscillator?

Exercise 10

Prove that, if the input of an LTI system is $x(t) = A\cos(2\pi f_0 t + \varphi)$, then the output must be $y(t) = \mid H(f_0) \mid A\cos(2\pi f_0 t + \varphi + \theta(f_0))$.

Exercise 11

Show that the resonance peak of a harmonic oscillator with low damping is $f_0 \approx \frac{1}{2\pi}\sqrt{\frac{k}{m}}$.

Hint: in the limit of low damping, the resonance frequency is the frequency at which the real part of the denominator of $H(f)$ vanishes.

Exercise 12

Assuming $\Delta f \ll f_0$, where f_0 is the resonance peak, prove that the quality factor of the harmonic oscillator is $Q = \frac{\sqrt{m \cdot k}}{\gamma}$.

Exercise 13

In this chapter, we made a close analogy with the Euclidean space. However, the relation:

Equation (1.89)

$$t_{mn} = \langle m|T|n \rangle$$

Giving the coefficients of the operator has not appeared in our description. What is the analogue of this relation in the function space?

Hint: write $y(t) = \int_{-\infty}^{+\infty} h(t - t_0)x(t_0)dt_0$ as a matrix equation. The coefficients of the matrix are $h(t - t_0)$, where t sets the row and t_0 the column. Then prove that $h(t_1 - t_2) = \langle \delta(t_1)|T|\delta(t_2) \rangle$.

Exercise 14

Suppose you have an LTI system designed to treat audio signals. The highest frequency humans can hear is about 20 kHz. You want to characterize your system by measuring the impulse response. How short should your input pulse be so that it can be treated as an impulse?

Exercise 15

Suppose that the input signal of the RC circuit of Figure 3.6a is the square wave of Figure 2.25. Find an expression for the output in terms of the Fourier series of this square wave.

Hint: the integral in Equation ((3.86)) reduces to a sum involving the coefficients of Equation (1.181).

Exercise 16

An AM modulator can be regarded as a system $T\{\}$ acting on the input $x(t)$ as:

$$T\{x(t)\} = x(t) \cdot [2 \cdot \cos(2\pi f_0 t)] \tag{3.96}$$

The modulator is a system that does not conserve frequency. After all, it is designed to change the frequency of the input signal (see Figure 2.18). If it does not conserve frequency, then the system as defined in Equation (3.96) is either nonlinear, or time variant, or both. Which one is it?

Exercise 17

Cavities are structures designed to store energy, and they play an important role in engineering and physics.

Typically, the rate of energy leakage from a cavity is proportional to the stored energy U. Mathematically, that means that:

$$\frac{dU}{dt} = -\frac{1}{\tau}U$$

where dU/dt is the rate of energy leakage and $1/\tau$ is the proportionality factor (τ is the cavity 'lifetime', and it has units of time). The negative sign means that the cavity is losing energy (dU/dt is negative). Indeed, the solution to this differential equation is:

$$U = U_0 e^{-\frac{t}{\tau}}$$

which implies that the energy decays exponentially in time.

In electromagnetic cavities, the energy U is proportional to the magnitude square of the electric field $U \propto |E|^2$. That entails that the electromagnetic field must be of the form:

$$E = E_0 e^{-\frac{t}{2\tau}} e^{i2\pi f_0 t}$$

where f_0 is the cavity resonance frequency. This solution implies that $|E|^2 = |E_0|^2 e^{-\frac{t}{\tau}}$, which indeed agrees with $U = U_0 e^{-\frac{t}{\tau}}$.

(a) Suppose that the cavity is initially empty, and then we inject a burst of energy into it. It follows that, after the burst, the field E decays exponentially, as $E = E_0 e^{-\frac{t}{2\tau}} e^{i2\pi f_0 t}$. The burst of energy is the physical realization of an impulse, so this exponentially decaying field is the impulse response of the cavity (strictly speaking, the field is proportional to the impulse response). Find the frequency response of the cavity and sketch its magnitude as a function of frequency. This form of the magnitude is called a 'Lorentzian curve', or 'Lorentzian line shape', and it appears all the time in engineering and physics.

(b) A key parameter of a cavity is the quality factor (Q-factor). We have seen that the Q-factor is defined as $Q = f_0/\Delta f$ (Equation (3.71)), where Δf is the FWHM of the magnitude of the frequency response of the cavity.

(c) Prove that:

$$\frac{f_0}{\Delta f} = 2\pi f_0 \tau$$

(d) Prove that another equivalent definition of Q is:

$$Q = 2\pi f_0 \frac{U}{|dU/dt|}$$

This last equality entails that Q can be defined as:

$$Q = 2\pi f_0 \frac{Energy\ stored}{rate\ of\ energy\ leakage}$$

4

Electric Circuits as LTI Systems

Learning Objectives

Analysis of electric circuits is based on the concepts of phasors and impedance. In this chapter, we will connect these concepts with the theory of representation of signals and systems. The main objective is to show that phasors and impedances provide a short-cut to find the frequency response of a circuit directly, without having to find the time domain representation.

4.1 Electric Circuits as LTI Systems

An electric circuit involves three basic elements: resistors, capacitors, and inductors. The voltage versus current relationship in these three elements are:

$$v_R(t) = Ri_R(t) \qquad i_C(t) = C\frac{dv_C(t)}{dt} \qquad v_L(t) = L\frac{di_L(t)}{dt} \qquad (4.1)$$

$$\underset{\text{Resistor}}{} \qquad \underset{\text{Capacitor}}{} \qquad \underset{\text{Inductor}}{}$$

where $v(t)$ and $i(t)$ are, respectively, voltage and current, and R, C, and L are the resistance, capacitance, and inductance of the respective elements.

The relationships of Equation (4.1) can be treated as relating output and input signals of an operator. Thus:

$$v_R(t) = T_R\{i_R(t)\} \qquad i_C(t) = T_C\{v_C(t)\} \qquad v_L(t) = T_L\{i_L(t)\} \qquad (4.2)$$

where:

$$T_R\{\ \} = R\cdot \qquad T_C\{\ \} = C\frac{d}{dt} \qquad T_L\{\ \} = L\frac{d}{dt} \qquad (4.3)$$

Notice that these three operators are linear. Furthermore, if R, C, and L are constants, then they are also time invariant (these parameters may vary in time due

---

Essentials of Signals and Systems, First Edition. Emiliano R. Martins.
© 2023 John Wiley & Sons Ltd. Published 2023 by John Wiley & Sons Ltd.
Companion website: www.wiley.com/go/martins/essentialsofsignalsandsystems

to nonideal conditions: for example, the resistance can change if the resistor gets too hot). If all elements of an electric circuit are described by LTI operators, then a combination of elements, that is, the circuit itself, must also be described by LTI operators. Thus, the theory of Chapter 3 (Representation of Systems) is applicable to electric circuits.

4.2   Phasors, Impedances, and the Frequency Response

To motivate the ideas of phasors and impedance, let us use the simplest possible circuit that already encapsulates all essential concepts: the RC circuit, as shown in Figure 3.6a. Our goal is to find the relationship between the output voltage $v_{out}(t)$, taken across the capacitor (see Figure 3.6a) and the input voltage $v_{in}(t)$, applied between one terminal of the resistor and one of the capacitor (see Figure 3.6a).

We can find this relationship using Kirchhoff's law. If $v_R(t)$ is the voltage across the resistor, and $v_{out}(t)$ is the voltage across the capacitor, then, according to Kirchhoff's voltage law, we have:

$$v_{in}(t) = v_R(t) + v_{out}(t)$$

But $v_R(t) = Ri_R(t)$, thus:

$$v_{in}(t) = Ri_R(t) + v_{out}(t)$$

Since the resistor and capacitor are connected in series, the current is the same in both elements, which entails that $i_R(t) = i_C(t) = Cdv_{out}(t)/dt$. Thus:

$$v_{in}(t) = RC\frac{dv_{out}(t)}{dt} + v_{out}(t) \tag{4.4}$$

According to Equation (4.4), the time domain representation of the operator describing the RC system of Figure 3.6a is:

$$A\{\} = RC\frac{d}{dt} + 1 \tag{4.5}$$

Comparing Equation (4.5) with Equation (3.13), we identify: $a_2 = 0$, $a_1 = RC$, and $a_0 = 1$. Thus, according to Equation (3.35) the spectrum of eigenvalues of the RC operator is:

$$\lambda(f) = iRC2\pi f + 1 \tag{4.6}$$

Notice that, according to Equation (4.4), we have a relationship of the form $input(t) = A\{output(t)\}$, which entails that the frequency response is the inverse of the spectrum of eigenvalues of $A\{\ \}$. Thus, we conclude that the frequency response of the RC system is:

$$H(f) = \frac{1}{iRC2\pi f + 1} \tag{4.7}$$

Incidentally, recall that we used this equation without proof in Worked Exercise: Charge and Discharge in an RC Circuit, of Chapter 3.

As explained in Chapter 3 (Representation of Systems), the relationship between input and output is given in terms of the frequency response through Equation (3.64):

$$v_{out}(t) = \int_{-\infty}^{+\infty} V_{in}(f)H(f)e^{i2\pi ft}df = \int_{-\infty}^{+\infty} \frac{V_{in}(f)}{iRC2\pi f + 1}e^{i2\pi ft}df \qquad (4.8)$$

Since the frequency response is a full description of an LTI system, with which a general relationship between input and output can be found, as in Equation (4.8), then the main goal of the analysis of an electric circuit is to find its frequency response. In our example, we first found the time domain representation of the operator (Equation (4.5)), and from it we found its frequency response. We did not have much trouble in finding the time domain representation of the operator because I chose a simple circuit as an example. But it is not hard to see that this approach soon gets impractical as the circuit gets more complex. Thus, we are led to the question: is it possible to find the frequency response straight away, without having to find the time domain representation of the operator?

To answer this question, recall that the frequency response is just a spectrum of eigenvalues. So, the real job is to find the spectrum of eigenvalues without having to find the time domain representation. The strategy to do it is not difficult. Recall that the eigenvalues are the proportion we get when we plug in an eigenvector in the input of the operator. So, let us say we plug in the eigenvector $Ve^{i2\pi ft}$, where V is a complex number describing the coordinate, or 'amplitude', of the eigenvector. Thus:

$$T\{Ve^{i2\pi ft}\} = \lambda(f)Ve^{i2\pi ft} \qquad (4.9)$$

Since different eigenvectors can have different amplitudes, it is better to allow for a variation on the coordinate of different eigenvectors by treating V as a function of the frequency. Thus, we rewrite Equation (4.9) as:

$$T\{V(f)e^{i2\pi ft}\} = \lambda(f)V(f)e^{i2\pi ft} \qquad (4.10)$$

I emphasize that, in Equation (4.10), $T\{\ \}$ is the time domain representation of the operator, so it acts on functions of time; so f in Equation (4.10) is treated as a fixed parameter.

Now, think of this: if we plug an eigenvector in the input, then, according to Equation (4.9), to find the eigenvalue with respect to that eigenvector, all we need to do is to find the relationship between the coordinates of the input and output. For example, say we have the relationship $v_{out}(t) = T\{v_{in}(t)\}$. If we plug an eigenvector in the input, that is, if we assume $v_{in}(t) = V_{in}(f)e^{i2\pi ft}$, then $v_{out}(t) = \lambda(f)V_{in}(f)e^{i2\pi ft}$, that is, the output is also an eigenvector, with the same frequency, but with the amplitude $V_{out}(f) = \lambda(f)V_{in}(f)$. Nothing surprising here: we learned in the previous chapter that LTI systems conserve frequency, and this is just a manifestation of frequency conservation. Thus, we can obtain $\lambda(f)$ by finding the ratio between $V_{out}(f)$ and $V_{in}(f)$. The only difference is that our system is of the form $v_{in}(t) = T\{v_{out}(t)\}$, so the ratio between $V_{out}(f)$ and $V_{in}(f)$ gives the inverse of the eigenvalues, that is, it gives the frequency response. Therefore, we can obtain the frequency response by finding the ratio between $V_{out}(f)$ and $V_{in}(f)$.

But what is the advantage of obtaining the frequency response by finding the ratio between the coordinates $V_{out}(f)$ and $V_{in}(f)$? Oh, my friend, there is a beautiful advantage: if the input is an eigenvector, then it is guaranteed that the voltage and current in every single element of the circuit is also an eigenvector with the same frequency. After all, there will be an LTI operator relating the input with the voltage or current in any element of the circuit. So, we know a priory that the voltage and current in all elements are of the form 'complex number times exponential', that is, of the form $v(t) = V(f)e^{i2\pi ft}$ and $i(t) = I(f)e^{i2\pi ft}$.

This a priori knowledge simplifies the analysis tremendously because, if the voltage and current have the same time dependence, then their ratio is not time dependent. Let us check this ratio for the three elements.

Beginning with a resistor: if the voltage drop across a resistor is $v_R(t) = V_R(f)e^{i2\pi ft}$ and the current is $i_R(t) = I_R(f)e^{i2\pi ft}$, then Equation (4.1) entails that:

$$\frac{V_R(f)e^{i2\pi ft}}{I_R(f)e^{i2\pi ft}} = R$$

which entails that:

$$\frac{V_R(f)}{I_R(f)} = R \tag{4.11}$$

Now let us check for the capacitor. Denoting the voltage and current, respectively, as $v_C(t) = V_C(f)e^{i2\pi ft}$ and $i_C(t) = I_C(f)e^{i2\pi ft}$, then Equation (4.1) entails that:

$$I_C(f)e^{i2\pi ft} = C\frac{dV_C(f)e^{i2\pi ft}}{dt} = i2\pi fCV_C(f)e^{i2\pi ft}$$

Thus, the ratio between voltage and current is:

$$\frac{V_C(f)e^{i2\pi ft}}{I_C(f)e^{i2\pi ft}} = \frac{1}{i2\pi fC}$$

which entails that:

$$\frac{V_C(f)}{I_C(f)} = \frac{1}{i2\pi fC} \tag{4.12}$$

Now for the inductor. Denoting the voltage and current, respectively, as $v_L(t) = V_L(f)e^{i2\pi ft}$ and $i_L(t) = I_L(f)e^{i2\pi ft}$, Equation (4.1) entails that:

$$V_L(f)e^{i2\pi ft} = L\frac{dI_L(f)e^{i2\pi ft}}{dt} = i2\pi fLI_L(f)e^{i2\pi ft}$$

Thus, the ratio between voltage and current is:

$$\frac{V_L(f)e^{i2\pi ft}}{I_L(f)e^{i2\pi ft}} = i2\pi fL$$

which entails that:

$$\frac{V_L(f)}{I_L(f)} = i2\pi fL \tag{4.13}$$

As expected, the ratios between voltage and current in all three elements do not depend on time. I emphasize that this is only true for eigenvectors, because only for eigenvectors the voltage and current have the same time dependence. If the input was not an eigenvector, then the voltage would have a general form $v(t)$, the current would have another general form $i(t)$, and their ratio would be some function of time.

Now, notice something quite interesting: if the current and voltages have the same time dependence, then, as illustrated in Figure 4.1, it follows that Kirchhoff's laws must also apply to the coordinates $V(f)$ and $I(f)$.

So, on the one hand, $V(f)$ and $I(f)$ obey Kirchhoff's laws. On the other hand, according to Equation (4.11), Equation (4.12), and Equation (4.13), their ratios in all elements are just numbers (for fixed f), not functions of time. But if neither $V(f)$ nor $I(f)$, nor their ratios, are functions of time, and they obey Kirchhoff's laws, then we can use the same techniques of circuit analysis that apply to circuits

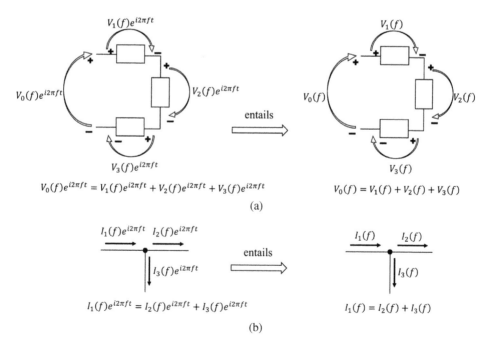

Figure 4.1 If the voltage and currents have the same time dependence, then Kirchhoff's voltage law (a) and current law (b) must apply to the coordinates $V(f)$ and $I(f)$.

without time dependent sources (that is, to DC circuits) in circuits that are time dependent (that is, to AC circuits). All we need to do is to treat the capacitors and inductors as if they were 'resistances' whose values are imaginary numbers according to Equation (4.12) and Equation (4.13), respectively. This generalization of the concept of resistance is called impedance. So, Equation (4.12) gives the impedance of a capacitor, while Equation (4.13) gives the impedance of an inductor. Thus, the ratio between the coordinates $V_{out}(f)$ and $V_{in}(f)$ – and consequently the frequency response – can be found by replacing each element with its impedance, and applying the same techniques of DC circuits.

To exemplify the idea, let us find the frequency response of the RC circuit again (see Figure 3.6a), but using the concept of impedances. Let us call the capacitor impedance Z_C. The capacitor is in series with the resistor, so the equivalent impedance is $Z_C + R$. Consequently, the current $I(f)$ is given by:

$$I(f) = \frac{V_{in}(f)}{Z_C + R}$$

We want the voltage $V_{out}(f)$, taken across the capacitor (see Figure 3.6a). Since the voltage is current times impedance, we have:

$$V_{out}(f) = I(f)Z_C = V_{in}(f)\frac{Z_C}{Z_C + R} = V_{in}(f)\frac{1}{\dfrac{R}{Z_C} + 1}$$

Thus, we conclude that the frequency response is:

$$H(f) = \frac{V_{out}(f)}{V_{in}(f)} = \frac{1}{\dfrac{R}{Z_C} + 1}$$

But, according to Equation (4.12), $Z_C = 1/(i2\pi fC)$. Therefore:

$$H(f) = \frac{V_{out}(f)}{V_{in}(f)} = \frac{1}{\dfrac{R}{Z_C} + 1} = \frac{1}{i2\pi fCR + 1}$$

Of course, this is the same result we had obtained in Equation (4.7). But now we obtained it straight away, without having to find the representation of the operator in the time domain.

You may have already guessed that 'phasor' is just yet another term for the coordinates $V(f)$ and $I(f)$. These coordinates are also frequently called 'complex amplitudes'. Regardless of how they are called, they are just the Fourier transforms of the time domain representation of the signals.

A wise old man once asked why engineers are obsessed with voltages and currents of the form $e^{i2\pi ft}$. After all, these functions cannot even represent something physical: they are complex, and they never begin, and never end. The wise old man was puzzled because he had not realized that, in assuming inputs of this funny $e^{i2\pi ft}$ form, engineers were, in fact, looking for information about the system, that is, for its frequency response.

Worked Exercise: An RLC Circuit as a Harmonic Oscillator

Consider an RLC circuit, as shown in Figure 4.2, where R is the resistance, Z_L is the impedance of an inductor, and Z_C is the impedance of a capacitor. Notice that this circuit is similar to the circuit of Figure 3.6a, but now with an inductor in series with the resistance. Find the frequency response of the RLC circuit and compare it with the frequency response of a harmonic oscillator (see Equation (3.69)).

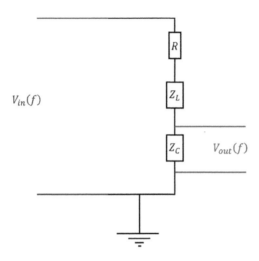

Figure 4.2 An RLC circuit.

Solution:

We have already found the frequency response of the RC circuit (Equation (4.7)). The only difference between the RC circuit and the RLC circuit is that the equivalent resistance R of the RC circuit must be replaced with the equivalent impedance $R + Z_L$ in the RLC circuit. Thus, making the substitution $R \to R + Z_L$ in Equation (4.7), we find:

$$H(f) = \frac{V_{out}(f)}{V_{in}(f)} = \frac{1}{i2\pi fC(R + Z_L) + 1}$$

According to Equation (4.13), $Z_L = i2\pi fL$. Thus, the frequency response reads:

$$H(f) = \frac{V_{out}(f)}{V_{in}(f)} = \frac{1}{i2\pi fC(R + i2\pi fL) + 1}$$

Rearranging, we find:

$$H(f) = \frac{1}{-LC(2\pi f)^2 + iRC2\pi f + 1} \tag{4.14}$$

Comparing the frequency response of the RLC circuit (Equation (4.14)) with the frequency response of the harmonic oscillator (Equation (3.69)), we find that they have the same form, and become identical with the following identifications:

$$m \rightarrow LC$$

$$k \rightarrow 1$$

$$\gamma \rightarrow RC \tag{4.15}$$

Thus, an RLC circuit is a resonator, behaving as a harmonic oscillator, with a resonance frequency given by (see Equation (3.70)):

$$f_0 \approx \frac{1}{2\pi}\sqrt{\frac{k}{m}} = \frac{1}{2\pi}\sqrt{\frac{1}{LC}} \tag{4.16}$$

Notice that the product RC plays the role of damping in the RLC circuit. Thus, the impulse response of the RLC circuit is a sinusoidal oscillation at the resonance frequency, with an exponential decay governed by the product RC (see discussion on the impulse response of a harmonic oscillator in Worked Exercise: Impulse and Frequency Responses of a Harmonic Oscillator, of Chapter 3).

4.3 Exercises

Exercise 1
Suppose you are working with an electric circuit, and then you notice that extraneous frequencies are being created by the system. What is most likely to be happening?

Exercise 2
What happens to the impedance of a capacitor at low frequencies? And at high frequencies? What about the impedance of an inductor?

Exercise 3
Obtain the equivalent impedance of two elements connected in parallel.

Exercise 4
Find the Q factor of the RLC resonator of Figure 4.2.

Exercise 5
Obtain an expression for the equivalent impedance of an inductor in series with a capacitor. What happens at the resonance frequency of Equation (4.16)? Interpret your result in light of the RLC resonator of Figure 4.2.

Exercise 6
Find an expression for the equivalent impedance of an inductor in parallel with a capacitor. What happens at the resonance frequency of Equation (4.16)?

Filters

Learning Objectives

Filtering is a quintessential application of LTI systems. In fact, any LTI system can be regarded as a filter. In this chapter, we will learn the essential concepts of filters: what they are, what they do, and what their main types are. We will also see examples of simple circuits acting as different types of filters.

5.1 Ideal Filters

In common language, a filter is a system that removes parts of something, while allowing other parts to go through unaffected. For example, a water filter is supposed to remove dirt from water. But we also do not want our filter to add anything to the water (maybe we do want new chemicals in the water, but that is not the job of the filter itself). Likewise, in the context of signals and systems, a filter is a system that selectively removes frequency components of a signal, while allowing a chosen part of the spectrum to go through as unaffected as possible.

One key challenge of filter design is precisely to allow the chosen part of the frequency components to go through as unaffected as possible, while blocking as much as possible the part that is to be filtered out. For example, let us suppose $x_1(t)$ is a beautiful song that got corrupted by another signal $x_2(t)$. This other signal could be anything: it could be noise, or part of another song that should not be there, or whatever. So, we have a signal $x(t) = x_1(t) + x_2(t)$, and we would like to filter $x_2(t)$ out of our song.

For the sake of illustration, suppose that the magnitudes of the Fourier transforms of these signals are as represented in Figure 5.1a. As with any LTI system, the frequency domain representation of the output, call it $Y(f)$, is given by the product of the filter's frequency response and the input, that is: $Y(f) = H(f)X(f)$.

Essentials of Signals and Systems, First Edition. Emiliano R. Martins.
© 2023 John Wiley & Sons Ltd. Published 2023 by John Wiley & Sons Ltd.
Companion website: www.wiley.com/go/martins/essentialsofsignalsandsystems

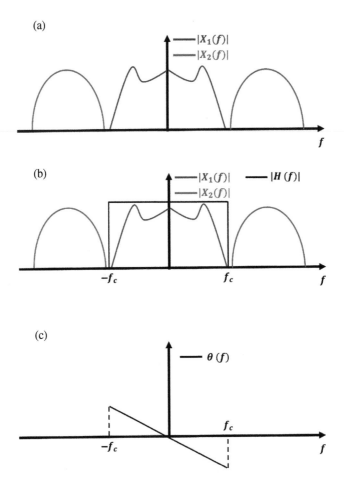

Figure 5.1 Example of an ideal low-pass filter. (a) Illustration of a signal consisting of the superposition of two signals $X_1(f)$ and $X_2(f)$. (b) If the objective is to filter out $X_2(f)$ without distorting $X_1(f)$, then the frequency response of the system must have a constant amplitude within the region where $|X_1(f)|$ is different from zero, but transit sharply to zero outside this region. The frequency delimitating the transition is called the cut-off frequency, and it is denoted by f_c. (c) To prevent distortion, the phase of the frequency response must be constant or linear.

Our goal is to filter out the signal $x_2(t)$ from $x(t)$. But we want to do that without distorting $x_1(t)$. That means that we want the output to be $x_1(t)$, that is, we want $Y(f) = X_1(f)$. But to get $Y(f) = X_1(f)$, we need $H(f) = 1$ within the spectral region where $X_1(f)$ is different from zero, and we also want $H(f) = 0$ outside this region to cut off the unwanted $X_2(f)$ part. The frequency that marks the transition is called the cut-off frequency, and it is represented as f_c in Figure 5.1b, alongside the magnitude of the ideal $H(f)$. In fact, the condition $H(f) = 1$ is not necessary to keep $x_1(t)$ undistorted. What is necessary is that the magnitude of $H(f)$ be constant up to the cut-off frequency. If the magnitude is constant but different from one, that means $x_1(t)$ got amplified or dampened, but not really distorted.

What about the phase of $H(f)$? Well, in principle, we want the phase to be constant in the region of interest (for frequencies up to the cut-off frequency). But that is also overkill because, as we have learned in Section 2.12.4 (Time Shifting), the imposition of a linear phase does not change the signal: it only translates it in time. Thus, the phase $\theta(f)$ of the frequency response $H(f)$ can be a linear function within the region of interest, and that only delays the signal, but does not distort it. How much it delays depends on the slope according to Equation (2.119). For example, if $\theta(f) = -\alpha f$, then, according to Equation (2.119), the time delay is $t_0 = \alpha/2\pi$. Notice that I am assuming a negative slope (positive α) in Figure 5.1c. In the list of exercises, you will be asked why positive slopes are not allowed.

The frequency response illustrated in Figure 5.1b and Figure 5.1c is an example of an ideal frequency response, which is characterized by constant amplitude and linear (or constant) phase within the region of interest, and a discontinuity in the transition frequency. Ideal filters are called ideal because they do the best possible job: they do not distort the signal in the spectral region of interest, while blocking all components outside the region of interest (in this example, frequencies above the cut-off frequency).

There are three main types of filters: low-pass filter, high-pass filter, and band-pass filter. The frequency response sketched in Figure 5.1b and Figure 5.1c is an example of an ideal low-pass filter: it lets the components with frequency lower than the cut-off to pass, and it blocks the frequencies higher than the cut-off. A high-pass filter, as illustrated in Figure 5.2b, does the opposite: it allows the frequencies higher than the cut-off to pass and the frequencies lower than the cut-off are blocked. Finally, a band-pass filter allows frequencies around a specific frequency to pass. This frequency is sometimes called central frequency, or resonance frequency. Notice that a harmonic oscillator (see Worked Exercise: Impulse and Frequency Responses of a Harmonic Oscillator, of Chapter 3) and an RLC circuit (see Worked Exercise: An RLC Circuit as a Harmonic Oscillator, of Chapter 4) are examples of band-pass filters.

The main objective of the engineer is to design a filter with a frequency response as close as possible to the ideal. Filter design is a science on its own, and as such it is outside the scope of this book (if you want to pursue this topic further, a good start is to look up Butterworth filters and Chebyshev filters). In Sections 5.2–5.4, we will see examples, as simple as possible, of each kind of filter. We begin with the low-pass filter, of which our good old RC circuit is an example.

5.2 Example of a Low-pass Filter

The RC circuit of Figure 3.6a is one of the simplest examples of a low-pass filter. It relies on the frequency dependence of the capacitor impedance to achieve the low-pass filtering: at low frequencies, the capacitor has a high impedance, so the signal from the input is mostly transferred to the capacitor. At high frequencies, however, the capacitor has a low impedance; consequently, the input signal is mostly transferred to the resistor and, as such, is cut off the output. Of course, such a simple circuit is far from ideal. For example, we have seen that the frequency response of the RC circuit is:

Equation (4.7)

$$H(f) = \frac{1}{iRC2\pi f + 1}$$

Therefore, its magnitude is:

$$|H(f)| = \sqrt{\frac{1}{(RC2\pi f)^2 + 1}} \tag{5.1}$$

According to Equation (5.1), the magnitude of the frequency response is not constant, as required by ideal filters, but decays monotonically with the frequency. In such case, we need a criterion for how low $|H(f)|$ must be to define the cut-off frequency. One common criterion is to require that $|H(f)|$ falls at the cut-off frequency by $1/\sqrt{2}$ of its maximum value. In the case of the RC circuit, $|H(f)|$ is maximum at $f = 0$ (it is, after all, a low-pass filter). Furthermore, the maximum value is $|H(0)| = 1$, as can be seen directly from Equation (5.1). Thus, imposing that:

$$|H(f_c)| = \frac{|H(0)|}{\sqrt{2}} = \frac{1}{\sqrt{2}} \tag{5.2}$$

We find that:

$$(RC2\pi f_c)^2 = 1$$

And, therefore:

$$f_c = \frac{1}{2\pi RC} \tag{5.3}$$

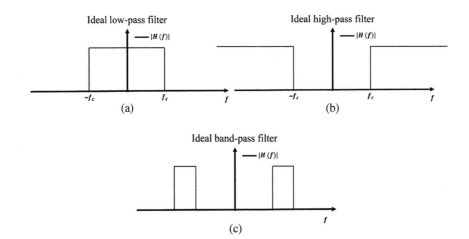

Figure 5.2 Illustration of the magnitude of the frequency response of (a) ideal low-pass filter, (b) ideal high-pass filter, and (c) ideal band-pass filter.

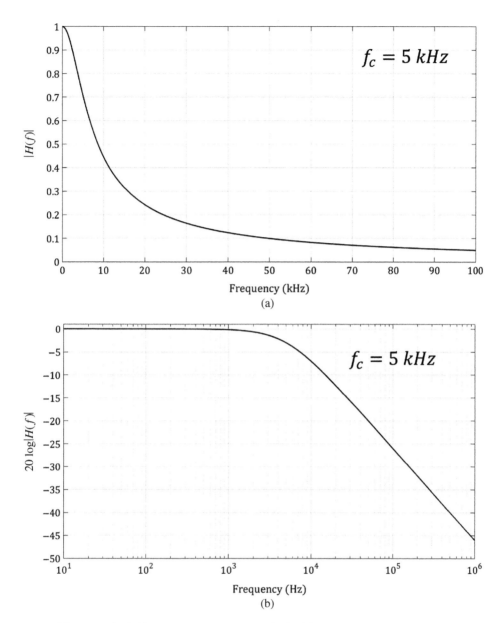

Figure 5.3 (a) Plot of $|H(f)|$ as defined in Equation (5.1). (b) Plot of frequency response in dB – $20\log(|H(f)|)$ – against frequency in a logarithmic scale. This is an example of a Bode plot.

With the help of Equation (5.3), Equation (5.1) can be re-expressed as:

$$|H(f)| = \sqrt{\frac{1}{\left(\frac{f}{f_c}\right)^2 + 1}} \qquad (5.4)$$

Since the magnitude of the frequency response of a physical system is always an even function, it does not make sense to plot the negative frequency part of the spectrum, since it contains the same information of the positive frequency part. Thus, it is much more usual to plot only the positive frequency part. Such a plot, assuming the frequency response of Equation (5.4) with $f_c = 5\ kHz$, is shown in Figure 5.3a. As expected, the magnitude decreases monotonically with the frequency. Notice the soft 'kink' at $f \sim 4\ f_c = 20$ kHz. Beyond this kink, the magnitude decreases slowly, and as such it requires a plot involving frequencies much higher than the cut-off frequency to see the magnitude dropping to low values. Consequently, such a plot 'hides' the region of the passing band, that is, the frequencies lower than the cut-off frequency.

A much nicer plot can be obtained when the frequency axis is plotted in logarithm scale and the magnitude is plotted in decibels, as shown in Figure 5.3b. This is an example of a Bode plot, and it is by far the most common way frequency responses of filters are plotted. Notice that the value of the magnitude at the cut-off frequency in decibel is $20 \cdot \log(1/\sqrt{2}) = -20 \cdot \log(\sqrt{2}) \approx -3\ dB$.

As you may have already noticed, the phase of $H(f)$ is also not ideal since it is not linear. However, it is approximately linear for frequencies lower than the cut-off frequency, as you will be asked to prove in the exercise list.

5.3 Example of a High-pass Filter

A high-pass filter can be obtained with the RC circuit by taking the output across the resistor. Another option is to replace the capacitor with an inductor, thus forming an RL circuit, and taking the output across the inductor. The idea is again to take advantage of the frequency dependence of the impedance: at low frequencies, the inductor has a low impedance, so the input is mostly transferred to the resistor and, as such, is cut off the output; at high frequencies, the inductor has a high impedance, and so the input is mostly transferred to the inductor.

The frequency response can be found straightforwardly. Noticing that the output voltage is the current times the inductor impedance:

$$V_{out}(f) = I(f)Z_L$$

And that the current is the input voltage divided by the total impedance, we obtain:

$$V_{out}(f) = V_{in}(f)\frac{Z_L}{Z_L + R}$$

Therefore, the frequency response is:

$$H(f) = \frac{Z_L}{Z_L + R} = \frac{i2\pi fL}{i2\pi fL + R} \tag{5.5}$$

Notice that $H(0) = 0$, as expected for a high-pass filter. Furthermore, notice that $H(f) \Rightarrow 1$ as $f \Rightarrow \infty$, which is also characteristic of a high-pass filter. Such

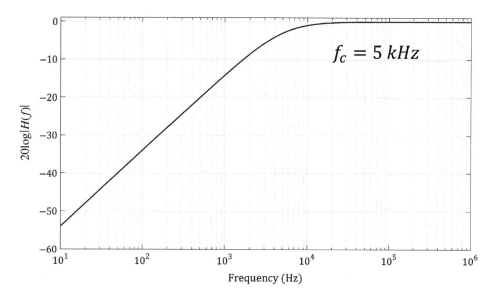

Figure 5.4 Example of Bode plot of an RL high pass filter assuming a cut-off frequency of 5 kHz.

characteristics are nicely captured by a Bode plot of Equation (5.5), as shown in Figure 5.4. In the exercise list, you will be asked to find the cut-off frequency of the RL high-pass filter.

5.4   Example of a Band-pass Filter

Any resonating system acts as a band-pass filter and we have already seen an example in the RLC resonator of Figure 4.2. Another option is to design a band-pass filter using the so-called 'tank circuit', which consists of an inductor in parallel with a capacitor. An example of a band-pass filter using a tank circuit is shown in Figure 5.5.

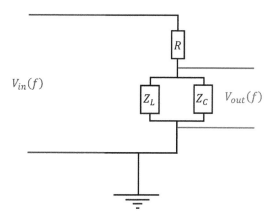

Figure 5.5 A band-pass filter based on a tank circuit.

The operation of the band-pass filter based on a tank circuit can be qualitatively described as follows: if, on the one hand, the frequency is too low, then the inductor impedance becomes too low, thus essentially short-circuiting the output; if, on the other hand, the frequency is too high, then the capacitor impedance becomes too low, thus again short-circuiting the output. Therefore, both too low and too high frequencies are cut off, and only a certain range of frequencies – in which the impedances of both the inductor and the capacitor are high – can pass.

In the exercise list of Chapter 4 (Electric Circuits as LTI Systems), you were asked to calculate the equivalent impedance of a capacitor in parallel with an inductor, that is, the equivalent impedance of a tank circuit. Denoting this equivalent impedance as $Z_{tank}(f)$, you found that:

$$Z_{tank}(f) = \frac{i2\pi fL}{1 - (2\pi f)^2 LC} \tag{5.6}$$

As you have also found in the same exercise, at the resonance frequency $f_0 = (1/2\pi)(1/\sqrt{LC})$, the impedance of the tank circuit goes to infinity. Thus, at the resonance frequency, the tank circuit acts as an open circuit. Consequently, the input is fully transferred to the output at the resonance frequency. In other words, the resonance frequency of the tank circuit coincides with the resonance frequency of the band-pass filter.

In the exercise list, you will be asked to find the frequency response of the band-pass filter based on the tank circuit, its two cut-off frequencies, and its Q-factor.

5.5 Exercises

Exercise 1
Why cannot the phase of the frequency response of an ideal filter, as illustrated in Figure 5.1, have a positive slope?

Exercise 2
The frequency response $H(f) = |H(f)|e^{i\theta(f)}$ of an ideal low-pass filter satisfies:

$$|H(f)| = 1 \quad for - f_c < f < f_c$$

And:

$$\theta(f) = -\alpha f$$

Find the impulse response of the ideal filter. Is it a causal system?
Hint: find the impulse response assuming $\alpha = 0$ and then use the property of time shifting to find the impulse response for $\alpha \neq 0$.

Exercise 3
Prove that the phase of the frequency response of an RC circuit is approximately linear for frequencies lower than the cut-off frequency.

Exercise 4
Find the cut-off frequency of the RL high-pass filter.

Exercise 5
Express the magnitude of the frequency response of the RL high-pass filter in terms of the cut-off frequency and compare it with Equation (5.4).

Exercise 6
Find the frequency response of the band-pass filter of Figure 5.5 and evaluate its magnitude at the resonance frequency.

Exercise 7
Find the two cut-off frequencies of the band-pass filter of Figure 5.5. Use these two cut-off frequencies to find an expression for the width of the resonance and evaluate what happens to the width when the resistance is increased. Explain the width dependence on the resistance qualitatively.

Exercise 8
Find the Q-factor of the band-pass filter of Figure 5.5.

Exercise 9
In the exercise list of Chapter 6, you will be asked to prove that the impulse response of the RC low-pass filter is:

$$h(t) = \frac{1}{RC} e^{-\frac{t}{RC}}$$

Since this is a low-pass filter, the convolution of $h(t)$ with a cosine having a frequency much higher than the cut-off frequency (Equation (5.3)) results in a 'weak' signal (a signal with low amplitude), whereas the convolution with a cosine with a frequency much lower that the cut-off frequency does not change the cosine significantly. Provide a qualitative explanation for this effect.

Exercise 10
In the exercise list of Chapter 6, you will be asked to prove that the impulse response of the RL high-pass filter is:

$$h(t) = \delta(t) - \frac{R}{L} e^{-\frac{R}{L} t}$$

Since this is a high-pass filter, the convolution of $h(t)$ with a cosine having a frequency much higher than the cut-off frequency (see Exercise 4 of Chapter 5) almost does not modify the cosine, whereas the convolution with a cosine having a frequency much lower than the cut-off frequency results in a weak signal. Provide a qualitative explanation for this effect.

6

Introduction to the Laplace Transform

Learning Objectives

The Laplace transform is a generalization of the Fourier transform, especially required for functions that do not have Fourier transforms. In this chapter, we will learn what a Laplace transform is, its main characteristics, in what conditions it exists, and how it is related to the Fourier transform. We will also learn how to deal with initial conditions by means of unilateral transforms.

6.1 Motivation: Stability of LTI Systems

The motivation to generalize the Fourier transform to the Laplace transform is related to the notion of stability of LTI systems. A system (in general, not only LTI) is stable if its output does not diverge. For example, consider a system consisting of a U-shaped structure (like a skate ramp) with a ball inside, as shown in Figure 6.1a. The input is the force applied to the ball, and the output is the ball's height. If the input is bounded, that is, if the force is finite, then of course the height of the ball is also bounded. Thus, the system of Figure 6.1a is an example of a stable system. The system of Figure 6.1b, however, is unstable, because the output is not bounded: even if we just flick the ball, it will go down forever and ever, which means that the output (the height) is unbounded: it goes to negative infinity.

Thus, a system is stable when all bounded inputs (any bounded input, without exception) result in bounded outputs. But, if there is a bounded input which results in an unbounded output, then the system is unstable.

In Chapter 3 (Representation of Systems), we learned that an LTI system is fully characterized by its impulse response. Thus, it must be possible to find out whether

---

Essentials of Signals and Systems, First Edition. Emiliano R. Martins.
© 2023 John Wiley & Sons Ltd. Published 2023 by John Wiley & Sons Ltd.
Companion website: www.wiley.com/go/martins/essentialsofsignalsandsystems

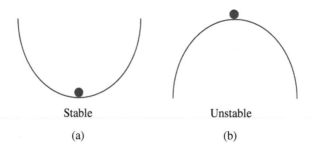

Stable   Unstable

(a)   (b)

Figure 6.1 (a) Example of a stable system. (b) Example of an unstable system.

an LTI system is stable or not by inspection of its impulse response. The main job of this section is to find what properties of the impulse response characterize an LTI system as stable or unstable.

We have just seen that a system is stable if bounded inputs lead to bounded outputs. A signal is bounded when it is never infinite. That means there exists a finite number, call it A, which is the highest value the signal ever reaches. Thus, mathematically, a signal $x(t)$ is bounded if there exists a finite number A such that:

$$|x(t)| \leq A \quad \text{for all } t \tag{6.1}$$

As we learned in Chapter 3 (Representation of Systems), if the system is LTI, then its input $x(t)$ is related to the output $y(t)$ through the convolution with the impulse response $h(t)$:

Equation (3.60)
$$y(t) = h(t) * x(t)$$

Thus, to test whether an LTI system is stable or not, we can check whether $y(t)$ is bounded when $x(t)$ is bounded. In other words, we assume that Equation (6.1) is true, and then we check if there exists a finite number B such that:

$$|y(t)| \leq B \quad \text{for all } t \tag{6.2}$$

We begin by writing the integral of convolution explicitly:

$$y(t) = h(t) * x(t) = \int_{-\infty}^{-\infty} h(t_0)x(t - t_0)dt_0$$

And now take the magnitude of the output:

$$|y(t)| = \left| \int_{-\infty}^{-\infty} h(t_0)x(t - t_0)dt_0 \right| \tag{6.3}$$

An integral is just a fancy sum of terms. So, on the right-hand side of Equation (6.3), we have the magnitude of a sum of terms. But the magnitude of the sum can never be larger than the sum of the magnitudes. For example, if two numbers involved in the sum are -5 and 2, then the sum is -3 and the magnitude of the sum is 3. But the sum of the magnitudes is $5 + 2 = 7$. Thus, it is guaranteed that:

$$\left| \int_{-\infty}^{-\infty} h(t_0)x(t - t_0)dt_0 \right| \leq \int_{-\infty}^{-\infty} |h(t_0)||x(t - t_0)|dt_0 \tag{6.4}$$

where the right-hand side of Equation (6.4) is the sum of the magnitudes.

We are assuming that the input is bounded, so the largest value $|x(t - t_0)|$ can take is A. Thus, it follows that:

$$\int_{-\infty}^{-\infty} |h(t_0)||x(t - t_0)|dt_0 \leq A \int_{-\infty}^{-\infty} |h(t_0)|dt_0$$

Consequently:

$$|y(t)| \leq A \int_{-\infty}^{-\infty} |h(t_0)|dt_0 \tag{6.5}$$

According to Equation (6.5), if the integral $\int_{-\infty}^{-\infty} |h(t_0)|dt_0$ does not blow up (that is, if the integral is not infinite), then it is guaranteed that the output is bounded when the input is bounded. Thus, we have found the condition of stability in LTI systems: an LTI system is stable when the magnitude of its impulse response is integrable. In other words, if there exists a finite number C such that:

$$\int_{-\infty}^{-\infty} |h(t)|dt \leq C \tag{6.6}$$

Then the LTI system is stable.

Unstable systems play an important role in engineering, especially in control engineering. But unstable systems pose a challenge: often they do not have a frequency response, because the direct Fourier transform of the impulse response blows up: it does not converge. And without a frequency response we lose the beauty and simplicity of the representation in the frequency domain, as amply discussed in Chapter 3 (Representation of Systems). So, we need a solution to this problem, and that is where the Laplace transform comes in.

Before introducing the Laplace transform, however, it is instructive to take a closer look at this assertion that unstable systems may not have frequency responses. We already know that the frequency response $H(f)$ is the Fourier transform of the impulse response:

$$H(f) = \int_{-\infty}^{+\infty} h(t)e^{-i2\pi ft}dt \tag{6.7}$$

Taking the magnitude of the frequency response, we find:

$$|H(f)| = \left| \int_{-\infty}^{+\infty} h(t)e^{-i2\pi ft}dt \right| \leq \int_{-\infty}^{+\infty} |h(t)||e^{-i2\pi ft}|dt \tag{6.8}$$

But $|e^{-i2\pi ft}| = 1$, so:

$$|H(f)| \leq \int_{-\infty}^{+\infty} |h(t)| dt \tag{6.9}$$

According to Equation (6.9), it is only guaranteed that the frequency response exists, that is, $|H(f)|$ is finite, if the magnitude of the impulse response is integrable. But this is exactly the condition of stability in LTI system (Equation (6.6)). Thus, since the magnitudes of the impulse responses of unstable systems are not integrable, they often do not have Fourier transforms, which means that their frequency responses do not exist. As mentioned earlier, without the frequency response we lose the simplicity of the frequency domain representation.

So, how can we circumvent this problem? To find an answer, we need to recall two pieces of information. The first piece is that the frequency domain representation is an especially simple representation because the functions $e^{i2\pi ft}$ are eigenvectors. The second piece is that the condition of existence of $H(f)$ – as shown in Equation (6.9) – only coincides with the integrability of $|h(t)|$ because $|e^{i2\pi ft}| = 1$. Thus, we are led to the question: can we find eigenvectors (to have a simple representation) whose magnitudes differ from one (so that the representation may exist even for unstable systems)? If we can, then we may have a representation in a basis of eigenvectors, even when $|h(t)|$ is not integrable.

Fortunately, the answer to this question is affirmative. Recall, from Section 3.6: Eigenvectors and Eigenvalues of LTI Operators, that any exponential of the form $e^{\alpha t}$ is an eigenvector. So, to find an eigenvector with magnitude different from one, we just need to plug in a real part in the argument of the exponential, turning $e^{i2\pi ft}$ into $e^{\sigma t + i2\pi ft}$, where σ is a real number. The function $e^{\sigma t + i2\pi ft}$ is still an eigenvector of LTI operators (since its time derivatives are proportional to it), but its magnitude is no longer one. Indeed: $|e^{\sigma t + i2\pi ft}| = |e^{\sigma t} e^{i2\pi ft}| = |e^{\sigma t}||e^{i2\pi ft}| = |e^{\sigma t}|$.

We have now a motivation to find a representation in a non-orthonormal basis, formed by functions of the type $e^{\sigma t + i2\pi ft}$, where each value of σ and f defines a different eigenvector. Thus, the direct Fourier transform is generalized to:

$$H(f, \sigma) = \int_{-\infty}^{+\infty} h(t) e^{-(\sigma t + i2\pi ft)} dt \tag{6.10}$$

Notice that $H(f, \sigma)$ is now a function of two variables: f and σ. That makes sense: since each pair f and σ indexes a different eigenvector, then we need the same pair to index the scalar product with these eigenvectors.

With this generalization, the magnitude of the transform reads:

$$|H(f, \sigma)| = \left| \int_{-\infty}^{+\infty} h(t) e^{-(\sigma t + i2\pi ft)} dt \right| \leq \int_{-\infty}^{+\infty} |h(t)||e^{-\sigma t}| dt \tag{6.11}$$

Comparing Equation (6.11) with Equation (6.9), we notice that $|H(f, \sigma)|$ may be finite even if $\int_{-\infty}^{+\infty} |h(t)| dt$ does not converge, because the extra term $|e^{-\sigma t}|$ may allow the convergence of $\int_{-\infty}^{+\infty} |h(t)||e^{-\sigma t}| dt$. Thus, by using eigenvectors of the type $e^{\sigma t + i2\pi ft}$, we may obtain a representation in a basis of eigenvectors – which, as we have seen, greatly simplifies the analysis of systems – even if the system is unstable, and as such does not have a frequency response.

Equation (6.10) defines the direct Laplace transform of $h(t)$. Notice that the Laplace transform includes the Fourier transform: the Fourier transform is the Laplace transform for the particular case $\sigma = 0$. We will further discuss the connection between Laplace and Fourier transform in Section 6.2, which discusses the Laplace transform pair.

6.2 The Laplace Transform as a Generalization of the Fourier Transform

Although there is nothing wrong about Equation (6.10), usually the Laplace transform is presented in terms of a complex variable s, defined as:

$$s = \sigma + i2\pi f \tag{6.12}$$

With the help of this definition, we can recast Equation (6.10) as:

$$H(s) = \int_{-\infty}^{+\infty} h(t)e^{-st}dt \tag{6.13}$$

Even though this form is more compact than Equation (6.10), it is not my favourite because it hides the fact that $H(s)$ is a function of two variables: σ and f. But that is the most likely way you will find the direct Laplace transform defined in other textbooks.

To find the inverse Laplace transform, we note that the Laplace transform of $h(t)$ is the Fourier transform of $h(t)e^{-\sigma t}$. Indeed:

$$H(s) = \int_{-\infty}^{+\infty} h(t)e^{-st}dt = \int_{-\infty}^{+\infty} (h(t)e^{-\sigma t})e^{-i2\pi ft}dt \tag{6.14}$$

Thus, if we have $H(s)$ but want $h(t)$, we can first inverse Fourier transform $H(s)$ to find $h(t)e^{-\sigma t}$:

$$h(t)e^{-\sigma t} = \int_{-\infty}^{+\infty} H(s)e^{i2\pi ft}df = \int_{-\infty}^{+\infty} H(\sigma, f)e^{i2\pi ft}df \tag{6.15}$$

Notice that σ is a fixed parameter in the integral in Equation (6.15). Now we can easily find $h(t)$ multiplying both sides of Equation (6.15) by $e^{\sigma t}$. Thus:

$$h(t) = e^{\sigma t} \int_{-\infty}^{+\infty} H(\sigma, f)e^{i2\pi ft}df$$

Since the integral is over f, we can bring the time-dependent exponential into it:

$$h(t) = e^{\sigma t} \int_{-\infty}^{+\infty} H(\sigma, f)e^{i2\pi ft}df = \int_{-\infty}^{+\infty} H(\sigma, f)e^{\sigma t + i2\pi ft}df \tag{6.16}$$

Finally, using again the definition of s (Equation (6.12)), we can re-express Equation (6.16) as:

$$h(t) = \int_{-\infty}^{+\infty} H(s)e^{st}df \tag{6.17}$$

Equation (6.13) and Equation (6.17) form a Laplace transform pair.

Laplace Transform Pair

DIRECT LAPALACE TRANSFORM

$$X(s) = \int_{-\infty}^{+\infty} x(t)e^{-st}dt$$

INVERSE LAPLACE TRANSFORM

$$x(t) = \int_{-\infty}^{+\infty} X(s)e^{st}df \tag{6.18}$$

Notice that, in Equation (6.18), I replaced $h(t)$ and $H(s)$ with $x(t)$ and $X(s)$. I replaced them to emphasize that, even though we have used the impulse response $h(t)$ to motivate the Laplace transform, it can be applied to any function.

According to the inverse Laplace transform (see Equation (6.16)), a general function $h(t)$ can be obtained as a linear combination of a group of functions of the type $e^{\sigma t + i2\pi ft}$. Furthermore, notice that σ is a fixed parameter in the integral in Equation (6.16). That means the group of functions involved in the linear combination is the group formed by all functions of the type $e^{\sigma t + i2\pi ft}$, with σ fixed and f ranging from $-\infty$ to $+\infty$. Therefore, since these functions, with fixed σ, can be used to compose a general function $h(t)$, then they form a basis. We can thus think of each value of σ as indexing a different basis, whereas f indexes a different eigenvector within the basis indexed by σ. In particular, the basis indexed by $\sigma = 0$ is the basis of the Fourier transform. Thus, while the Fourier transform of a signal gives the coordinates of this signal in one basis, the Laplace transform gives the coordinates of a signal in a group of many bases, where each σ indexes a different basis. Notice that, since σ is a real number, then there are infinitely many bases. In this sense, the Laplace transform is a kind of 'overkill' representation, as it is giving the coordinates not only in one basis (as in the Fourier transform) but also in a group with infinitely many bases, each basis with infinitely many eigenvectors.

6.3   Properties of Laplace Transforms

The properties of Laplace transforms and their proofs are so similar to the properties of Fourier transforms that it is not worth going through them in detail, as we did in Section 2.12 (Properties of Fourier Transforms). After all, as we have just seen, the Laplace transform of a function $x(t)$ is identical to the Fourier transform of the function $x(t)e^{-\sigma t}$, so the properties of Fourier transforms must be extendable to Laplace transforms. Furthermore, a property of Laplace transforms must coincide with the corresponding property of Fourier transforms in the particular case $\sigma = 0$. The property of product in the time domain, however, takes a somewhat awkward form, as shown in Appendix A – Laplace Transform Property of Product in the Time Domain. Nevertheless, most of the properties can be straightforwardly obtained replacing $i2\pi f$ with s (see Appendix B – List of Properties of Laplace Transforms). Thus, for example, the time-shifting property (Equation (2.119)) becomes:

if:

$$g_2(t) = g(t - t_0)$$

then:

$$G_2(s) = e^{-st_0} G(s) \tag{6.19}$$

And the property of differentiation becomes:

if:

$$g_2(t) = \frac{dg(t)}{dt}$$

then:

$$G_2(s) = sG(s) \tag{6.20}$$

The proofs of Equation (6.19) and Equation (6.20) follow the same logic of the proofs of Equation (2.119) and Equation (2.132), respectively, and are left as exercises.

In particular, the property of convolution in the time domain (which corresponds to product in the frequency domain – Equation (2.144) for Fourier transforms) leads to the representation between input and output of LTI systems in the s domain. Indeed, if $x(t)$ and $y(t)$ are the input and output of an LTI system whose impulse response is $h(t)$, then applying the direct Laplace transform to both sides of the relation $y(t) = h(t) * x(t)$ leads to:

$$Y(s) = H(s)X(s) \tag{6.21}$$

The function $H(s)$, which is the Laplace transform of the impulse response, is called 'the transfer function' of the LTI system. We learned that the frequency response fully specifies an LTI system. Likewise, the transfer function is a full specification of an LTI system. More precisely, each set of coordinates $H(\sigma, f)$ with fixed σ is a full specification of an LTI system. Thus, $H(\sigma, f)$ contains lots of full specifications of LTI systems: each value of σ for which $H(\sigma, f)$ exists indexes a different full specification.

6.4   Region of Convergence

In Section 6.1 (Motivation: Stability in LTI Systems), we saw that using an eigenvector of the type $e^{\sigma t + i2\pi f t}$ MAY allow convergence of the integral $\int_{-\infty}^{+\infty} x(t) e^{-(\sigma t + i2\pi f t)} dt$ even when the magnitude of $x(t)$ is not itself integrable. But that is not guaranteed. Maybe $x(t)$ is such a bad-boy that $\int_{-\infty}^{+\infty} x(t) e^{-(\sigma t + i2\pi f t)} dt$ never converges, and thus $x(t)$ has no Laplace transform. But these cases are rare. Usually, we encounter functions for which $\int_{-\infty}^{+\infty} x(t) e^{-(\sigma t + i2\pi f t)}$ converges for a certain range of values of σ. These ranges define the region of convergence (ROC) of the Laplace transform of a function $x(t)$.

It is convenient to classify functions into right-sided functions and left-sided functions. Thus, functions that have a beginning, but not an end, are called right-sided functions, whereas functions that never begin, but do have an end, are called left-sided functions. The motivation for the terminology is evident: right-sided functions extend all the way to positive infinite time, that is, to the right side of the time axis, whereas left-sided functions extend all the way to negative infinite time, that is, to the left side of the time axis.

Right-sided and left-sided functions can be specified with the help of the Heaviside step function $u(t)$, defined as:

$$u(t) = 1 \; for \; t \geq 0$$

$$u(t) = 0 \; for \; t < 0 \tag{6.22}$$

The ROC has some interesting properties that are nicely illustrated using exponential functions as examples. Thus, suppose that we want to find the ROC of the function $x(t) = e^{-\alpha t} u(t)$, where, for the sake of simplicity, we assume that α is a real number. Notice that this is a right-sided function: it begins at time $t = 0$ (the Heaviside step function ensures that $x(t) = 0$ for negative time), and it extends all the way to positive infinite time. To find the ROC of this function we can evaluate its Laplace transform explicitly. Thus:

$$X(s) = \int_{-\infty}^{+\infty} x(t) e^{-st} dt = \int_{-\infty}^{+\infty} e^{-\alpha t} u(t) e^{-st} dt$$

Since $u(t) = 0$ for $t < 0$, then the integral is nonzero only over positive time. Therefore:

$$X(s) = \int_{-\infty}^{+\infty} e^{-\alpha t} u(t) e^{-st} dt = \int_{0}^{+\infty} e^{-\alpha t} e^{-st} dt$$

$$= \int_{0}^{+\infty} e^{-(\alpha+s)t} dt = \int_{0}^{+\infty} e^{-(a+\sigma)t} e^{-i2\pi ft} dt$$

The integral above extends all the way to positive infinite time. As such, it can only converge if $e^{-(a+\sigma)t}$ goes to zero when t goes to infinity. In other words, the integral only converges if $a + \sigma$ is a positive number, that is, if $a + \sigma > 0$. For σ inside the ROC (that is, for $\sigma > -\alpha$), we find:

$$X(s) = \int_{0}^{+\infty} e^{-(\alpha+s)t} dt = -\left. \frac{e^{-(\alpha+s)t}}{\alpha+s} \right|_{-}^{\infty} = -\left. \frac{e^{-(a+\sigma)t} e^{-i2\pi ft}}{\alpha+s} \right|_{0}^{\infty}$$

$$= -\frac{e^{-(a+\sigma)\infty} e^{-i2\pi f\infty}}{\alpha+s} + \frac{e^{-(a+\sigma)0} e^{-i2\pi f0}}{\alpha+s} = -\frac{0 \cdot e^{-i2\pi f\infty}}{\alpha+s} + \frac{1}{\alpha+s} = \frac{1}{\alpha+s}$$

Thus, we have concluded that:

if:

$$x(t) = e^{-\alpha t} u(t)$$

then:

$$X(s) = \frac{1}{s+a}, \quad ROC : \sigma > -\alpha \tag{6.23}$$

The ROC can be graphically represented in a plot of the s plane, which features the real part of s – that is, $Re\{s\} = \sigma$ – along the horizontal axis and the imaginary part – that is, $Im\{s\} = 2\pi f = \omega$ – along the vertical axis. The ROC of Equation (6.23) is shown as the shaded region in Figure 6.2a. Since it depends only on σ, the ROC is delimited by a vertical line, excluding the point $\sigma = -\alpha$. Moreover, the ROC extends all the way to the right side of the point $\sigma = -\alpha$. This is a universal feature of ROC of right-sided functions: they extend from a certain vertical line all the way to the right, towards positive infinite σ. That makes sense: if the Laplace transform of a right-sided function converges for a certain σ_0, then it certainly converges for a larger σ (that is, for $\sigma > \sigma_0$), because $\sigma > \sigma_0$ guarantees faster convergence in a function that is extending towards positive infinite time (recall that σ enters the direct Laplace transform with a minus sign – due to the term e^{-st} – so, in a right-sided function, larger positive σ gives faster convergence than lower positive σ).

Likewise, the region of convergence of a left-sided function must be the left-sided region delimited by a vertical line. As an example, the ROC of the left-sided function

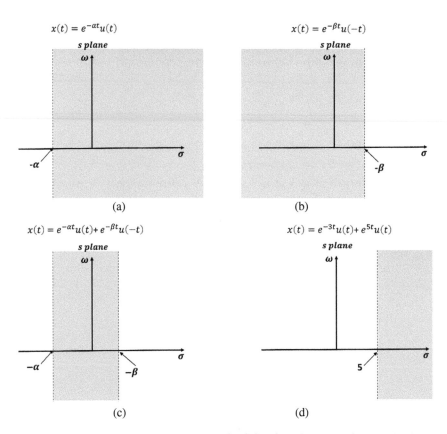

$x(t) = e^{-\alpha t}u(t)$

$x(t) = e^{-\beta t}u(-t)$

(a)   (b)

$x(t) = e^{-\alpha t}u(t) + e^{-\beta t}u(-t)$

$x(t) = e^{-3t}u(t) + e^{5t}u(t)$

(c)   (d)

Figure 6.2 Examples of graphical representation of ROCs of Laplace transforms. The frequency along the vertical axis is $\omega = 2\pi f$. (a) A right-sided function, assuming positive α. (b) A left-sided function, assuming negative β. (c) A combination of right-sided and left-sided functions assuming positive α and negative β. (d) A combination of two right-sided functions.

$x(t) = e^{-\beta t}u(-t)$ is shown in Figure 6.2b. Notice that the negative sign in the argument of $u(-t)$ ensures that it is zero for positive time, and one for negative time; consequently, $x(t) = e^{-\beta t}u(-t)$ is a left-sided function.

When two or more functions are added together, like in $x(t) = x_1(t) + x_2(t) + x_3(t) + \ldots$, then the ROC of the new function $x(t)$ is the overlap between the ROC of each of its composing functions $x_1(t)$, $x_2(t)$, $x_3(t)$, ... After all, for $x(t)$ to converge, it is necessary that all its composing functions $x_1(t)$, $x_2(t)$, $x_3(t)$, ... converge. If only one of them diverges, then $x(t)$ also diverges. For example, as shown in Figure 6.2c, the function $x(t) = e^{-\alpha t}u(t) + e^{-\beta t}u(-t)$ converges in the region $-\alpha < \sigma < -\beta$. Of course, this is only possible if $\alpha > \beta$. If, however, $\alpha < \beta$, then the ROC is null.

In all examples seen so far, the ROC includes the vertical axis crossing the origin (that is, the axis defined by $\sigma = 0$). Thus, these functions also have a Fourier transform. That is not necessarily the case, however. For example, the ROC of the function $x(t) = e^{-3t}u(t) + e^{5t}u(t)$ does not include the vertical axis crossing the origin (see Figure 6.2d), which means that this function does not have a Fourier transform.

Simple inspection of the ROC of the transfer function $H(s)$ of an LTI system can provide important information about the system. For example, by checking if the ROC includes the vertical axis crossing the origin (the axis $\sigma = 0$), we can tell whether the impulse response $h(t)$ has a frequency response or not. Thus, if the ROC includes the vertical axis crossing the origin, then $h(t)$ has a frequency response, which means that the system is stable, as discussed in Section 6.1 (Motivation: Stability in LTI Systems). If the ROC does not include the vertical axis crossing the origin, then the system is unstable. That is quite a nice party trick: one can tell whether the system is stable or not by just checking the ROC. Another cool party trick is the information on causality. We saw in Chapter 3 (Representation of Systems) that the impulse response of a causal system must be zero for negative time. Consequently, the impulse response of a causal system is a right-sided function. Thus, the ROC of the transfer function $H(s)$ of a causal LTI system must extend to the right of a vertical boundary (as in Figure 6.2a and Figure 6.2d). So, we can also tell whether the system is causal by simple inspection of the transfer function's ROC.

6.5   Inverse Laplace Transform by Inspection

One useful strategy to find inverse Laplace transforms is by inspection. One usually proceeds by rewriting the Laplace transform as a sum of terms that are promptly recognized as being the Laplace transforms of known functions. But notice that we need to know the ROC to find the inverse Laplace transform, since more than one function can have the same expression for the Laplace transform, but with a different ROC. One such example is given in the exercise list. Thus, keep in mind that the inverse Laplace transform is only uniquely specified when both its expression and the ROC are given.

Worked Exercise: Example of Inverse Laplace Transform by Inspection

Find the inverse Laplace transform of:

$$X(s) = \frac{1}{(s-5)(s+3)}, ROC : 5 > \sigma > -3$$

by inspection.

Solution:

The first part of the job is to express $X(s)$ as a sum of terms of the form $a/(s + \beta)$, from which we can readily identify the exponential term contributing to each term of the sum. For example, we have seen in Equation (6.23) that the Laplace transform of $e^{-\alpha t}u(t)$ is $1/(s+\alpha)$, $ROC : \sigma > -\alpha$ is. As another example, in the exercise list you will prove that the Laplace transform of $e^{-\beta t}u(-t)$ is $-1/(s+\beta)$, $ROC : \sigma < -\beta$.

We can express $X(s)$ as a sum of terms of the form $a/(s+\beta)$ using partial fraction decomposition. In our example, the expansion involves two terms:

$$X(s) = \frac{a}{(s-5)} + \frac{b}{(s+3)}$$

To find the coefficients a and b, we note that:

$$(s-5)X(s) = a + \frac{(s-5)b}{(s+3)}$$

which entails that:

$$(s-5)X(s)|_{s=5} = a + \frac{(5-5)b}{(5+3)} = a$$

Therefore:

$$a = (s-5)X(s)|_{s=5} = (s-5)\frac{1}{(s-5)(s+3)}\bigg|_{s=5} = \frac{1}{(s+3)}\bigg|_{s=5} = \frac{1}{8}$$

Likewise:

$$b = (s+3)X(s)|_{s=-3} = (s+3)\frac{1}{(s-5)(s+3)}\bigg|_{s=-3} = \frac{1}{(s-5)}\bigg|_{s=-3} = -\frac{1}{8}$$

Thus:

$$X(s) = \frac{1}{(s-5)(s+3)} = \frac{1}{8}\frac{1}{(s-5)} - \frac{1}{8}\frac{1}{(s+3)}, ROC : 5 > \sigma > -3 \qquad (6.24)$$

To complete the task, we need to consider the ROC. It involves the overlap between two regions, so it is of the form exemplified in Figure 6.2c. As such, one of the functions must be left-sided and the other must be right-sided. To find which one is which, we notice that σ extends to the right-side of -3. Thus, using the notation of Figure 6.2c, we find that $\alpha = 3$. That means the contribution $1/(s+3)$ is a right-sided function. Thus, the inverse Laplace transform of this contribution must be $e^{-3t}u(t)$. Likewise, we notice that the ROC extends to the left-side of 5, so, again using the notation of Figure 6.2c, we find that $\beta = -5$. Thus, the inverse Laplace transform of the contribution $1/(s-5)$ must be the left-sided function $-e^{+5t}u(-t)$ (as you will be asked to calculate in the exercise list, the direct Laplace transform of $e^{+5t}u(-t)$ is $-1/(s-5)$; hence the negative sign in $-e^{+5t}$). Thus, we conclude that the inverse Laplace transform of $X(s)$ is:

$$x(t) = \frac{1}{8}[-e^{+5t}u(-t) - e^{-3t}u(t)] \qquad (6.25)$$

Worked Exercise: Impulse Response of a Harmonic Oscillator

Find the inverse Fourier transform of the frequency response of a harmonic oscillator.

Solution:

The frequency response of the harmonic oscillator is:
Equation (3.69)

$$H(f) = \frac{1}{-m(2\pi f)^2 + i\gamma 2\pi f + k}$$

In the exercise list, you will be asked to prove that, if $x(t) = e^{-\alpha t}\sin(2\pi f_0 t)u(t)$, then its Laplace transform $X(s)$ is:
Equation (6.59)

$$X(s) = \frac{2\pi f_0}{(s + \alpha)^2 + (2\pi f_0)^2}, \ ROC : \sigma > -\alpha$$

Notice that the ROC extends to the right side of $-\alpha$. Thus, if $\alpha > 0$, then the ROC includes the axis $\sigma = 0$. Consequently, if $\alpha > 0$, then the Fourier transform of $e^{-\alpha t}\sin(2\pi f_0 t)u(t)$ exists, and can be obtained by replacing s with $i2\pi f$ in Equation (6.59). Thus, if $\alpha > 0$, then the Fourier transform of $x(t) = e^{-\alpha t}\sin(2\pi f_0 t)u(t)$ is:

$$X(f) = \frac{2\pi f_0}{(i2\pi f + \alpha)^2 + (2\pi f_0)^2} \tag{6.26}$$

Comparing Equation (6.26) with Equation (3.69), we notice that both denominators feature a second-order polynomial (that is, both equations are quadratic in f), whereas both numerators feature zero-order polynomials (that is, a constant). Thus, the inverse Fourier transform of Equation (3.69) must be of the form $e^{-\alpha t}\sin(2\pi f_0 t)u(t)$. We just need to find α and f_0 explicitly in terms of m, γ, and k. To cast Equation (3.69) in a form that can be explicitly compared with Equation (6.26), we notice that:

$$H(f) = \frac{1}{-m(2\pi f)^2 + i\gamma 2\pi f + k} = \frac{1}{m} \frac{1}{(i2\pi f)^2 + i\frac{\gamma}{m}2\pi f + \frac{k}{m}}$$

$$= \frac{1}{m} \frac{1}{\left(i2\pi f + \frac{\gamma}{2m}\right)^2 + \frac{k}{m} - \left(\frac{\gamma}{2m}\right)^2}$$

The last expression has almost in the same form of Equation (6.26): we just need the term in the numerator. Thus, we recast $H(f)$ as:

$$H(f) = \frac{1}{m\sqrt{\frac{k}{m} - \left(\frac{\gamma}{2m}\right)^2}} \frac{\sqrt{\frac{k}{m} - \left(\frac{\gamma}{2m}\right)^2}}{\left(i2\pi f + \frac{\gamma}{2m}\right)^2 + \frac{k}{m} - \left(\frac{\gamma}{2m}\right)^2} \tag{6.27}$$

Comparing Equation (6.27) with Equation (6.26), we find that:

$$\alpha = \frac{\gamma}{2m}$$

And:

$$2\pi f_0 = \sqrt{\frac{k}{m} - \left(\frac{\gamma}{2m}\right)^2} = \sqrt{\frac{k}{m}} \sqrt{1 - \frac{\gamma^2}{4mk}} \tag{6.28}$$

Therefore, with the help of Equation (6.26), we conclude that the inverse Fourier transform of $H(f)$ is:

$$h(t) = \frac{1}{\sqrt{mk\left(1 - \frac{\gamma^2}{4mk}\right)}} e^{-\frac{\gamma}{2m}t} \sin\left(\sqrt{\frac{k}{m}} \sqrt{1 - \frac{\gamma^2}{4mk}}\, t\right) u(t) \tag{6.29}$$

Thus, we have found that the impulse response of a harmonic oscillator consists of a sinusoidal damped oscillation. The frequency of the sinusoidal oscillation is given by Equation (6.28), which reduces to Equation (3.70) in the limit of low damping ($\gamma \ll \sqrt{4mk}$). The coefficient governing the amplitude decay is $\alpha = \gamma/2m$. Finally, notice that our solution assumes real f_0, so it is only valid for $\gamma^2/4mk \leq 1$ (see Equation (6.28) – if this condition is not satisfied, we get a negative number inside the root square, and hence an imaginary f_0).

A plot of Equation (6.29) assuming $m = k = 1$ and $\gamma = 0.01$ is shown in Figure 3.4d.

6.6   Zeros and Poles

In Section 3.2 (Operators Representing Linear and Time Invariant systems), we found that the most general input–output relationship of LTI systems must be of the form $y(t) = \sum_{n=1}^{\infty} a_n \frac{d^n x(t)}{dt^n} + a_0 \cdot x(t)$, but that $y(t)$ may involve derivatives of the physical output. Writing these derivatives explicitly, the most general relationship reads:

$$\sum_{n=1}^{\infty} b_n \frac{d^n y(t)}{dt^n} + b_0 \cdot y(t) = \sum_{n=1}^{\infty} a_n \frac{d^n x(t)}{dt^n} + a_0 \cdot x(t) \tag{6.30}$$

Taking the Laplace transform of both sides of Equation (6.30), with the help of the property of differentiation (Equation (6.20)), we find:

$$\sum_{n=1}^{\infty} b_n s^n Y(s) + b_0 \cdot Y(s) = \sum_{n=1}^{\infty} a_n s^n X(s) + a_0 \cdot X(s)$$

Therefore:

$$\frac{Y(s)}{X(s)} = \frac{\sum\limits_{n=1}^{\infty} a_n s^n + a_0}{\sum\limits_{n=1}^{\infty} b_n s^n + b_0}$$

But $Y(s)/X(s)$ is the transfer function $H(s)$ of the LTI system (see Equation 6.21). Thus:

$$H(s) = \frac{\sum\limits_{n=1}^{\infty} a_n s^n + a_0}{\sum\limits_{n=1}^{\infty} b_n s^n + b_0} \qquad (6.31)$$

According to Equation (6.31), the transfer function of an LTI system is a ratio of polynomials. The values of s for which the polynomial in the numerator is null are called the 'zeros' of the transfer function. Likewise, the values of s for which the polynomial in the denominator is null are called the 'poles' of the transfer function. Zeros and poles play a significant role in control engineering. Notice that the transfer function goes to infinity when it is evaluated in one of its poles. Therefore, the ROC cannot include any pole.

Worked Exercise: Finding the Zeros and Poles

Suppose that the transfer function of an LTI system is given by:

$$H(s) = \frac{(s - 3)}{s^2 - 5s + 4}$$

Find its zeros and poles.

Solution:

There is only one zero, and by straightforward inspection we find it to be $s_0 = 3$. The polynomial in the denominator is of second degree, so there must be two poles. The poles are the solutions to the equation:

$$s^2 - 5s + 4 = 0$$

Solving the equation using the quadratic formula, we find $s_1 = 1$ and $s_2 = 4$, which are thus the poles of $H(s)$.

Worked Exercise: Poles of a Harmonic Oscillator

The relationship between the input $x(t)$ and output $y(t)$ of a harmonic oscillator is (see Equation (3.65)):

$$x(t) = m\frac{d^2y(t)}{dt^2} + \gamma\frac{dy(t)}{dt} + ky(t)$$

where $x(t)$ is the applied force, $y(t)$ is the mass position, m is the mass, γ is the damping coefficient, and k is the spring constant (see Figure 3.3)

Find the transfer function, the zeros and the poles of the harmonic oscillator, assuming $m = k = 1$ and $\gamma = 0.01$.

Solution:

Taking the Laplace transform of both sides of the equation, we find:

$$X(s) = ms^2Y(s) + \gamma sY(s) + kY(s)$$

Therefore, the transfer function is:

$$H(s) = \frac{Y(s)}{X(s)} = \frac{1}{ms^2 + \gamma s + k} \tag{6.32}$$

Since the numerator is a constant, there are no zeros in the transfer function. The two poles are found with the quadratic formula:

$$s = \frac{-\gamma \pm \sqrt{\gamma^2 - 4mk}}{2m} = \frac{-\gamma \pm i\sqrt{4mk - \gamma^2}}{2m}$$

If γ^2 is small compared to $4mk$, then the square root can be approximated to $\sqrt{4mk}$. Thus, for small γ, we get:

$$s \approx \frac{-\gamma \pm i\sqrt{4mk}}{2m} = -\frac{\gamma}{2m} \pm i\sqrt{\frac{k}{m}} \tag{6.33}$$

Plugging in the numbers, we find the two poles:

$$s_1 = \frac{-0.01}{2} + i$$

And:

$$s_1 = \frac{-0.01}{2} - i$$

It is instructive to reflect on how these results relate to the analysis of the frequency response of a harmonic oscillator (see Worked Exercise: Impulse and

Frequency Responses of a Harmonic Oscillator, of Chapter 3, and Worked Exercise: Impulse Response of a Harmonic Oscillator).

Recall that the Fourier transform is a particular case of the Laplace transform, that is, it is the Laplace transform with $\sigma = 0$. Consequently, $H(s)$ in Equation (6.32) reduces to $H(f)$ in Equation (3.69) by imposing $s = i2\pi f$ (check that this is indeed the case). According to our result, is it possible that $H(f)$ goes to infinity? Since $H(f)$ coincides with $H(s)$ when s is a purely imaginary number (in other words, when $\sigma = 0$), then $H(f)$ can only go to infinity if at least one of the poles is a purely imaginary number (see Equation (6.12)). But, according to Equation (6.33), the requirement for a purely imaginary pole is $\gamma = 0$. Thus, the frequency response only goes to infinity if there is no damping in the harmonic oscillator. If there is no damping, then the poles are:

$$s = \pm i\sqrt{\frac{k}{m}}$$

Since $s = i2\pi f$, these poles correspond to the frequencies: $2\pi f = \pm\sqrt{k/m}$, which are the positive and negative resonance frequencies of a harmonic oscillator (see Equation (6.28)). In other words, the frequency response of a harmonic oscillator without damping is infinite at the resonance frequency.

Thus, we have concluded that, if a force oscillating at the resonance frequency is applied to a harmonic oscillator without damping, then the amplitude of the mass oscillation (the output) goes to infinity, since the frequency response blows up at the resonance frequency. With damping, however, the amplitude does not go to infinity, but it may reach very high values, as the frequency is 'near' a pole.

6.7 The Unilateral Laplace Transform

In the analysis of signals and systems, often we do not need to know, or cannot know, the behaviour of the signals in all times. Often, we are happy with knowledge of only a part of the time window. For example, suppose that the system is a harmonic oscillator (see Figure 3.3) and we want to know how the mass moves when a certain force is applied. In this example, we do not care about how the mass was moving before the force was applied. It may be that we cannot even know how the mass was moving before the force was applied. So, we want to find a way of describing how the mass moves when the force is applied, without having to know how it moved before the force was applied. More precisely, we want to be able to find the mass movement only with knowledge of its initial conditions, that is, the position and velocity of the mass at the instant we begun applying the force. This is the motivation for the unilateral Laplace transform.

For convenience, we define the initial time of interest as $t = 0$. Thus, we assume we want to obtain the output of a system for positive time, without having to know how it behaved for negative time. More precisely, we want to describe its behaviour for positive time using only the initial conditions at time $t = 0$. In our example of the harmonic oscillator, we would set the origin of the time coordinates to coincide with the instant the force of interest is applied.

Mathematically, we can 'delete' the information for negative time by means of the step function $u(t)$. For example, suppose that $y(t)$ is the mass position of the harmonic oscillator. If we want to know how it behaves for positive time, and we do not know or do not care about how it behaved for negative time, then we can create a new function $y(t)u(t)$, which represents the mass position with the negative time erased out by the function $u(t)$. So, we will be studying functions of the type 'function times step function'.

These types of functions, with the negative time erased out, motivate the definition of the unilateral Laplace transform. Mathematically, the unilateral Laplace transform is just the Laplace transform of a function multiplied by the step function. For example, suppose that we have a function $g_2(t) = g(t)u(t)$. The BILATERAL Laplace transform (the bilateral Laplace transform is the normal one we have been studying) is:

$$G_2(s) = \int_{-\infty}^{+\infty} g_2(t)e^{-st}dt \tag{6.34}$$

This is just the good old direct Laplace transform, as in Equation (6.18). Now we use the fact that the step function erased the negative time:

$$G_2(s) = \int_{-\infty}^{+\infty} g_2(t)e^{-st}dt = \int_{-\infty}^{+\infty} g(t)u(t)e^{-st}dt = \int_{0}^{+\infty} g(t)e^{-st}dt$$

The last equality is the UNILATERAL Laplace transform of $g(t)$. Thus, by definition, the unilateral Laplace transform of $g(t)$ is:

$$G(s) = \int_{0}^{+\infty} g(t)e^{-st}dt \tag{6.35}$$

Of course, $G_2(s) = G(s)$. But, please, pay attention to the terminology, because it may be slippery. We say that $G(s)$ in Equation (6.35) is the unilateral Laplace transform of $g(t)$, while $G_2(s)$ in Equation (6.34) is the bilateral Laplace transform (or just Laplace transform) of $g_2(t) = g(t)u(t)$. Notice that, although $G_2(s) = G(s)$, $g_2(t) \neq g(t)$.

This is the key bit that often leads to confusion. The function $g(t)$ and the function $g_2(t) = g(t)u(t)$ are not the same function. But the bilateral Laplace transform of $g_2(t)$ is identical to the unilateral Laplace transform of $g(t)$. We need to keep this in mind, because it leads to a subtlety in the property of differentiation of the unilateral Laplace transform. I will first show you this property for the unilateral Fourier transform, and then we extend it to the unilateral Laplace transform.

6.7.1 The Differentiation Property of the Unilateral Fourier Transform

In Section 6.7, we defined the unilateral Laplace transform (Equation (6.35)). This definition entails that the unilateral Fourier transform is:

$$G(f) = \int_0^{+\infty} g(t)e^{-i2\pi ft}dt \tag{6.36}$$

And we have also seen that the unilateral Fourier transform of $g(t)$ coincides with the Fourier transform of $g_2(t) = g(t)u(t)$. Indeed:

$$G_2(f) = \int_{-\infty}^{+\infty} g_2(t)e^{-i2\pi ft}dt = \int_{-\infty}^{+\infty} g(t)u(t)e^{-i2\pi ft}dt = \int_0^{+\infty} g(t)e^{-i2\pi ft}dt \tag{6.37}$$

Thus, $G_2(f) = G(f)$.

Now, define $g_d(t)$ as the time derivative of $g(t)$:

$$g_d(t) = \frac{dg(t)}{dt} \tag{6.38}$$

In this section, we want to find the relationship between the unilateral Fourier transform of $g(t)$ and the unilateral Fourier transform of $g_d(t)$.

You may be tempted to think that the unilateral Fourier transform of $g_d(t)$, call it $G_d(f)$, must be $G_d(f) = i2\pi fG(f)$, but that is only part of the story. There is an important difference: on the one hand, since $G(f)$ coincides with the Fourier transform of $g(t)u(t)$, then it follows that $i2\pi fG(f)$ is the Fourier transform of the derivative of $g(t)u(t)$; on the other hand, $G_d(f)$ is the unilateral Fourier transform of $g_d(t)$, and as such it coincides with the Fourier transform of $g_d(t)u(t)$. But the derivative of $g(t)u(t)$ does not coincide with $g_d(t)u(t)$: there is an extra term. Therefore, since $i2\pi fG(f)$ is the Fourier transform of the derivative of $g(t)u(t)$, and $G_d(f)$ is the Fourier transform of $g_d(t)u(t)$, and these are not the same function, then it follows that $i2\pi fG(f) \neq G_d(f)$. To find the correct relation, we need to take that extra term into account.

To find the correct relation, let us go back to charted waters, and use the differentiation property of the Fourier transform (the good old one, not the unilateral one). So, if $g_2(t) = g(t)u(t)$, then the differentiation property of Fourier transforms entails that the Fourier transform of $g_{2d}(t)$, defined as:

$$g_{2d}(t) = \frac{dg_2(t)}{dt} \tag{6.39}$$

is:

$$G_{2d}(f) = i2\pi fG_2(f) \tag{6.40}$$

We can use Equation (6.40) to find the differentiation property of the unilateral Fourier transform. To see how they are connected, let us work out the derivative of Equation (6.39) explicitly:

$$g_{2d}(t) = \frac{dg_2(t)}{dt} = \frac{d[g(t)u(t)]}{dt} = \frac{dg(t)}{dt}u(t) + g(t)\frac{du(t)}{dt} \tag{6.41}$$

So, the derivative of $g_{2d}(t)$ involves two terms. The first term, involving the derivative of $g(t)$, is the term whose Fourier transform we want to find (recall that we want the unilateral Fourier transform of dg/dt, which coincides with the Fourier transform of $u(t)dg/dt$), and the second term, involving the derivative of the step function, is the extra term.

In the exercise list, you will prove that the derivative of the step function is the impulse function:

$$\frac{du(t)}{dt} = \delta(t) \tag{6.42}$$

Therefore:

$$g_{2d}(t) = \frac{d[g(t)u(t)]}{dt} = \frac{dg(t)}{dt}u(t) + g(t)\delta(t) \tag{6.43}$$

Thus, the Fourier transform of $g_{2d}(t)$ involves two terms:

$$G_{2d}(f) = \int_{-\infty}^{+\infty} g_{2d}(t)e^{-i2\pi ft}dt = \int_{-\infty}^{+\infty} \frac{dg(t)}{dt}u(t)e^{-i2\pi ft}dt + \int_{-\infty}^{+\infty} g(t)\delta(t)e^{-i2\pi ft}dt$$

The last integral can be evaluated using the sifting property of the delta function:

$$\int_{-\infty}^{+\infty} g(t)e^{-i2\pi ft}\delta(t)dt = g(0)e^{-i2\pi f0} = g(0) \tag{6.44}$$

Therefore:

$$G_{2d}(f) = \int_{-\infty}^{+\infty} g_{2d}(t)e^{-i2\pi ft}dt = \int_{-\infty}^{+\infty} \frac{dg(t)}{dt}u(t)e^{-i2\pi ft}dt + g(0)$$

Using Equation (6.40), we conclude that:

$$\int_{-\infty}^{+\infty} \frac{dg(t)}{dt}u(t)e^{-i2\pi ft}dt = i2\pi fG_2(f) - g(0) \tag{6.45}$$

We have found Equation (6.45) using the differentiation property of the Fourier transform. Now let us see how it can be used to establish the differentiation property of the unilateral Fourier transform.

Just to recap, our problem is to find the relationship between the unilateral Fourier transform of $g(t)$ and the unilateral Fourier transform of its time derivative $g_d(t)$. Let us write the latter unilateral Fourier transform explicitly:

$$G_d(f) = \int_0^{+\infty} g_d(t)e^{-i2\pi ft}dt = \int_0^{+\infty} \frac{dg(t)}{dt}e^{-i2\pi ft}dt \tag{6.46}$$

Notice that the last integral in Equation (6.46) coincides with the integral on the left-hand side of Equation (6.45). Indeed:

$$\int_{-\infty}^{+\infty} \frac{dg(t)}{dt}u(t)e^{-i2\pi ft}dt = \int_0^{+\infty} \frac{dg(t)}{dt}e^{-i2\pi ft}dt$$

Thus, using Equation (6.45) in Equation (6.46), we find:

$$G_d(f) = \int_0^{+\infty} \frac{dg(t)}{dt} e^{-i2\pi ft} dt = i2\pi f G_2(f) - g(0)$$

But $G_2(f)$ is the Fourier transform of $g_2(t) = g(t)u(t)$. Therefore, $G_2(f) = G(f)$, where $G(f)$ is the UNILATERAL Fourier transform of $g(t)$. Thus:

$$G_d(f) = \int_0^{+\infty} \frac{dg(t)}{dt} e^{-i2\pi ft} dt = i2\pi f G(f) - g(0) \tag{6.47}$$

We have concluded that, if $G(f)$ is the UNILATERAL Fourier transform of $g(t)$, then the UNILATERAL Fourier transform of $g_d(t) = dg/dt$ is $G_d(f) = i2\pi f G(f) - g(0)$.

Thus, the differentiation property of the unilateral Fourier transform involves a new term: $g(0)$. This term is the initial condition.

To sum up, recall that the term $g(0)$ appears because $i2\pi f G(f)$ is the Fourier transform of the derivative of $g(t)u(t)$. But we do not want the Fourier transform of the derivative of $g(t)u(t)$. Instead, we want the Fourier transform of $g_d(t)u(t)$ – that is, the unilateral Fourier transforms of $g_d(t)$. However, the derivative of $g(t)u(t)$ involves two terms. One term is $g_d(t)u(t)$, which is what we want, and the other term is $g(t)\delta(t)$. Since we want the Fourier transform of $g_d(t)u(t)$, then we need to remove the Fourier transform of the term $g(t)\delta(t)$ from $i2\pi f G(f)$. That is why we need to subtract $g(0)$ from $i2\pi f G(f)$.

Worked Exercise: Differentiation Property of the Unilateral Fourier Transform Involving Higher Order Derivatives

If $G(f)$ is the unilateral Fourier transform of $g(t)$, find the unilateral Fourier transform of:

$$g_{dd}(t) = \frac{d^2 g(t)}{dt^2}$$

in terms of $G(f)$ and the initial conditions.

Solution:

Define:

$$g_d(t) = \frac{dg(t)}{dt}$$

Then:

$$g_{dd}(t) = \frac{dg_d(t)}{dt}$$

According to the property of differentiation of the unilateral Fourier transform (Equation (6.47)):

$$G_{dd}(f) = i2\pi f G_d(f) - g_d(0)$$

where $G_{dd}(f)$ is the unilateral Fourier transform of $g_{dd}(t)$, $G_d(f)$ is the unilateral Fourier transform of $g_d(t)$, and:

$$g_d(0) = \frac{dg(t)}{dt}\bigg|_{t=0}$$

But, still according to Equation (6.47):

$$G_d(f) = i2\pi f G(f) - g(0)$$

Therefore:

$$G_{dd}(f) = i2\pi f G_d(f) - g_d(0) = i2\pi f[i2\pi f G(f) - g(0)] - g_d(0)$$

We have thus concluded that:

$$G_{dd}(f) = \int_0^{+\infty} \frac{d^2 g(t)}{dt^2} e^{-i2\pi f t} dt = i2\pi f[i2\pi f G(f) - g(0)] - \frac{dg(t)}{dt}\bigg|_{t=0} \qquad (6.48)$$

The property for higher order derivatives can be found in a similar way. For example, the unilateral Fourier transform of $d^3 g/dt^3$ can be found by applying the differentiation property on the function $d^2 g/dt^2$ and using Equation (6.48). Thus:

$$G_{ddd}(f) = \int_0^{+\infty} \frac{d^3 g(t)}{dt^3} e^{-i2\pi f t} dt$$

$$= i2\pi f\left[i2\pi f[i2\pi f G(f) - g(0)] - \frac{dg(t)}{dt}\bigg|_{t=0}\right] - \frac{d^2 g(t)}{dt^2}\bigg|_{t=0} \qquad (6.49)$$

Worked Exercise: Example of Differentiation Using the Unilateral Fourier Transform

Assuming α is a positive real number, find the unilateral Fourier transforms of $g(t) = e^{-\alpha t}$ and $g_d(t) = dg/dt$ by direct integration. Then find the unilateral Fourier transform of $g_d(t)$ using the property of differentiation and compare with the result obtained by direct integration.

Solution:

We begin by evaluating the unilateral Fourier transform of $g(t) = e^{-\alpha t}$:

$$G(f) = \int_0^{+\infty} g(t) e^{-i2\pi f t} dt = \int_0^{+\infty} e^{-\alpha t} e^{-i2\pi f t} dt = \int_0^{+\infty} e^{-(\alpha + i2\pi f)t} dt$$

Thus:

$$G(f) = \int_0^{+\infty} e^{-(\alpha + i2\pi f)t} dt = - \left. \frac{e^{-(\alpha + i2\pi f)t}}{\alpha + i2\pi f} \right|_0^{+\infty}$$

$$= -\frac{e^{-\alpha\infty} e^{-i2\pi f\infty}}{\alpha + i2\pi f} + \frac{e^{-\alpha 0} e^{-i2\pi f 0}}{\alpha + i2\pi f}$$

$$= -\frac{0 \cdot e^{-i2\pi f\infty}}{\alpha + i2\pi f} + \frac{1}{\alpha + i2\pi f} = \frac{1}{\alpha + i2\pi f} \tag{6.50}$$

Now, notice that:

$$g_d(t) = \frac{dg(t)}{dt} = -\alpha e^{-\alpha t} = -\alpha g(t)$$

Thus, from Equation (6.50), it follows immediately that:

$$G_d(f) = -\frac{\alpha}{\alpha + i2\pi f} \tag{6.51}$$

where $G_d(f)$ is the unilateral Fourier transform of $g_d(t)$.

Now we calculate $G_d(f)$ using the property of differentiation:

$$G_d(f) = i2\pi f G(f) - g(0) = i2\pi f \frac{1}{\alpha + i2\pi f} - e^{-\alpha 0}$$

$$= \frac{i2\pi f}{\alpha + i2\pi f} - 1 = \frac{i2\pi f - \alpha - i2\pi f}{\alpha + i2\pi f} = -\frac{\alpha}{\alpha + i2\pi f} \tag{6.52}$$

Because math is math, Equation (6.52) agrees with Equation (6.51).

Worked Exercise: Discharge of an RC Circuit

As an example of application of unilateral Fourier transforms, consider the RC circuit of Figure 3.6a. Suppose that the input voltage is short-circuited at $t = 0$. Find the output after the input is short-circuited.

Solution:

The time domain relationship between input and output is given by: Equation (4.4)

$$v_{in}(t) = RC\frac{dv_{out}(t)}{dt} + v_{out}(t)$$

We do not have complete information about the input $v_{in}(t)$ itself. All we know about the input is that the circuit is short-circuited at $t = 0$, which means that

$v_{in}(t)u(t) = 0$. So, since we do not know anything about $v_{in}(t)$ at negative times, we had better use the unilateral Fourier transform. Applying it on both sides of Equation (4.4), and using the property of differentiation (Equation (6.47)), we find:

$$V_{in}(f) = RC[i2\pi f V_{out}(f) - v_{out}(0)] + V_{out}(f)$$

where $v_{out}(0)$ is the voltage across the capacitor at time $t = 0$.
  Since $v_{in}(t)u(t) = 0$, it follows that $V_{in}(f) = 0$. Thus:

$$V_{out}(f) = \frac{RCv_{out}(0)}{1 + RCi2\pi f} = v_{out}(0)\frac{1}{\frac{1}{RC} + i2\pi f} \tag{6.53}$$

Comparing Equation (6.53) with Equation (6.50), we identify that $V_{out}(f)$ is the unilateral Fourier transform of $v_{out}(0)e^{-\frac{1}{RC}t}$. Therefore:

$$v_{out}(t) = v_{out}(0)e^{-\frac{t}{RC}} \tag{6.54}$$

This is a well-known result in circuit theory: the voltage in a capacitor drops exponentially as it discharges through a resistor, and the time constant of the discharge is RC. We have also discussed capacitor discharge in Worked Exercise: Charge and Discharge in an RC Circuit, of Chapter 3. Finally, notice that Equation (6.54) only applies to positive time, since we obtained it by means of the unilateral Fourier transform.

6.7.2   Generalization to the Unilateral Laplace Transform

The proof of the differentiation property of the unilateral Laplace transform follows the same logic of the proof of the differentiation property of the unilateral Fourier transform. As such, it can be found by replacing $i2\pi f$ with s in Equation (6.47). Thus, if $G_d(s)$ is the unilateral Laplace transform of dg/dt, and $G(s)$ is the unilateral Laplace transform of $g(t)$, then:

$$G_d(s) = \int_0^{+\infty} \frac{dg(t)}{dt}e^{-st}dt = sG(s) - g(0) \tag{6.55}$$

The higher order derivatives can also be found in a similar manner. Thus, Equation (6.48) becomes:

$$G_{dd}(s) = \int_0^{+\infty} \frac{d^2g(t)}{dt^2}e^{-st}dt = s[sG(s) - g(0)] - \frac{dg(t)}{dt}\Big|_{t=0} \tag{6.56}$$

And Equation (6.49) becomes:

$$G_{ddd}(s) = \int_0^{+\infty} \frac{d^3g(t)}{dt^3}e^{-st}dt = s\left[s[sG(s) - g(0)] - \frac{dg(t)}{dt}\Big|_{t=0}\right] - \frac{d^2g(t)}{dt^2}\Big|_{t=0} \tag{6.57}$$

6.8 Exercises

Exercise 1

We saw in Chapter 4 (Electric Circuits as LTI Systems) that the impedance of capacitors and inductors is the ratio between the amplitudes of their voltage and current when their time dependence is of the form $e^{i2\pi ft}$. Generalize the concept of impedance to currents and voltages of the form e^{st}, and find the impedance of capacitor and inductors in terms of s.

Exercise 2

Prove the time-shifting property of the Laplace transform (Equation (6.19)).

Exercise 3

Prove the property of differentiation (Equation (6.20)).

Exercise 4

In the derivation leading to Equation (6.23), identify where the condition $a + \sigma > 0$ was used, and explain what would have happened if $a + \sigma \le 0$.

Exercise 5

Find the Laplace transform and ROC of the left-sided function $x_1(t) = e^{-\beta t}u(-t)$. Compare your result with the Laplace transform of the right-sided function $x_2(t) = -e^{-\beta t}u(t)$. Notice that their Laplace transform expressions are identical: the only difference is the ROC.

Exercise 6

Prove that the Laplace transform of the signal $x(t) = e^{-\alpha t}\cos(2\pi f_0 t)u(t)$ is:

$$x(t) = e^{-\alpha t}\cos(2\pi f_0 t)u(t) \Longleftrightarrow X(s) = \frac{s + \alpha}{(s + \alpha)^2 + (2\pi f_0)^2}, ROC : \sigma > -\alpha \quad (6.58)$$

What values of s maximize $X(s)$? Interpret you result in light of the decomposition of $x(t)$ as a linear combination of eigenvectors.

Exercise 7

Prove that the Laplace transform of the signal $x(t) = e^{-\alpha t}\sin(2\pi f_0 t)u(t)$ is:

$$x(t) = e^{-\alpha t}\sin(2\pi f_0 t)u(t) \Longleftrightarrow X(s) = \frac{2\pi f_0}{(s + \alpha)^2 + (2\pi f_0)^2}, ROC : \sigma > -\alpha \quad (6.59)$$

Exercise 8

Obtain the impulse response of the RC low pass filter through inverse Fourier transformation of the frequency response (Equation (4.7)) by inspection.
    Hint: use Equation (6.23).

Exercise 9
Find the inverse Laplace transform of:

$$X(s) = \frac{1}{(s-5)(s+3)}, \quad ROC : \sigma > 5$$

by inspection.

Exercise 10
Find the inverse Laplace transform of:

$$X(s) = \frac{1}{(s-5)(s+3)}, \quad ROC : \sigma < -3$$

by inspection.

Exercise 11
The transfer function of an LTI system is given by:

$$H(s) = \frac{1}{(s - i80\pi)(s + i80\pi)}$$

If you apply a sinusoidal frequency in the input of this LTI system, is there any frequency that would make the output go to infinity?

Exercise 12
Is it possible for:

$$H(s) = \frac{1}{(s+i5)}, \quad ROC\ includes\ \sigma = 0$$

to be the transfer function of a physical LTI system?

Exercise 13
The ROC of the transfer function of system A is $\sigma < -3$ and of system B is $\sigma > 3$. Only one of them is a physical system. Which one? And is it a stable or unstable system?

Exercise 14
Find the transfer function of the band-pass filter of Figure 5.5, its zeros and poles.

Exercise 15
Prove that $du/dt = \delta(t)$, where $u(t)$ is the Heaviside step function (Equation (6.22)).
    Hint: first prove that $du/dt|_{t\neq0} = 0$, then prove $du/dt|_{t=0} = \infty$ and finally prove that $\int_{-\infty}^{+\infty}(du/dt)dt = 1$ (see Equation (2.18)).

Exercise 16

Using Equation (6.23) and the property of differentiation of the BILATERAL Laplace transform, find the inverse Laplace transform of:

$$X(s) = \frac{s}{s + \alpha}, \quad ROC : \sigma > -\alpha$$

And confirm your result by calculating the direct Laplace transform of the function you found by direct integration.

Exercise 17

Obtain the impulse response of the RL high-pass filter through inverse Fourier transformation of its frequency response (Equation (5.5)).

Exercise 18

Using the property of differentiation of the UNILATERAL Laplace transform, calculate the UNILATERAL Laplace transform of:

$$x(t) = \frac{de^{-\alpha t}}{dt}$$

Compare your result with the result of Exercise 16 of Chapter 6 and explain the difference.

Exercise 19

Find the UNILATERAL Laplace transform of the mass position $y(t)$ of a harmonic oscillator in terms of the unilateral Laplace transform of the applied force $x(t)$, and the initial position and velocity of the mass.

Exercise 20

Consider the harmonic oscillator of Figure 3.3. Suppose that you pick the mass with your fingers, pull it to the position y_0, hold it, and then release it at time $t = 0$. Find the time dependence of the mass oscillation. Compare your result with the impulse response of the harmonic oscillator (Equation (6.29)).

   Hint: use Equation ((6.58)) and Equation ((6.59)) to obtain the inverse Laplace transform by inspection.

Exercise 21

Find the general solution (that is, a solution for general initial conditions) for the positive time dependence of the mass position in a harmonic oscillator assuming that no force is applied at positive time.

Interlude: Discrete Signals and Systems: Why do we Need Them?

Hitherto, we have been considering the theory of representation of 'continuous time' signals and systems. That is just a weird way of saying that we have been treating time for what it is: a continuous variable. But now we need to consider the theory of representation of 'discrete time' signals and systems. A discrete time signal is just a discrete signal, that is, a discrete function. But, in the context of signals and systems, we use the term 'discrete time' because almost always we are interested in discrete functions that were obtained by sampling a continuous time signal.

A discrete signal is just a sequence of numbers. The sequence $5.76, 10, 9i, 10 + 8.9874i$ is an example of a (boring) discrete signal. A discrete system is just a mathematical operation that turns a sequence of numbers (a discrete signal) into another sequence of numbers (another discrete signal).

Maybe you are under the impression that studying properties of sequences of numbers makes as much sense as ordering a pineapple pizza. After all, why would we be interested in sequences of numbers, or in lumping dinner and dessert together? The latter question is unfathomable, but the answer to the former is quite simple: even though we are ultimately interested in analyzing physical (and, as such, continuous time) signals and systems, in practice we do that using computers. For example, if we want to know the Fourier transform of a song, how on earth could we evaluate its Fourier transform integral by hand? No way Jose, we need a computer. As another example, engineers and physicists often need to analyze systems – be it a new device or a law of physics – and they often do that through a mathematical representation of the system in a computer. But, of course, computers can only handle discrete functions. Thus, to take advantage of computers, we need to extend the theory we have developed so far to include discrete signals and systems. Fortunately, the extension of the theory to discrete functions is not too taxing: often it only involves replacing an integral with a sum. But there are some peculiarities and surprises that deserve spelling out.

Since, in practice, we are interested in physical (continuous time) signals and systems, but we want to study them using computers, a major skill of engineers and physicists is to be able to relate the discrete functions stored in their computers to their continuous time counterparts. Thus, a major goal of the following two chapters is to show you in detail how discrete signals and their Fourier transforms are related to their continuous time versions.

7

The Sampling Theorem and the Discrete Time Fourier Transform (DTFT)

Learning Objectives

The Discrete Time Fourier Transform (DTFT) is the Fourier transform of a discrete function. We need a Fourier representation of discrete functions for the same reason we need it for continuous functions: it is a representation in a basis of eigenvectors, which greatly simplifies the analysis of systems because it diagonalizes their operators. The main objective of this chapter is to learn how the DTFT of a discrete function relates to the Fourier transform of its corresponding continuous time function. Along the way, we will stumble on an astonishing, not to say supernatural, theorem: the sampling theorem.

7.1 Discrete Signals

In Chapter 3, we learned that the representation of signals and systems in the frequency domain greatly facilitates their analysis because the operators become diagonal. Thus, the main goal of this chapter is to find a Fourier transform suitable for discrete signals, and to investigate how it relates to the Fourier transform of their continuous time counterparts. Before we do so, however, we need to say a few words about discrete signals.

A discrete signal, or discrete function, is a sequence of numbers. We represent these sequences of numbers by a general discrete function $x[n]$, where n is an integer number (we can say that n is the discrete time variable). We use [ ] instead of ( ) to differentiate between discrete and continuous functions. And we also allow n to range from $-\infty$ to $+\infty$.

Let us see an example. Say that the sequence of numbers is $8, \pi, \sqrt{2}, i3$. We can represent this sequence with a function $x[n]$, by indexing each value of the

Essentials of Signals and Systems, First Edition. Emiliano R. Martins.
© 2023 John Wiley & Sons Ltd. Published 2023 by John Wiley & Sons Ltd.
Companion website: www.wiley.com/go/martins/essentialsofsignalsandsystems

sequence with the variable n. Thus, say we choose to index the first number of the sequence with $n = -1$. Then, we have: $x[-1] = 8$, $x[0] = \pi$, $x[1] = \sqrt{2}$, $x[2] = i3$. Furthermore, since we allow n to range from $-\infty$ to $+\infty$, then it is implied that $x[n] = 0$ for all other values of n.

We can also represent discrete functions with explicit mathematical expressions, for example, $x[n] = e^{-n^2}$. Importantly, we are often interested in discrete functions that were obtained by sampling continuous functions at intervals of time T_S. This interval of time T_S is called the 'sampling period', and it is a central parameter in our study. If $x[n]$ is obtained by sampling $x(t)$ at intervals of time T_S, then it follows that:

$$x[n] = x(nT_S) \tag{7.1}$$

The equality above should be interpreted in the sense that, for all n, the number $x[n]$ is equal to the number $x(nT_S)$.

For example, say that we have the continuous function:

$$x(t) = e^{-\left(\frac{t}{\sigma}\right)^2} \tag{7.2}$$

And we want to create a discrete function by sampling the continuous time function at intervals T_S. Then, according to Equation (7.1):

$$x[n] = e^{-\left(\frac{nT_S}{\sigma}\right)^2} \tag{7.3}$$

Unless stated otherwise, from now on we will assume that our discrete signal was obtained from a continuous time signal, and that our interest is to study the continuous time signal through the discrete time signal.

One interesting feature of discrete time functions is that it is possible to sample a periodic continuous time function and obtain a nonperiodic discrete time function. Thus, suppose that we have a periodic function $x(t)$, specified as:

$$x(t) = \cos(2\pi f t) \tag{7.4}$$

And now we sample it following the recipe of Equation (7.1):

$$x[n] = x(nT_S) = \cos(2\pi f n T_S) \tag{7.5}$$

Question: how can we tell whether $x[n]$ is periodic or not? For $x[n]$ to be periodic, there must exist an integer number N for which $x[n + N] = x[n]$ for all n. Let us check if that is true by evaluating $x[n + N]$:

$$x[n + N] = \cos(2\pi f T_S(n + N))$$
$$= \cos(2\pi f T_S n)\cos(2\pi f T_S N) - \sin(2\pi f T_S n)\sin(2\pi f T_S N) \tag{7.6}$$

According to Equation (7.6), the condition $x[n + N] = x[n]$ requires that $\cos(2\pi f T_S N) = 1$ and $\sin(2\pi f T_S N) = 0$. These two conditions can be simultaneously satisfied only if $2\pi f T_S N$ is a multiple of 2π. In other words, these two conditions require that $fT_S N$ be an integer number. Thus, we have concluded that $x[n]$ is periodic only if:

$$fT_S N = m \tag{7.7}$$

where m is an integer number. In other words, for $x[n]$ to be periodic, it is necessary that the product fT_S reduces to a ratio of two integer numbers:

$$fT_S = \frac{m}{N} \tag{7.8}$$

Recall that the ratio of two integers defines a 'rational number'. So, for $x[n]$ to be periodic, the sampling period T_S must be chosen so that its product with f is a rational number. If it is not, then $x[n]$ is nonperiodic, though it has been sampled from a periodic function.

7.2 Fourier Transforms of Discrete Signals and the Sampling Theorem

Quite often, engineers and physicists need to calculate Fourier transforms of continuous time signals using only samples of the signal (usually stored digitally). We denote the continuous time signal by $g(t)$, whereas its samples constitute the discrete signal $g[n]$. In this section, we will learn how to obtain the Fourier transform of $g(t)$ through the samples $g[n]$, and in what conditions it is possible to recover $g(t)$ from $g[n]$.

Our job is essentially to find a Fourier representation of $g[n]$ that gives information about the Fourier transform of $g(t)$. The connection between the two representations can be conveniently established by means of an intermediary continuous time function, which we call $g_d(t)$.

The intermediary function $g_d(t)$ is a continuous time function that is as close as possible to the discrete function $g[n]$. To elucidate the meaning of $g_d(t)$, consider as an example a signal $g(t)$, as sketched on the left-hand side of Figure 7.1. Suppose this signal passes through an ideal analog-to-digital (A/D) converter. An ideal A/D converter is essentially an electronic switch that captures the amplitude of the input signal at intervals of time T_S. We can conceptualize an ideal A/D converter as transforming the signal $g(t)$ into another signal $g_d(t)$, which is always zero, except at instants multiples of the sampling period T_S (that is, except at $t = nT_S$). So, $g_d(t)$ picks up the value of $g(t)$ at $t = nT_S$, but it loses the information at $t \neq nT_S$. Thus, $g_d(t)$ looks like a discrete function in the sense that, whereas a discrete function is only specified for integers values of n, the intermediary function is only nonzero at instants $t = nT_S$. Once the values of $g(nT_s)$ are stored in a computer, then we have a sequence of numbers, that is, then we have $g[n]$. Thus, whereas $g(t)$ and $g_d(t)$ are representations of continuous time signals, $g[n]$ is the representation of a sequence of numbers.

The intermediary function $g_d(t)$ carries almost the same information of the discrete function $g[n]$ (hence the subindex d in $g_d(t)$). I said 'almost' because there is

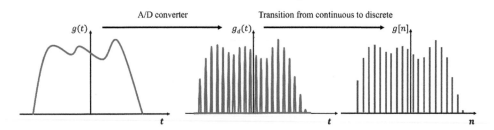

Figure 7.1 Illustration of the relationship between a continuous time signal $g(t)$, the intermediary signal $g_d(t)$, and the discrete signal $g[n]$. A sketch of an example signal $g(t)$ is shown on the left-hand side of the figure. As shown in the middle figure, the information at $t \neq nT_s$ is lost when the signal passes through an A/D converter. Thus, the intermediary signal $g_d(t)$ is zero at $t \neq nT_s$. The discrete signal $g[n]$ – illustrated on the right-hand side of the figure – is obtained when the data carried by $g_d(t)$ is turned into a sequence of numbers, for example by storing them in a computer.

a significant difference between the information carried by $g_d(t)$ and by $g[n]$: since $g_d(t)$ is still a continuous time function, it still has information about time. Thus, if you look at a certain 'spike' in the sketch of $g_d(t)$ in the middle of Figure 7.1, you will be able to 'locate' the spike in time: you know when it happened. However, if you look at the corresponding 'spike' in the sketch of $g[n]$ on the right-hand side of Figure 7.1, you will not be able to know when the spike occurred, because $g[n]$ does not carry information about time: after all, it is just a sequence of numbers (it is a function of n, not of t). But we can recover information about time if we know the sampling period T_S. I emphasize, however, that T_S is an 'external' datum: it is not in $g[n]$ (we cannot infer T_S from $g[n]$).

We need to define a bit of notation. Later, we will use the letter G to denote the DTFT of $g[n]$. So, to avoid confusion, from now on we use $G_c(f)$ to denote the Fourier transform of $g(t)$. The subscript c is a reminder that $G_c(f)$ is the Fourier transform of a continuous time signal $g(t)$.

Now let us consider our main problem, which is to obtain the Fourier transform of $g(t)$ through its samples $g[n]$. Recall that the Fourier transform of $g(t)$ is:
Equation (2.81)

$$G_c(f) = \int_{-\infty}^{+\infty} g(t)e^{-i2\pi f t}dt$$

Since we are assuming that $g[n]$ was obtained by sampling $g(t)$ at intervals $t = nT_S$, then it makes sense to convert the integral in Equation (2.81) into a sum involving the samples taken at $t = nT_S$. Such a conversion reads:

$$\int_{-\infty}^{+\infty} g(t)e^{-i2\pi f t}dt \Rightarrow \sum_{n=-\infty}^{+\infty} g(nT_S)e^{-i2\pi f n T_S} = \sum_{n=-\infty}^{+\infty} g[n]e^{-i2\pi f n T_S} \quad (7.9)$$

where we used $g[n] = g(nT_S)$. The sum depends on the frequency f, so we can associate it with a function of f, which we denote by $G_d(f)$. Thus:

$$G_d(f) = \sum_{n=-\infty}^{+\infty} g[n]e^{-i2\pi fnT_s} \qquad (7.10)$$

Recall that our goal is to obtain $G_c(f)$ from $g[n]$. The sum in Equation (7.10) is connected to $G_c(f)$ in the sense of Equation (7.9): it is like the Fourier transform of $g(t)$, but with the integral replaced with a sum. Thus, Equation (7.10) is kind of the best we can do to compute $G_c(f)$ from $g[n]$. But is it good enough? In other words, can we infer $G_c(f)$ from $G_d(f)$? And, if we can, how are they related?

To answer these questions, let us invoke the intermediary function $g_d(t)$. Recall that it is always zero, except at instants $t = nT_S$. How can we represent such a function mathematically? Well, we do know a function that is always zero, except at an instant of time: the impulse function. The problem is that the impulse function is undefined at the instant whereat it is different from zero (it goes to infinity). But we can circumvent this problem if we use its 'area', because the 'area' is well defined. So, we can represent the values of $g_d(t)$ that are different from zero by modulating the area of the impulse function. For example, suppose that we want a function that is always zero, except at $t = 0$, and that somehow stores the value of $g(t)$ at $t = 0$. The function $g(t)\delta(t)$ does this job: it is always zero, except at $t = 0$; and its 'area' (or, more precisely, its integral around the origin) is equal to $g(0)$ – if you forgot why, it is due to the sifting property of $\delta(t)$: $\int_{-\infty}^{+\infty} \delta(t)g(t)dt = g(0)$. If we want another function that is always zero, except at $t = T_S$, and that stores the value of $g(t)$ at $t = T_S$, then the function $g(t)\delta(t - T_S)$ does the job: it is always zero, and its 'area' is equal to $g(T_S)$. Since we want a function that is always zero, except at $t = nT_S$, for any integer value n, then we can take the product of $g(t)$ with the impulse train $imt(t)$. Recall that:

Equation (2.163)

$$imt(t) = \sum_{n=-\infty}^{\infty} \delta(t - nT)$$

Setting the period of the impulse train to $T = T_S$, the mathematical representation of $g_d(t)$ reads:

$$g_d(t) = g(t) \cdot imt(t) = \sum_{n=-\infty}^{\infty} g(t)\delta(t - nT_S) \qquad (7.11)$$

So, $g_d(t)$ can be thought of as an impulse train modulated by the function $g(t)$, where the height of each arrow is given by the value of $g(t)$ in the corresponding instant, just like in Figure 2.10 (in Figure 2.10, the red line represents $g(t)$ and the blue arrows represent $g_d(t)$ – each arrow represents one element of the sum in Equation (7.11)).

The beauty of $g_d(t)$ is that, on the one hand, it is still a continuous time function, and as such it has a Fourier transform integral; but, on the other hand, its Fourier transform integral reduces to the sum in Equation (7.10) (that is why I chose to call the sum $G_d(f)$). We can prove that this is true by straightforward evaluation of the Fourier transform of $g_d(t)$:

$$G_d(f) = \int_{-\infty}^{+\infty} g_d(t)e^{-i2\pi ft}\, dt = \int_{-\infty}^{+\infty} \sum_{n=-\infty}^{\infty} g(t)\delta(t-nT_S)e^{-i2\pi ft}\, dt$$

Swapping the order of integral and sum:

$$G_d(f) = \sum_{n=-\infty}^{\infty} \int_{-\infty}^{+\infty} g(t)\delta(t-nT_S)e^{-i2\pi ft}\, dt$$

Using the sifting property of the impulse function:

$$G_d(f) = \sum_{n=-\infty}^{\infty} g(nT_S)e^{-i2\pi fnT_S}$$

Since $g(nT_S) = g[n]$, we conclude that:

$$G_d(f) = \sum_{n=-\infty}^{\infty} g[n]e^{-i2\pi fnT_S} \tag{7.12}$$

Thus, we have proved that the sum in Equation (7.10) is the Fourier transform of $g_d(t)$. That is wonderful, because now we can easily connect $G_d(f)$ with $G_c(f)$ through Equation (7.11) and the properties of Fourier transforms. More specifically, since $g_d(t)$ is the product between $g(t)$ and $imt(t)$, it follows from the property of product in the time domain (Equation (2.152)) that $G_d(f)$ is the convolution between $G_c(f)$ and the Fourier transform of $imt(t)$. Recall that the Fourier transform of $imt(t)$ is an impulse train in the frequency domain:
Equation (2.164)

$$IMT(f) = \frac{1}{T} \sum_{k=-\infty}^{\infty} \delta\left(f - k\frac{1}{T}\right)$$

Thus, setting $T = T_S$ and using the property of product in the time domain (Equation (2.152)), we get:

$$G_d(f) = G_c(f) * IMT(f) = G(f) * \left[f_S \sum_{k=-\infty}^{\infty} \delta(f - kf_S) \right] \tag{7.13}$$

where f_S is the 'sampling frequency', defined as:

$$f_S = \frac{1}{T_S} \tag{7.14}$$

Recall that the convolution is a linear operation, which means that the convolution with a sum of functions is the sum of the convolutions. Thus:

$$G_d(f) = G_c(f) * \left[f_S \sum_{k=-\infty}^{\infty} \delta(f - kf_S) \right] = f_S \sum_{k=-\infty}^{\infty} G_c(f) * \delta(f - kf_S) \qquad (7.15)$$

According to Equation (7.15), $G_d(f)$ is the sum of convolutions of $G_c(f)$ with impulse functions. We have already proved that the convolution of a function with an impulse function results in the function shifted to the 'place' where the impulse function is 'located' (see Equation (2.26)). Thus, $G_c(f) * \delta(f - kf_S) = G_c(f - kf_S)$. Consequently, Equation (7.15) reduces to:

$$G_d(f) = f_S \sum_{k=-\infty}^{\infty} G_c(f - kf_S) \qquad (7.16)$$

Putting the pieces together, we have concluded that:

$$\sum_{n=-\infty}^{\infty} g[n]e^{-i2\pi f n T_S} = f_S \sum_{k=-\infty}^{\infty} G_c(f - kf_S) \qquad (7.17)$$

Forget about spiritual journeys in search of your inner self: Equation (7.17) is one of the most important things you will learn in your life.

According to Equation (7.17), the sum $\sum_{n=-\infty}^{\infty} g[n]e^{-i2\pi f n T_S}$ results in infinitely many 'copies' of $G_c(f)$, each copy centred on a multiple of f_S. Thus, it is possible to obtain $G_c(f)$ from $g[n]$, as long as there is no overlap between these copies. For example, suppose that the highest frequency of $G_c(f)$ is f_0, as represented in Figure 7.2a (for simplicity, in this illustration I assume that $G_c(f)$ is a real function, but recall from Chapter 2 that in general $G_c(f)$ is a complex function). If, on the one hand, the sampling frequency f_S is higher than $2f_0$, then there is no overlap between the copies. This situation is illustrated in Figure 7.2b. If, on the other hand, the sampling frequency f_S is lower than $2f_0$, then there is overlap between the copies. This latter situation is illustrated in Figure 7.2c.

Thus, we have concluded that it is possible to perfectly recover $G_c(f)$ from $G_d(f)$ if:

$$f_S > 2f_0 \qquad (7.18)$$

where f_0 is the highest frequency of the signal.

Notice the astonishing weirdness of this result. If $f_S > 2f_0$, then we can recover $G_c(f)$ from $G_d(f)$; but $G_d(f)$ is a representation of the samples $g[n]$, whereas $G_c(f)$ is a representation of $g(t)$. Thus, the assertion that we can recover $G_c(f)$ from $G_d(f)$ is logically equivalent to the assertion that we can recover $g(t)$ from $g[n]$. But how can that be? How can we know the values of $g(t)$ that were not stored in $g[n]$? Recall that $g(t)$ is a continuous signal, so it is defined by an infinity of coordinates (as we saw in Chapter 2 [Representation of Signals], each value of $g(t)$ is a different coordinate, indexed by t). But $g[n]$ is a finite sequence of numbers. So, how can we deduce an infinite number of coordinates from a finite number of coordinates? That is bizarre, but it is what Equation (7.17) is teaching.

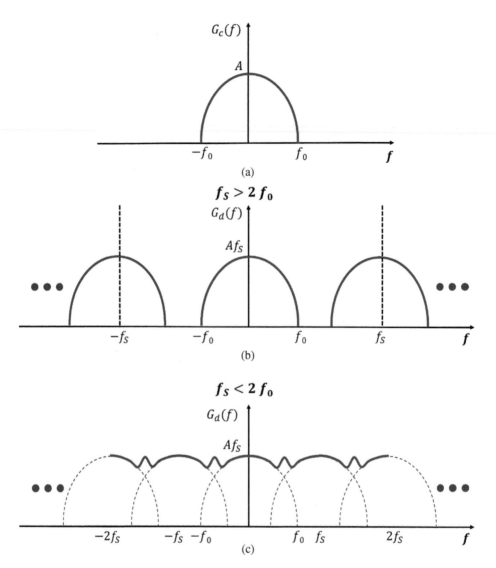

Figure 7.2 (a) Illustration of $G_c(f)$ with highest frequency f_0; for simplicity $G_c(f)$ is assumed to be real. $G_d(f)$ is formed by copies of $G_c(f)$. (b) If $f_S > 2f_0$, then there is no superposition between the copies, and hence $G_c(f)$ can be recovered from $G_d(f)$. If $f_S < 2f_0$, then there is superposition between the copies, and $G_c(f)$ cannot be recovered.

To gain more insights into this math from cuckoo land, let us write $g(t)$ explicitly in terms of $g[n]$. Of course, we are assuming that the condition $f_S > 2f_0$ has been satisfied, in which case Equation (7.16) entails that:

$$G_c(f) = \frac{1}{f_S}G_d(f) \quad for \quad \frac{-f_S}{2} \le f \le \frac{f_S}{2} \tag{7.19}$$

Equation (7.19) is just the mathematical assertion that $G_c(f)$ coincides with the central copy of $G_d(f)$ (that is, the copy centred at the origin, and which is indexed by $k = 0$ in the sum on the right-hand side of Equation (7.16)), with the amplitude factor f_S corrected for. Inverse Fourier transforming $G_c(f)$, and using Equation (7.19), we get:

$$g(t) = \int_{-\infty}^{+\infty} G_c(f)\, e^{i2\pi ft}\, df = \int_{-\frac{f_S}{2}}^{+\frac{f_S}{2}} \frac{1}{f_S} G_d(f)\, e^{i2\pi ft}\, df$$

Using Equation (7.12):

$$g(t) = \int_{-\frac{f_S}{2}}^{+\frac{f_S}{2}} \frac{1}{f_S} G_d(f)\, e^{i2\pi ft}\, df = \int_{-\frac{f_S}{2}}^{+\frac{f_S}{2}} \frac{1}{f_S}\left[\sum_{n=-\infty}^{\infty} g[n] e^{-i2\pi fnT_S}\right] e^{i2\pi ft}\, df$$

Swapping the order of the sum and integral, we obtain:

$$g(t) = \frac{1}{f_S}\sum_{n=-\infty}^{\infty} g[n]\int_{-\frac{f_S}{2}}^{+\frac{f_S}{2}} e^{i2\pi f(t-nT_S)}\, df \tag{7.20}$$

Evaluating the integral, we get:

$$g(t) = \frac{1}{f_S}\sum_{n=-\infty}^{\infty} g[n]\left[\frac{e^{i2\pi f(t-nT_S)}}{i2\pi(t-nT_S)}\Bigg|_{-\frac{f_S}{2}}^{\frac{f_S}{2}}\right]$$

$$= \frac{1}{f_S}\sum_{n=-\infty}^{\infty} g[n]\frac{e^{i2\pi\frac{f_S}{2}(t-nT_S)} - e^{-i2\pi\frac{f_S}{2}(t-nT_S)}}{i2\pi(t-nT_S)}$$

With a bit of rearrangement, the sum of exponentials can be combined into a sinc function (Equation (2.85)). Thus:

$$g(t) = \sum_{n=-\infty}^{\infty} g[n]\frac{1}{\pi f_S(t-nT_S)}\frac{e^{i2\pi\frac{f_S}{2}(t-nT_S)} - e^{-i2\pi\frac{f_S}{2}(t-nT_S)}}{2i}$$

$$= \sum_{n=-\infty}^{\infty} g[n]\frac{1}{\pi f_S(t-nT_S)}\sin\left(2\pi\frac{f_S}{2}(t-nT_S)\right)$$

The last step uses Equation (2.65).

Now we are ready to cast the result in terms of the sinc function (Equation (2.85)):

$$g(t) = \sum_{n=-\infty}^{\infty} g[n]\frac{\sin(\pi f_S(t-nT_S))}{\pi f_S(t-nT_S)} = \sum_{n=-\infty}^{\infty} g[n]\mathrm{sinc}(\pi f_S(t-nT_S))$$

Recall that $f_S = 1/T_S$ (Equation (7.14)). Thus, rearranging the sum above, we finally conclude that:

$$g(t) = \sum_{n=-\infty}^{\infty} g[n]\mathrm{sinc}\left(\pi\frac{(t-nT_S)}{T_S}\right) \tag{7.21}$$

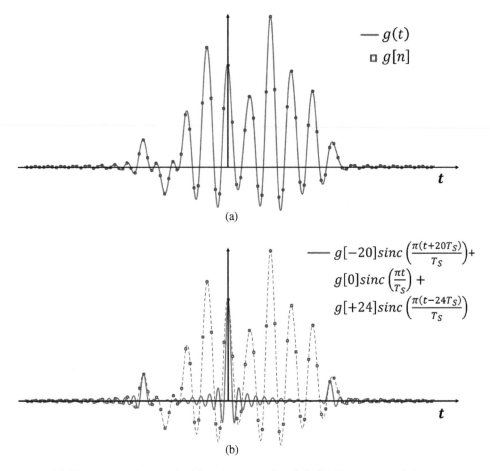

Figure 7.3 (a) Illustration of a signal $g(t)$ and its samples $g[n]$. (b) Illustration of the reconstruction of $g(t)$ through $g[n]$ by interpolating *sinc* functions. The figure shows the sum of three *sinc* functions. The others are omitted for better visualization. The plot assumes $T_S = 0.1$ s.

According to Equation (7.21), a signal $g(t)$ can be recovered from its samples $g[n]$ by interpolating *sinc* functions, each one centred at a multiple of T_S, and with amplitude set by $g[n]$. This process is illustrated in Figure 7.3. A signal $g(t)$ is shown in Figure 7.3a, together with samples taken at instants $t = nT_S$, marked with blue squares (the example assumes $T_S = 0.1$ s, but the numbers are omitted from the axes to avoid clutter). These samples constitute the discrete signal $g[n]$. If we add up a bunch of *sinc* functions, each one centred at an instant nT_S, and with amplitude $g[n]$, then we recover $g(t)$. Three of these *sinc* functions are shown explicitly in Figure 7.3b. The others are omitted to help visualizing the process.

We have concluded, again, that we can obtain $g(t)$ from $g[n]$: we just need to follow the recipe of Equation (7.21), and the *sinc* functions will fill in the 'gaps' between the samples.

Notice that, at each instant t, $g(t)$ is formed by the contribution of infinitely many *sinc* functions, each one centred at a multiple of T_S, and which add up to $g(t)$. How do these *sinc* functions 'know' how to arrange themselves so that when they are added up the result is exactly $g(t)$? That would already be astonishing for a single instant t, but according to Equation (7.21) they 'know' how to do it at all instants t! Sometimes math gets a bit too close to the supernatural.

And this surreal result is not the only one entailed by Equation (7.17). Recall that $G_d(f)$ is the Fourier transform of the 'digital' signal $g_d(t)$, and $G_c(f)$ is the Fourier transform of the continuous time original signal $g(t)$. So, if we want to physically recover $g(t)$ from $g_d(t)$ – for example, using a circuit – then Equation (7.17) (or, equivalently, Equation (7.16)) is teaching which circuit we need to use: we need to use a low-pass filter with a cut-off frequency at $f_S/2$. After all, by passing $g_d(t)$ through this low-pass filter, the 'extra' copies are filtered out, leaving only the central copy, which means that the output is $g(t)$ (apart from the f_S factor in the amplitude). In other words, Equation (7.17) (or Equation (7.16)) is teaching that a digital-to-analog converter is a low-pass filter.

That is a lot of teaching for such a tiny equation, do you agree?

We have seen that the Fourier transform of a discrete signal $g[n]$ coincides with the Fourier transform of the product of the continuous time signal with an impulse train, which product is represented by $g_d(t)$. Since product in the time domain leads to convolution in the frequency domain, and convolution with an impulse train makes a function periodic, then the Fourier transform of a discrete signal is a periodic function. But notice that this is not a new story: we have already seen it before, in Section 2.13 (The Fourier Series). The only difference is that the domains are swapped: in Section 2.13 (The Fourier Series), we considered a periodic function in the time domain, which led to discretization in the frequency domain, while here we discretized the time domain, leading to a periodic function in the frequency domain. In both cases, we used the feature that 'discretization in one domain leads to periodic functions in the other domain'. Keep this feature of Fourier transforms in mind: it is quite useful, and we will use it again in Chapter 8.

To conclude this section, we need to give credit where credit is due, and also to learn a bit of terminology. This business of recovering $g(t)$ from its samples $g[n]$ goes by the name of 'Nyquist–Shannon sampling theorem'. This name combines two giants of communication theory: Harry Nyquist and Claude Shannon. Incidentally, it was Shannon who figured out that entropy and information are essentially the same thing, arguably the most beautiful discovery falling within the scope of electrical engineering. Shannon also derived Equation (7.21), which is known as 'Whittaker–Shannon interpolation formula'. Finally, if the sampling frequency does not satisfy $f_S > 2f_0$ (in other words, if $f_S < 2f_0$), then the periods overlap and, consequently, $g(t)$ cannot be recovered from $g[n]$. This effect of loss of information due to poor sampling is known as 'aliasing'.

Now that we have figured out how to obtain $G_c(f)$ from $g[n]$, we can define an 'official' Fourier transform of discrete signals. But let us open a new section for that.

7.3 The Discrete Time Fourier Transform (DTFT)

In Section 7.2, we learned that the sum in Equation (7.12), which I rewrite below for convenience, results in copies of the Fourier transform of the continuous time signal of interest. As such, Equation (7.12) is a natural choice to extend Fourier transforms to discrete time signals, with a clear-cut connection with the Fourier transform of the continuous time counterpart, as we have just seen in Section 7.2. Equation (7.12)

$$G_d(f) = \sum_{n=-\infty}^{\infty} g[n] e^{-i2\pi f n T_S}$$

However, there is a snag: Equation (7.12) involves T_S, which is a parameter external to $g[n]$. It is a bit odd, not to say inconvenient, to define a transform of a signal that involves parameters that are not inherent to the signal. So, we need to get rid of T_S in Equation (7.12). There is a neat way to do it: we can define a new variable v as:

$$v = fT_S \qquad (7.22)$$

And recast Equation (7.12) in terms of v. Thus:

$$G(v) = \sum_{n=-\infty}^{\infty} g[n] e^{-i2\pi v n} \qquad (7.23)$$

In the form of Equation (7.23), Equation (7.12) no longer involves T_S, so it is a more convenient definition of a Fourier transform of discrete signals. The transformation of Equation (7.23) is called the Discrete Time Fourier Transform (DTFT). The main goal of this section is to spell out how the DTFT is related to the Fourier transform of the continuous time signal.

The first significant feature to notice in the DTFT is that v is a dimensionless frequency. Indeed, according to Equation (7.22):

$$v = fT_S = \frac{f}{f_S} \qquad (7.24)$$

Since both f and f_S have units of inverse of time (like Hertz, for example), then v is dimensionless. It makes sense that the 'frequency' of the DTFT is a dimensionless quantity: after all, $g[n]$ is just a sequence of numbers, so there is no notion of time in it. Recall, from Section 2.11 (The Physical Meaning of Fourier Transforms), that a signal with high-frequency components is a 'fast' signal: its amplitude changes a lot in a short period of time. But a sequence of numbers cannot be fast because there is no time in it. It is just a sequence of numbers. That is why its 'frequency' is dimensionless: a dimensionless frequency, on its own, does not give information about how fast a signal is. But we can recover information about time if we know T_S. Thus, for example, say that we are investigating two discrete signals $g_1[n]$ and $g_2[n]$, whose DTFTs are $G_1(v)$ and $G_2(v)$, respectively.

Now, suppose we pick a certain v_0 and ask: are $G_1(v_0)$ and $G_2(v_0)$ components of a high or of a low frequency? We cannot answer this question, because it is nonsense: we do not have a notion of time in $g_1[n]$ and $g_2[n]$. But we can recover it if we know T_S. So, suppose that $g_1[n]$ and $g_2[n]$ were generated by sampling $g_1(t)$ and $g_2(t)$ with sampling frequencies f_{S1} and f_{S2}, respectively. In this case, $G_1(v_0)$ and $G_2(v_0)$ are carrying information about the Fourier components $f_1 = v_0 f_{S1}$ and $f_2 = v_0 f_{S2}$ – see Equation (7.24). Now we can answer the question whether $G_1(v_0)$ and $G_2(v_0)$ are high- or low-frequency components: it depends on f_1 and f_2. For example, if f_1 is high and f_2 is low, then $G_1(v_0)$ is a high-frequency component and $G_2(v_0)$ is a low-frequency component. Notice that we reached two different conclusions for the same v_0, because the sampling frequencies f_{S1} and f_{S2} were different in our example.

Our decision to define the DTFT in terms of the dimensionless frequency v is relevant to the connection between the DFTF of $g[n]$, the Fourier transform of the intermediary function $g_d(t)$ and the Fourier transform of $g(t)$. This relationship is illustrated in Figure 7.4, where I assume $f_S > 2f_0$. As shown in Figure 7.4a and

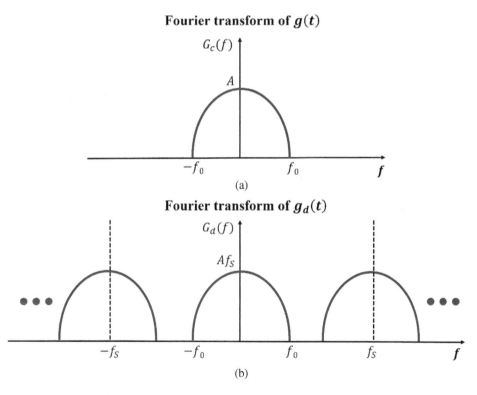

(a)

(b)

Figure 7.4 Illustration of the relationship between the Fourier transform of the signal of interest $g(t)$, the Fourier transform of the intermediary function $g_d(t)$, and the DTFT of the discrete signal $g[n]$. (a) Illustration of an example of Fourier transform. (b) The Fourier transform of $g_d(t)$ is formed by copies of the Fourier transform of $g(t)$, each copy centred at a multiple of f_S. Thus, the period of $G_d(f)$ is f_S. (c) The DTFT of $g[n]$ is a scaled version of $G_d(f)$, with scaling factor f_S (see Equation (7.24)). Thus, the period of $G(v)$ is 1.

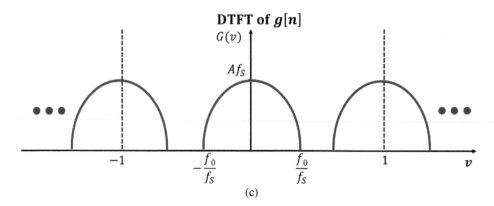

Figure 7.4 (*Continued*)

Figure 7.4b, the relationship between $G_d(f)$ and $G_c(f)$ is the same as before: $G_d(f)$ is a periodic function, with period f_S, and formed by copies of $G_c(f)$.

The DTFT of $g[n]$ is shown in Figure 7.4c. Comparing Equation (7.23) with Equation (7.12), and recalling that $f_S T_S = 1$ (see Equation (7.14)), we notice that $G(v)$ is obtained from $G_d(f)$ by making the substitution $f = v f_S$. Thus:

$$G(v) = G_d(v f_S) \tag{7.25}$$

Equation (7.25) entails that $G(v)$ is a scaled version of $G_d(f)$, with scaling factor set by f_S. Consequently, $G(v)$ is also a periodic function, but now its period is $v = f_S/f_S = 1$.

In practice, engineers and physicists use algorithms to calculate the DTFT of the sampled signal and, from the DTFT of $g[n]$, they get the Fourier transform of $g(t)$ (actually, the algorithm calculates the DFT, which is the subject of Chapter 8 – but the concept is the same). Thus, it is important to spell out how we can obtain $G_c(f)$ from $G(v)$. As this is better done by means of worked examples, we conclude this section with two such examples.

Worked Exercise: Example of a Matlab Routine to Calculate the DTFT

Write a Matlab routine to calculate the DTFT of the Gaussian function $g(t) = 5e^{-\left(\frac{t}{10}\right)^2}$.

Solution:

The first step is to discretize the function. For that, we need to define a time vector. It is good practice to begin by defining the number of discretization points. Later, in Chapter 8, we will learn how to use the function fft, which works best if the number of points is a power of 2. In this example, I choose the number of points to be $N_{pt} = 2^{10}$.

We also need a vector n with the integers indexing the samples. It is advisable to define the vector n in a such a way that we get $n(N_{pt}/2 + 1) = 0$ (this is again due to a peculiarity of the *fft*, to be discussed in Chapter 8). In our example, $N_{pt} = 1024$, so we want $n(513) = 0$. The lines of code defining N_{pt} and n are:

```
clear all
close all

Npt = 2^10; % number of points

n = (-Npt/2:1:Npt/2 -1); % definition of vector n
```

Now we need to choose a time step, that is, we need to choose the sampling period T_S. For that, we begin by choosing the edges of the time window. Since we are sampling a Gaussian function with width of 10 s, I choose a time window beginning in $t_i = -300$ and ending in $t_f = +300$. That gives plenty of 'space' for the Gaussian to go to zero. The time window, together with the number of points, set the time step $T_S = (t_f - t_i)/(N_{pt})$:

```
ti = -300; % beginning of the time window
tf = -ti; % end of the time window

Ts = (tf-ti)/(Npt); % sampling period
```

Now the time vector can be defined:

```
t = n*Ts; % definition of time vector
```

The next group of lines discretize the function $g(t)$

```
A = 5;     % amplitude of the gaussian
sigma = 10; % width of the gaussian

g = t*0;   % initialize vector g with the same
number of points of t.
g = A*exp(-(t./sigma).^2); % discretize the gaussian
```

Now we need to define the vector v containing the values of the dimensionless frequency for which we will calculate the DTFT. The vector v does not need to have the same length of the vector t, so I choose $N_{pt\text{-}v} = 2^{11}$ points. The following group of lines define the vector v following the same logic of the vector t. Notice that I chose the vector v to range from $v_i = -2$ to $v_f = 2$. This range includes three complete periods, and part of a fourth and fifth. So, this range suffices to visualize that the DTFT is indeed periodic.

```
Npt_v = 2^11; % number of points of vector v

vi = -2;  % beginning of the dimensionless frequency
window
vf = -vi; % end of the dimensionless frequency window

v_step = (vf-vi)/(Npt_v); % step

k = (-Npt_v/2:1:Npt_v/2 -1); % definition of an
auxiliary vector k
v = k*v_step;  % definition of v
```

Finally, the group of lines below calculate the DTFT vector. Notice that the command .* was used, and that this is the Matlab command for point by point vector multiplication. Make sure you understood that the lines below execute the sum in Equation (7.23).

```
G_DTFT = v*0; % initialize vector G_DTFT with the same
number of points of v

for u = 1:length(v)

  G_DTFT(u) = sum(g.*exp(-1i*2*pi*v(u)*n));
  end

figure(1)
plot(v,abs(G_DTFT)) % plot the DTFT.
xlabel('v')
ylabel('DTFT')
```

Notice that the plot function uses the command 'abs', which returns the magnitude of a complex number. In general, the DTFT is a complex function, so we need

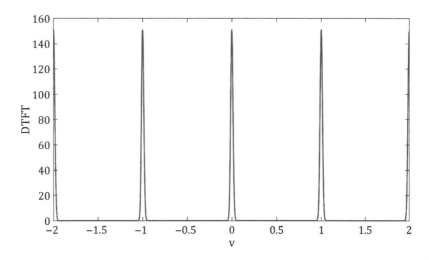

Figure 7.5 Plot of the DTFT of the Gaussian function.

to plot magnitude and phase separately. In this example, however, the DTFT is a real function, so the 'abs' command was not necessary. I just left it there because in general we do need it.

The resulting plot is shown in Figure 7.5. Notice that the DTFT is indeed periodic, and the period is indeed 1. Furthermore, notice the peak of the curves, which is at 151.2494. Let us check if that makes sense. According to Equation (2.183), the peak of the Fourier transform of a Gaussian is $A \cdot \sigma\sqrt{\pi}$. In our example, $A = 5$, $\sigma = 10$, which gives a peak of 88.6227. Thus, according to Equation (7.16) (or, if you prefer, according to Figure 7.4), the peak of the DTFT should be $88.6227 \cdot f_S$. In our example, $f_S = 1.706667$. Thus, the DTFT peak should be $88.6227 \times 1.706667 = 151.2494$, which agrees with the value shown in Figure 7.5.

Finally, notice that the periods are well spaced, which means that $f_S \gg 2f_0$. The closer f_S is to $2f_0$, the closer to each other the periods are. So, you can check if the condition $f_S > 2f_0$ has been satisfied by checking whether the periods overlap.

Worked Exercise: Fourier Transform from the DTFT

Calculate the DTFT of $g(t) = 5e^{-\left(\frac{t}{10}\right)^2}$, and then convert the DTFT vector into a vector storing samples of the Fourier transform of $g(t)$.

Solution:

The time discretization is the same as in the previous exercise. The definition of the vector v is also essentially the same as before, with just one difference: since our goal is to obtain the Fourier transform, we should define v only within the central period. Thus, we need to set $v_i = -0.5$ and $v_f = 0.5$. This is the only difference between the definition of the vector v to be used in this exercise and in the previous exercise. The code to calculate the DTFT vector is also the same as in the previous exercise.

Since we are interested in the Fourier transform, we need to generate a frequency vector according to Equation (7.24). Thus:

```
fs = 1/Ts;  % define sampling frequency
f = v*fs;   % define frequency vector
```

A vector storing the samples of the Fourier transform can be straightforwardly obtained from the DTFT vector by correcting the amplitude (see Equation (7.16) or Figure 7.4). Thus:

```
G_FT = G_DTFT./fs;  % create a vector storing
samples of the Fourier transform

plot(f,abs(G_FT))  % plot the FT
xlabel('f (Hz)')
ylabel('FT')
```

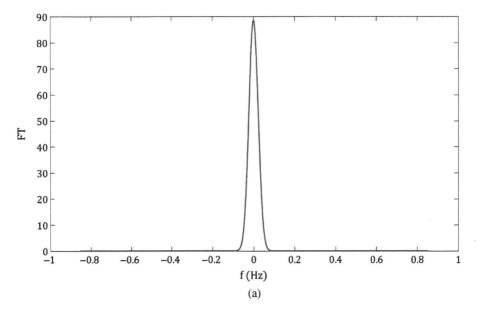

(a)

Figure 7.6 (a) Plot of the Fourier transform. (b) Plot of the DTFT, including only the central period.

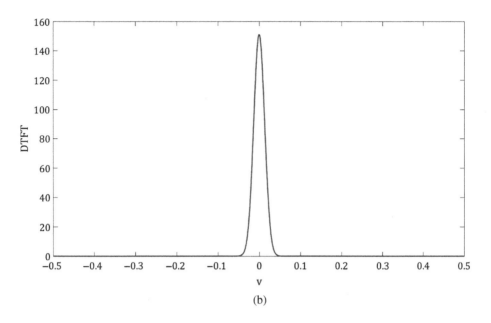

(b)

Figure 7.6 *(Continued)*

The plot generated by the command above is shown in Figure 7.6a. For comparison, a plot of the DTFT is shown in Figure 7.6b.

7.4   The Inverse DTFT

A transform is not of much use without its inverse, so we need to find the inverse of the DTFT. We begin with the inverse Fourier transform of $g(t)$:

$$g(t) = \int_{-\infty}^{+\infty} G_c(f) e^{i2\pi ft}\, df$$

Equation (7.16) entails that (see also Figure 7.4):

$$G_c(f) = \frac{1}{f_S} G_d(f) \quad for \quad \frac{-f_S}{2} \leq f \leq \frac{f_S}{2} \tag{7.26}$$

Thus:

$$g(t) = \int_{-\infty}^{+\infty} G_c(f) e^{i2\pi ft}\, df = \int_{-\frac{f_S}{2}}^{+\frac{f_S}{2}} \frac{1}{f_S} G_d(f) e^{i2\pi ft}\, df \tag{7.27}$$

Since $g[n] = g(nT_S)$, then:

$$g[n] = g(nT_S) = \int_{-\frac{f_S}{2}}^{+\frac{f_S}{2}} \frac{1}{f_S} G_d(f) e^{i2\pi fnT_S}\, df \tag{7.28}$$

To get the inverse DTFT, we need to recast Equation (7.28) in terms of the DTFT. First, we rewrite the integral in Equation (7.28) in terms of v. Recalling that $f = v f_S$ (see Equation (7.24)), we obtain:

$$g[n] = \int_{-\frac{f_S}{2}}^{+\frac{f_S}{2}} \frac{1}{f_S} G_d(f) e^{i2\pi f n T_S} df = \int_{-\frac{1}{2}}^{+\frac{1}{2}} \frac{1}{f_S} G_d(v f_S) e^{i2\pi v f_S n T_S} f_S \, dv$$

$$= \int_{-\frac{1}{2}}^{+\frac{1}{2}} G_d(v f_S) e^{i2\pi v n} \, dv \qquad (7.29)$$

Notice that we used $f_S T_S = 1$ in the argument of the exponential. Moreover, recall that, according to Equation (7.25), $G(v) = G_d(v f_S)$. Therefore, we conclude that:

$$g[n] = \int_{-\frac{1}{2}}^{+\frac{1}{2}} G_d(v f_S) e^{i2\pi v n} \, dv = \int_{-\frac{1}{2}}^{+\frac{1}{2}} G(v) e^{i2\pi v n} \, dv \qquad (7.30)$$

The last integral in Equation (7.30) is the inverse DTFT.

We conclude this section by combining the direct DTFT (Equation (7.23)) and the inverse DTFT in a box:

DTFT PAIR
DIRECT DTFT

$$G(v) = \sum_{n=-\infty}^{\infty} g[n] e^{-i2\pi v n}$$

INVERSE DTFT

$$g[n] = \int_{-0.5}^{+0.5} G(v) e^{i2\pi v n} \, dv \qquad (7.31)$$

7.5   Properties of the DTFT

We have learned that $G_d(f)$ is the Fourier transform of $g_d(t)$, which is the product of a signal with an impulse train (see Equation (7.11)). We have also learned that the DTFT is a scaled version of $G_d(f)$; in other words, that $G(v) = G_d(v f_S)$ – see Equation (7.25). Thus, it follows that $G(v)$ can be interpreted as being the Fourier transform of the product of a signal with an impulse train with impulses spaced of $T_S = 1$. Mind you, I am not saying that the DTFT requires that $T_S = 1$. In fact, usually $T_S \neq 1$. What I am saying is that, if T_S was 1, then the Fourier transform of $g_d(t)$ would coincide with the DTFT of $g[n]$. If $T_S \neq 1$, then the DTFT is a scaled version of the Fourier transform of $g_d(t)$, as affirmed by Equation (7.25). But, even if $T_S \neq 1$, there exists **another** signal $g_{2d}(t)$, with $T_S = 1$, whose Fourier transform coincides with the DTFT. Of course, this **other** signal $g_{2d}(t)$, with $T_S = 1$, is itself is a scaled version of the original $g_d(t)$, with $T_S \neq 1$.

The existence of this other scaled signal $g_{2d}(t)$, with $T_S = 1$, whose Fourier transform coincides with the DTFT of $g[n]$, implies that the DTFT can be interpreted as being a particular case of the Fourier transform. Such an interpretation is very useful, because it entails that the properties of Fourier transforms must also be applicable to the DTFT. In fact, many of the DTFT properties are identical to the equivalent Fourier transform property. There are some subtleties, however, arising from the fact that the DTFT involves discrete functions.

In this section, we survey the main properties of the DTFT, and see how they relate to the properties of the Fourier transform. The emphasis is on the properties that are a bit different from the Fourier transform. A few that are quite similar are left as exercises.

7.5.1 'Time' shifting

In Section 2.12.4 (Time Shifting), we learned that time shifting a function induces a phase modulation in its Fourier transform (Equation (2.119)). There is an analogous property of the DTFT, but we can no longer speak of a 'proper' time shifting in a discrete signal $g[n]$. Instead, we can loosely say that, if $g_2[n] = g[n - n_0]$, then $g_2[n]$ is a 'time' shifted version of $g[n]$.

The property of 'time' shifting relates the DTFT of $g_2[n]$ and the DTFT of $g[n]$. Its proof follows essentially the same logic of the proof of Equation (2.119), but using sums instead of integrals. Thus, evaluating the DTFT of $g_2[n]$, we get:

$$G_2(v) = \sum_{n=-\infty}^{\infty} g_2[n] e^{-i2\pi vn} = \sum_{n=-\infty}^{\infty} g[n - n_0] e^{-i2\pi vn}$$

Making the substitution $n' = n - n_0$:

$$G_2(v) = \sum_{n'=-\infty}^{\infty} g[n'] e^{-i2\pi v(n'+n_0)}$$

And isolating the term that does not depend on n', we get:

$$G_2(v) = e^{-i2\pi vn_0} \sum_{n'=-\infty}^{\infty} g[n'] e^{-i2\pi vn'} = e^{-i2\pi vn_0} G(v)$$

Thus, we have concluded that:

if:

$$g_2[n] = g[n - n_0]$$

then:

$$G_2(v) = e^{-i2\pi vn_0} G(v) \tag{7.32}$$

7.5.2 Difference

The property of difference is straightforward, but its relation to the continuous time domain is a bit subtle. We begin by establishing the property, and then we investigate how it relates to the continuous time domain.

If $g_2[n] = g[n] - g[n-1]$, then $g_2[n]$ is the difference of $g[n]$. Their relationship can be easily obtained: the DTFT of $g_2[n]$, which we denote by $G_2(v)$, is the difference between the DTFT of $g[n]$, which we denote by $G(v)$, and the DTFT of $g[n-1]$. But, according to Equation (7.32), the DTFT of $g[n-1]$ is $e^{-i2\pi v}G(v)$. Thus:

if:

$$g_2[n] = g[n] - g[n-1]$$

then:

$$G_2(v) = G(v) - e^{-i2\pi v}G(v) = (1 - e^{-i2\pi v})G(v) \tag{7.33}$$

Now let us see how it relates to the continuous time domain.

The property of difference is analogous to the property of differentiation (Equation (2.132)). Indeed, if we wanted to calculate the derivative of $g(t)$ numerically, we would take the difference of $g[n]$, and divide by the time step. Thus, if $g_3[n]$ is to contain samples of the derivative of $g(t)$ calculated numerically, then:

$$g_3[n] = \frac{g[n] - g[n-1]}{T_S} \tag{7.34}$$

According to Equation (7.33), the DTFT of $g_3[n]$ reads:

$$G_3(v) = \frac{(1 - e^{-i2\pi v})}{T_S}G(v) \tag{7.35}$$

This result does not connect smoothly with the time domain. For, if:

$$g_3(t) = \frac{dg(t)}{dt} \tag{7.36}$$

Then $g_3[n]$ should contain samples of $g_3(t)$. Actually, since $g_3[n]$ is the derivative calculated numerically (as in Equation (7.34)), and $g_3(t)$ is the exact derivative, then the precise statement is that $g_3[n]$ should approximate to samples of $g_3(t)$ as $T_S \to 0$.

But this conclusion does not seem to agree with Equation (7.35). Indeed, according to Equation (2.132), $G_3(f) = i2\pi f G(f)$, where $G_3(f)$ and $G(f)$ are the Fourier transforms of $g_3(t)$ and $g(t)$, respectively. Thus, since the central period of the

DTFT corresponds to a scaled version of the Fourier transform, we should expect that $G_3^{cp}(v) = i2\pi f G^{cp}(v)$, where $G_3^{cp}(v)$ and $G^{cp}(v)$ are the central periods of the DTFT of $g_3[n]$ and $g[n]$, respectively. But, according to Equation (7.35), instead of having $G_3^{cp}(v) = i2\pi f G^{cp}(v)$, we have:

$$G_3^{cp}(v) = \frac{(1 - e^{-i2\pi v})}{T_S} G^{cp}(v) \tag{7.37}$$

How can that be? Where is the term $i2\pi f$ in Equation (7.37)? Has math gone crazy? No, it has not. Fortunately, we can almost always trust math to make sense of life. This term is there, but it is hidden. Let us extract it. First, recall that we expect $i2\pi f$ to show up in the limit $T_S \to 0$. This limit corresponds to $f_S \to \infty$, which entails that $f_S \gg f_0$ (recall that f_0 is the maximum frequency of the signal). In this condition, $G^{cp}(v)$ is nonzero only for $|v| \ll 1$ (see Figure 7.4c). Consequently, we can expand the term $e^{-i2\pi v}$ in a Taylor series and retain only the first-order term (the higher order terms play a role for higher values of v, but that does not matter because $G^{cp}(v) = 0$ for higher values of v). Thus, we can make the approximation:

$$e^{-i2\pi v} \approx 1 - i2\pi v \tag{7.38}$$

Using this approximation in Equation (7.37), we get:

$$G_3^{cp}(v) \approx \frac{[1 - (1 - i2\pi v)]}{T_S} G^{cp}(v) = \frac{i2\pi v}{T_S} G^{cp}(v)$$

But $v/T_S = f$ (see Equation (7.22)). Thus, we conclude that:

$$G_3^{cp}(v) \approx i2\pi f G^{cp}(v) \tag{7.39}$$

The shorter T_S is, the better the approximation of Equation (7.39) is. So, Equation (7.37) agrees with the property of differentiation of continuous time functions if T_S is sufficiently short.

But how short is sufficiently short? We need to use the approximation in Equation (7.38), and we can only use it if $G^{cp}(v)$ is 'concentrated' in a region of small v (in other words, if $G^{cp}(v)$ is nonzero only for $|v| \ll 1$). But for $G^{cp}(v)$ to be 'concentrated' in a region of small v, we need $f_S \gg f_0$ (see Figure 7.4c). So, we need to choose a sampling frequency (f_S) that is much higher than the highest frequency of the signal (f_0). Thus, how short sufficiently short is depends on the highest frequency of the signal. The higher f_S with respect to f_0 is, the better the approximation of Equation (7.39) is (recall that high f_S entails short T_S).

The approximation of Equation (7.39) is related to another funny feature of the relationship between continuous and discrete time: as we will see in Chapter 9 (Discrete Systems), it is much more accurate to calculate derivatives in the 'frequency' domain than in the 'time' domain.

7.5.3 Sum

The property of sum is analogous to the property of integration. The sum signal $g_2[n]$ is defined as:

$$g_2[n] = \sum_{m=-\infty}^{n} g[m] \tag{7.40}$$

Notice that the sum is over m, and ends at $m = n$, which is analogous to the integral in Equation (2.133), which is over τ and ends at $\tau = t$. Following a similar logic to the one developed in Section 2.12.7 (Integration), we notice that Equation (7.40) entails that:

$$g_2[n] - g_2[n-1] = \sum_{m=-\infty}^{n} g[m] - \sum_{m=-\infty}^{n-1} g[m]$$

The two sums involve the same terms, except that the first sum stops at $m = n$, and the second sum stops at $m = n - 1$. Thus, all terms cancel each other, except the term $g[n]$, which appears only in the first sum. Therefore:

$$g_2[n] - g_2[n-1] = g[n] \tag{7.41}$$

According to Equation (7.40) and Equation (7.41), if $g_2[n]$ is the sum of $g[n]$, then the difference of $g_2[n]$ must be $g[n]$. This conclusion is analogous to the relationship between derivative and integral (if $g_2(t)$ is the integral of $g(t)$, then the derivative of $g_2(t)$ must be $g(t)$).

Using Equation (7.33) in Equation (7.41), we get:

$$(1 - e^{-i2\pi v})G_2(v) = G(v)$$

This relationship is almost correct, but like in the property of integral (Equation (2.135)), we need to include the contribution of the impulse function, which captures the DC level. The complete expression reads:

if:

$$g_2[n] = \sum_{m=-\infty}^{n} g[m]$$

then:

$$G_2(v) = \frac{G(v)}{(1 - e^{-i2\pi v})} + \frac{G(0)}{2} \sum_{k=-\infty}^{\infty} \delta(v - k) \tag{7.42}$$

where k is an integer number.

Notice that the single impulse function of Equation (2.135) became an impulse train in Equation (7.42). The impulse train appears because the DTFT is a periodic function with unit period, so if there is an impulse at $v = 0$, then necessarily there is also an impulse at $v = 1$, $v = -1$, $v = 2$, and so on.

7.5.4 Convolution in the 'Time' Domain

In Chapter 3 (Representation of Systems), we learned that the property of convolution in the time domain, described in Section 2.12.8 (Convolution in the Time Domain), lies at the heart of the analysis of continuous time LTI systems. Likewise, the property of convolution in the 'discrete time' domain lies at the heart of the analysis of discrete LTI systems, to be treated later in Chapter 9 (Discrete Systems).

Like its continuous time counterpart, the property of convolution of discrete signals affirms that the convolution between two discrete signals leads to the product between their DTFTs. The proof is similar to the proof of the continuous time property (Equation (2.144)). The main difference is the definition of the convolution of discrete time signals, which is essentially the sum version of the convolution integral. Thus, we define the convolution of discrete signals as:

$$g[n] = a[n] * b[n] = \sum_{m=-\infty}^{+\infty} a[m]b[n-m] \qquad (7.43)$$

Notice that Equation (7.43) is the discrete version of Equation (2.138), with the discrete variable m playing the role of the continuous variable t_0, and the discrete variable n playing the role of the continuous variable t.

The procedure to calculate the convolution between two discrete signals follows essentially the same steps of the continuous time: the axes are changed from n to m, then one of the signals is flipped (in the sum in Equation (7.43), the signal $b[n]$ is flipped) and the flipped signal is shifted to evaluate the result of the convolution for each value of n. Though there is nothing mysterious in the operation of convolution of discrete signals, it is helpful to do it by hand at least once, using simple and short signals. In the exercise list, you will be asked to evaluate the convolution between two simple discrete signals.

To prove that $G(v) = A(v)B(v)$, where $G(v)$, $A(v)$, and $B(v)$ are the DTFTs of $g[n]$, $a[n]$, and $b[n]$, respectively, we begin by writing $G(v)$ explicitly as the DTFT of $g[n]$, and substitute Equation (7.43). Thus:

$$G(v) = \sum_{n=-\infty}^{\infty} g[n]e^{-i2\pi vn} = \sum_{n=-\infty}^{\infty} \sum_{m=-\infty}^{+\infty} a[m]b[n-m]e^{-i2\pi vn}$$

Evaluating the sum over n, and using Equation (7.32), we get:

$$G(v) = \sum_{m=-\infty}^{\infty} a[m] \sum_{n=-\infty}^{+\infty} b[n-m]e^{-i2\pi vn} = \sum_{m=-\infty}^{\infty} a[m]e^{-i2\pi vm}B(v) \qquad (7.44)$$

Notice that the sum over m in the last expression of Equation (7.44) is the DTFT of $a[n]$. Therefore:

$$G(v) = \sum_{m=-\infty}^{\infty} a[m]e^{-i2\pi vm}B(v) = A(v)B(v) \qquad (7.45)$$

Thus, we have concluded that:

if:

$$g[n] = a[n] * b[n]$$

then:

$$G(v) = A(v)B(v) \tag{7.46}$$

We will use this property in Chapter 9 (Discrete Systems).

7.5.5   Product in the Time Domain

In this section, we will express the DTFT of $g[n]$ in terms of the DTFT of $a[n]$ and $b[n]$ when $g[n] = a[n]b[n]$. The proof is simple and similar to Section 2.12.9 (Product in the Time Domain). We begin by evaluating the DTFT of $g[n]$, and expressing $a[n]$ in terms of its inverse DTFT. Thus:

$$G(v) = \sum_{n=-\infty}^{\infty} g[n]e^{-i2\pi vn} = \sum_{n=-\infty}^{\infty} a[n]b[n]e^{-i2\pi vn}$$

$$= \sum_{n=-\infty}^{\infty} \left[\int_{-0.5}^{+0.5} A(v')e^{i2\pi v'n}dv' \right] b[n]e^{-i2\pi vn}$$

Now we invert the order of sum and integral:

$$G(v) = \int_{-0.5}^{+0.5} A(v') \left[\sum_{n=-\infty}^{\infty} b[n]e^{-i2\pi(v-v')n} \right] dv'$$

The sum is the DTFT of $b[n]$ evaluated at the frequency $v - v'$. Therefore:

$$G(v) = \int_{-0.5}^{+0.5} A(v')B(v - v')dv' \tag{7.47}$$

We have thus concluded that:

if:

$$g[n] = a[n]b[n]$$

then:

$$G(v) = \int_{-0.5}^{+0.5} A(v')B(v - v')dv' \tag{7.48}$$

Equation (7.48) is the property of product in the 'time' domain of DTFTs. It is similar to its continuous time counterpart (Equation (2.151)), but with an important difference: the integral in Equation (7.47) is a 'circular convolution'. Notice that, contrary to the convolution in Equation (2.151) (which, by the way, is called a 'linear convolution'), the integral in Equation (7.47) does not range from $-\infty$ to $+\infty$. Indeed, it ranges only over the central period of the DTFT. Circular convolutions are a type of convolution involving periodic functions, and its relationship with linear convolutions plays a crucial role in the Discrete Fourier Transform (DFT), as we will see in more details in Chapter 8.

7.5.6   The Theorem that Should not be: Energy of Discrete Signals

This section deals with Parseval's theorem for discrete signals. Our main goal is to establish a relationship between the energy of a discrete signal, and the energy of its continuous time counterpart. This relationship can be established in a simple way using the frequency domain representation. We will see that the comparison between the discrete and continuous time energies leads to a counter-intuitive conclusion.

We begin by defining the energy of a discrete signal. The energy E_d (the subscript d is a reminder that we are dealing with discrete signals) of a signal $g[n]$ is defined as:

$$E_d = \sum_{n=-\infty}^{\infty} |g[n]|^2 \tag{7.49}$$

Our first goal is to find an expression for E_d in terms of the DTFT of $g[n]$ (in other words, we want Parseval's theorem for discrete signals). We will need the following identity:

$$\sum_{n=-\infty}^{\infty} e^{i2\pi v n} = \sum_{k=-\infty}^{\infty} \delta(v - k) \tag{7.50}$$

Equation (7.50) is the discrete version of Equation (2.74) and the intuitions behind them are also analogous: if v is not zero (and also not an integer number), then the sum of exponentials in Equation (7.50) involves all possible phases, so for any term $e^{i2\pi v n_1}$ there will be a term $e^{i2\pi v n_2}$ which is π shifted with respect to it (that is, $e^{i2\pi v n_2} = -e^{i2\pi v n_1}$). Thus, the terms cancel each other out and the sum is zero. But, if v is an integer number, then the exponential is always 1 regardless of the value of n (that is, $e^{i2\pi v n} = 1$ if v is an integer number). In this case, we are adding the number 1 infinite times, which results in an infinite number. Therefore, the sum in n is a function of v, and this function is zero everywhere, except for integer values of v. As such, it is a train of impulses in v, with each impulse 'located' at an integer number. In the exercise list, you will be asked to prove Equation (7.50) using the DTFT pair.

To find an expression for E_d in terms of the DTFT, we use the inverse DTFT (see Equation (7.31)) to write $g[n]$ and $g^*[n]$ in terms of $G(v)$. Thus:

$$g[n] = \int_{-0.5}^{+0.5} G(v)e^{i2\pi vn} \, dv$$

$$g^*[n] = \int_{-0.5}^{+0.5} G^*(v)e^{-i2\pi vn} \, dv$$

which entails that:

$$|g[n]|^2 = \left[\int_{-0.5}^{+0.5} G(v)e^{i2\pi vn} \, dv\right]\left[\int_{-0.5}^{+0.5} G^*(v)e^{-i2\pi vn} \, dv\right]$$

Expressing the product of integrals as a double integral, we get:

$$|g[n]|^2 = \int_{-0.5}^{+0.5}\int_{-0.5}^{+0.5} G(v)G^*(v')e^{i2\pi(v-v')n} \, dv dv' \tag{7.51}$$

According to Equation (7.49), to find E_d we need to sum Equation (7.51) over n. Thus:

$$E_d = \sum_{n=-\infty}^{\infty} |g[n]|^2 = \sum_{n=-\infty}^{\infty}\int_{-0.5}^{+0.5}\int_{-0.5}^{+0.5} G(v)G^*(v')e^{i2\pi(v-v')n} \, dv dv'$$

The only term that depends on n is the exponential. So, it makes sense to evaluate the sum over n before evaluating the integrals. Thus:

$$E_d = \sum_{n=-\infty}^{\infty} |g[n]|^2 = \int_{-0.5}^{+0.5}\int_{-0.5}^{+0.5} G(v)G^*(v')\left[\sum_{n=-\infty}^{\infty} e^{i2\pi(v-v')n}\right] dv dv' \tag{7.52}$$

According to Equation (7.50), the sum in Equation (7.52) results in $\sum_{k=-\infty}^{\infty} \delta(v - v' - k)$. Therefore:

$$E_d = \sum_{n=-\infty}^{\infty} |g[n]|^2 = \int_{-0.5}^{+0.5}\int_{-0.5}^{+0.5} G(v)G^*(v')\left[\sum_{k=-\infty}^{\infty} \delta(v - v' - k)\right] dv dv' \tag{7.53}$$

Now notice that, since the limits of integration of v and v' are ± 0.5, then $-1 < v - v' < 1$. Therefore, it guaranteed that $\delta(v - v' - k) = 0$, except for $k = 0$, because the argument of $\delta(v - v' - k)$ is always nonzero, except for $k = 0$ (recall that k is an integer). For example, $\delta(v - v' - 10) = 0$, because there is no combination of v and v' within the limits of the integrals that results in $v - v' = 10$. Since this is true for all values of k different from zero, then it follows that $\delta(v - v' - k) = 0$ if $k \neq 0$. That allows us to retain only the term $k = 0$ of the sum. Thus, Equation (7.53) reduces to:

$$E_d = \sum_{n=-\infty}^{\infty} |g[n]|^2 = \int_{-0.5}^{+0.5}\int_{-0.5}^{+0.5} G(v)G^*(v')\delta(v - v') \, dv dv'$$

Integrating over v' and using the sifting property of impulse functions, we obtain:

$$E_d = \sum_{n=-\infty}^{\infty} |g[n]|^2 = \int_{-0.5}^{+0.5} G(v)G^*(v)dv = \int_{-0.5}^{+0.5} |G(v)|^2 dv$$

Thus, we have concluded that:

$$E_d = \sum_{n=-\infty}^{\infty} |g[n]|^2 = \int_{-0.5}^{+0.5} |G(v)|^2 dv \qquad (7.54)$$

Equation (7.54) is Parseval's theorem for discrete functions.

Now we turn our attention to the most interesting bit. We want to answer the question: if $g[n]$ has been sampled from $g(t)$, and the condition $f_S > 2f_0$ has been satisfied, then what is the relationship between the energy of $g[n]$ and the energy of $g(t)$? As a recap, according to Parseval's theorem in the continuous time domain (see Equation (2.162)):

$$E_c = \int_{-\infty}^{+\infty} |g(t)|^2 dt = \int_{-\infty}^{+\infty} |G_c(f)|^2 df \qquad (7.55)$$

where E_c is the energy of $g(t)$. So, we want to find a relationship between E_c and E_d. Mind you, this is not an idle curiosity: often we want to know the energy E_c of a signal of interest, but we only have its samples $g[n]$, which means that in practice we want to know E_c but can only compute E_d.

Suppose you write a routine to do the calculation $\sum_{n=-\infty}^{+\infty} |g[n]|^2$. So, you have E_d, but you want E_c. Looking at the expressions in the time domain, it seems to be impossible to work out E_c from E_d. But, if we look at the expressions in the frequency domain, then we can quite easily connect them, because we know the relationship between the DTFT $G(v)$ and the Fourier transform $G_c(f)$: the central period of $G(v)$ is a scaled version of $G_c(f)$, with an amplitude factor of f_S. So, we can express the integral $\int_{-\infty}^{+\infty} |G_c(f)|^2 df$ in terms of the DTFT $G(v)$, and thus connect E_d with E_c. Let us do it.

First, recall that f_0 is the highest frequency of $G_c(f)$, which means that $G_c(f) = 0$ for frequencies above f_0. So, the integral from $-\infty$ to ∞ reduces to an integral from $-f_0$ to f_0. It is more convenient, however, to set the limits of the integral between $-f_S/2$ and $f_S/2$ (recall that we are assuming $f_S > 2f_0$, which entails that $f_S/2 > f_0$). Thus:

$$E_c = \int_{-\infty}^{+\infty} |G_c(f)|^2 df = \int_{-\frac{f_S}{2}}^{+\frac{f_S}{2}} |G_c(f)|^2 df \qquad (7.56)$$

It is more convenient to express Equation (7.56) in terms of $G_d(f)$. Recall that $G_d(f)$ is formed of copies of $G_c(f)$, with a factor of f_S, that is:

Equation (7.16)

$$G_d(f) = f_S \sum_{k=-\infty}^{\infty} G_c(f - kf_S)$$

Since the limits of the integral in Equation (7.56) covers only the central period $(k = 0)$ of $G_d(f)$, we can make the substitution:

$$E_c = \int_{-\frac{f_S}{2}}^{+\frac{f_S}{2}} |G_c(f)|^2 df = \int_{-\frac{f_S}{2}}^{+\frac{f_S}{2}} \frac{|G_d(f)|^2}{f_S^2} df$$

Now we rewrite the last integral in terms of $v = f/f_S$. Thus:

$$E_c = \int_{-\frac{f_S}{2}}^{+\frac{f_S}{2}} |G_c(f)|^2 df = \int_{-\frac{1}{2}}^{+\frac{1}{2}} \frac{|G_d(vf_S)|^2}{f_S^2} f_S dv = \int_{-\frac{1}{2}}^{+\frac{1}{2}} \frac{|G_d(vf_S)|^2}{f_S} dv \qquad (7.57)$$

According to Equation (7.25), $G(v) = G_d(vf_S)$. Therefore, we conclude that:

$$E_c = \frac{1}{f_S} \int_{-\frac{1}{2}}^{+\frac{1}{2}} |G(v)|^2 dv \qquad (7.58)$$

But the integral in Equation (7.58) is the discrete energy (see Equation (7.54)). Therefore, we conclude that:

$$E_c = \frac{E_d}{f_S} \qquad (7.59)$$

Thus, if you want to know the energy of the continuous time signal, you can work out the energy of the discrete signal and divide it by the sampling frequency. So, Equation (7.59) is a quite useful result.

But it is not only that: it is an astonishing result! For, say you know nothing about signals and systems, and nothing about Equation (7.59), and you are given the task of evaluating E_c. What do you do? Well, the obvious: you compute the time integral in Equation (7.55) using a computer. In other words, you turn it into the Riemann sum:

$$\int_{-\infty}^{+\infty} |g(t)|^2 dt \longrightarrow \sum_{n=-\infty}^{\infty} |g[n]|^2 \Delta t \qquad (7.60)$$

But, as a good student of calculus, you know that the Riemann sum only converges to the integral in the limit $\Delta t \to 0$. So, according to calculus, $\sum_{n=-\infty}^{\infty} |g[n]|^2 \Delta t$ is only APPROXIMATELY equal $\int_{-\infty}^{+\infty} |g(t)|^2 dt$, but to know $\int_{-\infty}^{+\infty} |g(t)|^2 dt$ EXACTLY is impossible, because we would need $\Delta t \to 0$.

But that is not true. According to Equation (7.59), the Riemann sum $\sum_{n=-\infty}^{\infty} |g[n]|^2 \Delta t$ is EXACTLY equal to the integral $\int_{-\infty}^{+\infty} |g(t)|^2 dt$ if $f_S > 2f_0$. To see it, recall that Δt is the time step, that is, $\Delta t = T_S$, so the Riemann sum is $\sum_{n=-\infty}^{\infty} |g[n]|^2 \Delta t = \sum_{n=-\infty}^{\infty} |g[n]|^2 T_S$. But, according to Equation (7.59):

$$\int_{-\infty}^{+\infty} |g(t)|^2 dt = \frac{\sum_{n=-\infty}^{\infty} |g[n]|^2}{f_S}$$

where I used Equation (7.49) and Equation (7.55). Since $1/f_S = T_S$, we have:

$$\int_{-\infty}^{+\infty} |g(t)|^2 dt = \sum_{n=-\infty}^{\infty} |g[n]|^2 T_S \tag{7.61}$$

That is madness: Equation (7.59), recast as Equation (7.61), is affirming that the Riemann sum is EXACTLY equal to the integral for finite T_S. Contrary to one of the most sacred notions of calculus, we have concluded that T_S does not have to be infinitesimal for the sum to converge to the integral: it only has to be sufficiently small for the condition $f_S > 2f_0$ to be satisfied.

7.6 Concluding Remarks

We have covered the most important features of the theory of representation of discrete signals in the frequency domain. But maybe you noticed a loophole in our argument. The motivation for this theory is that we want to analyze signals and systems using a computer, so the signals must be discretized. That led to the development of the Fourier representation of discrete signals. But, surely, we also want to able to use computers to perform operations in the frequency domain. Alas, we cannot do that with the DTFT, because the DTFT is a continuous function (v is a continuous variable). Thus, to be able to do operations in the frequency domain using computers, we need to discretize the DTFT itself, so that both 'time' and 'frequency' domains are discretized. This is where the DFT comes in, which is the subject of Chapter 8.

7.7 Exercises

Exercise 1
Suppose we want to evaluate the Fourier transform of $g(t)$ using a computer, and we proceed by converting the integral into the Riemann sum:

$$\int_{-\infty}^{+\infty} g(t) e^{-i2\pi ft} \, dt \Rightarrow \sum_{n=-\infty}^{n=+\infty} g[n] e^{-i2\pi fnT_S} \, T_S$$

According to the fundamental theorem of calculus, the Riemann sum is only exactly equal to the integral when $T_S \rightarrow 0$. So, is it possible to evaluate exactly the Fourier transform of $g(t)$ using its samples $g[n]$, or only approximate results can be obtained?

Exercise 2
Show that the DTFT pair is mathematically identical to the Fourier series pair (Equation (2.176) and Equation (2.173)) with the condition $T = 1$, and with the roles of time and frequency reversed (more precisely, with t replaced with $-v$ and k replaced with n).

Exercise 3

Suppose that the signal $g(t) = A \cdot e^{-\left(\frac{t}{\sigma}\right)^2}$, whose Fourier transform is $G_c(f) = A \cdot \sigma\sqrt{\pi}e^{-(\pi\sigma f)^2}$, is sampled at the sampling frequency f_S. Write an explicit expression for the DTFT of the sampled signal. Then compare your result with the DTFT calculated numerically (as in the Worked Exercise: Example of a Matlab Routine to Calculate the DTFT).

   Hint: the explicit expression can be found using Equation (7.17) and Equation (7.25).

Exercise 4

Suppose that the Fourier transform of a certain signal is $G_c(f) = cos(\pi \cdot 0.1 \cdot f)$ for $-5 < f < 5$, and $G_c(f) = 0$ for f outside the range $-5 < f < 5$. The signal was sampled at intervals $T_S = 0.05$ s.

(a) Sketch the Fourier transform $G_c(f)$, the Fourier transform of the intermediary function $G_d(f)$ and the DTFT of the sampled signal. Has the condition $f_S > 2f_0$ been satisfied?

(b) Find the value of the DTFT evaluated at the dimensionless frequencies $v_1 = 0.2$, $v_2 = 0.6$, $v_3 = 0.9$ and $v_4 = 3.2$.

Exercise 5

Write two Matlab routines to calculate the DTFT of the Gaussian signal defined in Worked Exercise: Example of a Matlab Routine to Calculate the DTFT. In each routine, choose a different T_S (using the routine of the worked exercise, to change T_S you need to change either the number of points, or the limit ti). Compare the DTFTs you obtained with the two different T_S and comment on their differences.

Exercise 6

Suppose that a rectangular function (see Worked Exercise: The Fourier Transform of the Rectangular Function, of Chapter 2) has been sampled, thus generating the following discrete signal $w[n]$:

$$w[n] = 1 \quad for - n_0 \le n \le n_0$$

$$w[n] = 0 \quad for \ n \ outside \ the \ range - n_0 \le n \le n_0$$

(a) Prove that the DTFT of $w[n]$ is given by:

$$W(v) = \frac{sin[\pi v(1 + 2n_0)]}{sin(\pi v)}$$

(b) Show that, for small v and $n_0 \gg 1$, then $W(v) \approx 2n_0 sinc(2\pi v n_0)$.

(c) Consider a continuous time rectangular function, as given by Equation (2.7) with $c = n_0 + 1/2$. Suppose this function has been sampled with $T_S = 1$ (I chose $c = n_0 + 1/2$ to avoid a delta function overlapping with the discontinuity of the rectangular function). Find an equation for $W(v)$ in terms of the Fourier transform of the rectangular function (this Fourier transform is given in Equation (2.86)).

(d) Now write a Matlab code to calculate and plot the DTFT of $w[n]$. Compare your result with the $W(v)$ deduced in part (a), with the approximated function of part (b), and with the $W(v)$ obtained from the Fourier transform in part (c). Check how the curves compare for different values of n_0 and explain their behaviour.

Exercise 7
Starting from the direct DTFT, prove that $G(v) = G(v+1)$ for all v. This is a proof that $G(v)$ is periodic with period 1.

Exercise 8
Find the symmetries of the DTFT of a real signal $g[n]$.
   Hint: follow a logic analogous to the proof of Equation (2.112).

Exercise 9
If $g_2[n] = e^{i2\pi v_0 n} g[n]$, find the DTFT of $g_2[n]$ in terms of the DTFT of $g[n]$. This is the property of spectral shifting.

Exercise 10
In Section 7.5.2 (Difference), we learned that the property of difference approximates the property of differentiation in the limit of $T_S \to 0$. We have seen that the approximation connecting difference and differentiation is $1 - e^{-i2\pi v} \approx i2\pi v$. To get an idea of the maximum error involved in discrete time differentiation, evaluate the error defined as:

$$error = 100\% \cdot \frac{|i2\pi v_{max} - (1 - e^{-i2\pi v_{max}})|}{2\pi v_{max}}$$

where v_{max} is the maximum v_{max} for which $G(v) \neq 0$, at the three following conditions: $f_S = 2f_0$, $f_S = 5f_0$, and $f_S = 100f_0$.

Exercise 11
Work out $g[n] = a[n] * b[n]$, where $a[-1] = 1$, $a[0] = 2$, $a[1] = 3$, and $a[n] = 0$ otherwise; $b[-1] = 4$, $b[0] = 0$, $b[1] = -1$, and $b[n] = 0$ otherwise.

Exercise 12
Prove Equation (7.50) using the DTFT pair.
   Hint: define the function $S(v) = \sum_{n=-\infty}^{\infty} e^{i2\pi v n}$. This function is periodic in v with unit period: $S(v+1) = \sum_{n=-\infty}^{\infty} e^{i2\pi(v+1)n} = \sum_{n=-\infty}^{\infty} e^{i2\pi v n} e^{i2\pi n} = \sum_{n=-\infty}^{\infty} e^{i2\pi v n} = S(v)$. Thus, to prove that $S(v) = \sum_{k=-\infty}^{\infty} \delta(v-k)$, all we need to do is to prove that $S(v) = \delta(v)$ for $-0.5 < v < 0.5$. To prove the latter expression, we need to evaluate the integral $\int_{-0.5}^{0.5} X(v)S(v-v')dv$. If the result is $X(v')$, then we will have proved that $S(v) = \delta(v)$ for $-0.5 < v < 0.5$ and, consequently, that $S(v) = \sum_{k=-\infty}^{\infty} \delta(v-k)$.

Exercise 13

Prove that the result of the circular convolution of the function:

$$W(v) = \frac{sin[\pi v(1 + 2n_0)]}{sin(\pi v)}$$

with itself is itself.

Hint: see Exercise 6 of Chapter 7.

Exercise 14

Without touching pencil and paper, let alone a computer, evaluate the integral:

$$\int_{-0.5}^{+0.5} \left| \frac{sin[\pi v(1 + 2n_0)]}{sin(\pi v)} \right|^2 dv$$

8

The Discrete Fourier Transform (DFT)

Learning Objectives

In Chapter 7, we found a 'frequency' domain representation of discrete signals through the DTFT. The DTFT, however, is itself a continuous function, and as such cannot be implemented in a computer. Thus, to be able to perform operations in the frequency domain using a computer, we need to discretize the DTFT. This discretization leads to the Discrete Fourier Transform (DFT). In Chapter 7, we learned that the consequence of discretizing the time domain is that the frequency domain becomes periodic. In this chapter, we will learn that the consequence of discretizing the frequency domain is that the time domain becomes periodic. This is the key difference between the DFT and the DTFT, and it is crucial to keep it in mind when implementing the former. In this chapter, we will learn how this feature of periodicity in both domains affects the implementation of the DFT and how it connects to the Fourier transform. Importantly, we will learn how to use the function *fft*, which is a fast implementation of the DFT. With the content covered in this chapter, we will be able to perform operations in both time and frequency domains using a computer.

8.1 Discretizing the Frequency Domain

We have seen that a theory of discrete signals and systems is necessary to enable computational analysis. This necessity motivated the definition of a Fourier transform of discrete signals, which led to the DTFT. This transformation, however, goes only half-way, because the DTFT is itself a continuous function (v is a continuous variable). Consequently, we cannot use a computer to get the inverse DTFT, because the inverse DTFT is an integral, and not a sum. Thus, to be able to perform

Essentials of Signals and Systems, First Edition. Emiliano R. Martins.
© 2023 John Wiley & Sons Ltd. Published 2023 by John Wiley & Sons Ltd.
Companion website: www.wiley.com/go/martins/essentialsofsignalsandsystems

operations in the frequency domain, we need to discretize the DTFT itself, and turn the inverse DTFT into a sum.

In Chapter 7, we approached the problem of discretizing the time domain by investigating what happened to the Fourier transform when it was turned into a sum. We learned that the sum leads to a periodic function, in which each period is a copy of the Fourier transform. So, we concluded that the effect of discretizing the time domain is to make the frequency domain periodic. We can follow the same approach here and ask the question: what happens to the inverse DTFT if we replace the integral with a sum? The answer is that, unsurprisingly, the 'time' domain becomes periodic. Let us check if that is indeed true.

An intuitive way to understand what happens when we discretize the DTFT, turning the inverse DTFT into a sum, is to recall that the DTFT can be understood as a particular case of the Fourier transform: it is the Fourier transform of a signal sampled with $T_S = 1$. We also assume that the sampling theorem has been satisfied, which entails that $f_S > 2f_0$. Since we are assuming $f_S = 1$, we also need to assume that $f_0 < 0.5$. Thus, we focus attention on the Fourier transform of a signal whose highest frequency is less than 0.5.

Let us spell this out by supposing that we have a continuous time signal $g(t)$, whose highest frequency is less than 0.5. Recall that, by assuming a highest frequency less than 0.5, we are assuming that $G_c(f) = 0$ for frequencies above 0.5 (see Figure 7.4). Thus, the inverse Fourier transform integral can be limited to the range ± 0.5:

$$g(t) = \int_{-0.5}^{+0.5} G_c(f)e^{i2\pi ft}df \tag{8.1}$$

Suppose further that we are interested in evaluating $g(t)$ at integer times (like 5 s, 6 s, -4 s, for example). Thus, we want to evaluate $g(t)$ at times $t = n$, again taking n to be an integer number (the only difference between n of Chapter 7 and n here is that by equating $t = n$ we are giving it units of time, while in Chapter 7 it had no units). In this case, Equation (8.1) is turned into:

$$g(n) = \int_{-0.5}^{+0.5} G_c(f)e^{i2\pi fn}df \tag{8.2}$$

Comparing the inverse DTFT integral (see Equation (7.30), which I repeat below for convenience) with the integral in Equation (8.2), we notice that they have the same form (recall that both v in Equation (7.30) and f in Equation (8.2) are dummy indexes, so it is immaterial that one equation uses v and the other f).

Equation (7.30)

$$g[n] = \int_{-0.5}^{+0.5} G(v)e^{i2\pi vn}dv$$

These two equations have the same form because, by choosing to evaluate $g(t)$ at times $t = n$, we are essentially sampling it with $T_S = 1$, which makes $v = f$ (see Equation (7.24)).

If the integral in Equation (7.30) is mathematically identical to the integral in Equation (8.1) when the latter is evaluated at integer times, then whatever applies to the integral in Equation (8.1) must also apply to the integral in Equation (7.30). Thus, instead of investigating what happens if we discretize Equation (7.30), we can check what happens if we discretize Equation (8.1), because the conclusion necessarily also applies to Equation (7.30).

Maybe you are asking yourself what the advantage is of investigating the discretization of Equation (8.1) instead of Equation (7.30), since the latter is our final goal. The advantage is that, as it turns out, we have already done this job in Section 2.13 (The Fourier Series). Indeed, we already know what happens if we discretize Equation (8.1): we will be turning the inverse Fourier transform into a Fourier series, and thus $g(t)$ will be turned into a periodic function. As we learned all that in Section 2.13 (The Fourier Series), we just need to recap the main ideas, and then apply them to the inverse DTFT.

In Section 2.13 (The Fourier Series), we learned that the sum in Equation (2.173), which I repeat below for convenience (using $f_T = 1/T$), results in a periodic function $g_p(t)$, whose period is T.

Equation (2.173) (with $f_T = 1/T$)

$$g_P(t) = \sum_{k=-\infty}^{\infty} G[k]e^{i2\pi \frac{k}{T}t}$$

Furthermore, according to Equation (2.175) (repeated below), $G[k]$ are samples of the Fourier transform $G_c(f)$ (allowing for the amplitude correction $1/T$), and each sample is spaced by $1/T$ in the frequency domain – that is, the 'discretization step' in the frequency domain is $1/T$ (for example, $G[1]$ is the sample 'located' at $f = 1/T$, and $G[2]$ is the sample 'located' at $f = 2/T$, so the 'distance' between them is $\Delta f = 2/T - 1/T = 1/T$). Also, recall that $G_c(f)$ is the Fourier transform of the central period of $g_P(t)$, which we denoted by $g(t)$.

Equation (2.175) (with $f_T = 1/T$)

$$G[k] = \frac{1}{T}G_c\left(\frac{k}{T}\right)$$

Equation (2.173) is already teaching what happens when we discretize the Fourier transform: instead of getting the nonperiodic signal $g(t)$, which we would have gotten if we had done the integral, when we do the sum we get a periodic signal $g_P(t)$ formed by copies of $g(t)$ (see Figure 2.23). Since Equation (8.1) is just a particular case of the inverse Fourier transform, Equation (2.173) is also teaching what happens if we turn the integral in Equation (8.1) into a sum: we get a signal $g_P(t)$, which is formed by copies of $g(t)$.

Let us be a bit more specific. Suppose we want to convert the integral in Equation (8.1) into a sum involving N samples of $G(f)$ (also, suppose that N is an even integer number). What will be the 'space' between the samples – that is, the 'length' of the frequency discretization step? Well, the integral ranges from

$f = -0.5$ to $f = +0.5$, so we need to cover a 'distance' of $0.5 - (-0.5) = 1$. Thus, we need a step of $[0.5 - (-0.5)]/N = 1/N$ to cover this 'distance' with N samples. According to Equation (2.173) and Equation (2.175), if the frequency domain step is $1/T$, then the period in the time domain is T. Therefore, a step of $1/N$ entails a period N. Thus, N plays a double role: it is the number of samples in the frequency domain, and it is also the period in the time domain.

Rewriting Equation (2.173) with the identification $T = N$, we get:

$$g_P(t) = \sum_{k=-\infty}^{\infty} G[k] e^{i2\pi \frac{k}{N} t}$$

Since we only need to cover the 'distance' from $f = -0.5$ to $f = +0.5$, then we do not need the sum to go from $-\infty$ to ∞, but only from $-N/2$ to $N/2 - 1$ (for example, if $N = 10$, then k must range from -5 to 4, and we do not need the sample $k = 5$ because it already belongs to the next period of the DTFT: it is identical to the sample $k = -5$). Adapting the limits of the sum to the range of interest, we get:

$$g_P(t) = \sum_{k=-\frac{N}{2}}^{\frac{N}{2}-1} G[k] e^{i2\pi \frac{k}{N} t}$$

Finally, using Equation (2.175), we get:

$$g_P(t) = \frac{1}{N} \sum_{k=-\frac{N}{2}}^{\frac{N}{2}-1} G_c\left(\frac{k}{N}\right) e^{i2\pi \frac{k}{N} t} \tag{8.3}$$

The relationship between $g(t)$ in Equation (8.1) and $g_P(t)$ in Equation (8.3) is illustrated in Figure 8.1.

Now that we know what happens to $g(t)$ when we discretize $G_c(f)$ in Equation (8.1), we can easily answer the question of what happens if we discretize the inverse DTFT (Equation (7.30)). Recall that the inverse DTFT is mathematically identical to Equation (8.1) when the latter is evaluated at integer multiples of time. Look at Figure 8.1d: the same integer values of $g(t)$ are also there in the central period of $g_p(t)$. So, if we evaluate the sum at an instant of time that belongs to the central period of $g_p(t)$, then we get the same result we get with the integral. In other words:

If n is in the central period of $g_p(t)$, then:

$$g[n] = \int_{-0.5}^{+0.5} G(v) e^{i2\pi v n} dv = \frac{1}{N} \sum_{k=-\frac{N}{2}}^{\frac{N}{2}-1} G\left(\frac{k}{N}\right) e^{i2\pi \frac{k}{N} n} \tag{8.4}$$

$$g(t) = \int_{-0.5}^{+0.5} G_c(f)e^{i2\pi ft}\,df = \int_{-0.5}^{+0.5} G(v)e^{i2\pi vt}\,dv$$

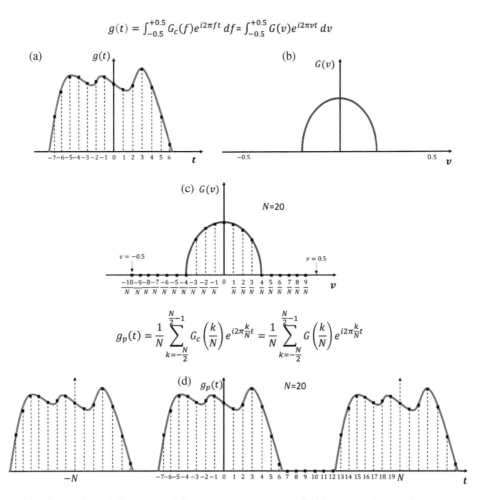

$$g_p(t) = \frac{1}{N}\sum_{k=-\frac{N}{2}}^{\frac{N}{2}-1} G_c\left(\frac{k}{N}\right)e^{i2\pi\frac{k}{N}t} = \frac{1}{N}\sum_{k=-\frac{N}{2}}^{\frac{N}{2}-1} G\left(\frac{k}{N}\right)e^{i2\pi\frac{k}{N}t}$$

Figure 8.1 Illustration of the effect of discretizing the DTFT. (a) An example signal. (b) Schematical illustration of the Fourier transform of an example signal (this is not an actual Fourier transform: it is only an illustration). (c) Same as (b) but showing the 'position' of the N samples of the Fourier transform $G_c(f)$. The example assumes $N = 20$. (d) According to the Fourier series pair, if the integral is turned into a sum, the result is a periodic version of the original signal.

But:

If n is NOT in the central period of $g_p(t)$, then:

$$g[n] = \int_{-0.5}^{+0.5} G(v)e^{i2\pi vn}\,dv \neq \frac{1}{N}\sum_{k=-\frac{N}{2}}^{\frac{N}{2}-1} G\left(\frac{k}{N}\right)e^{i2\pi\frac{k}{N}n} \qquad (8.5)$$

The integral in Equation (8.5) is different from the sum because, whereas the integral gives $g(n)$, the sum gives $g_p(n)$, and outside the central period these functions are not the same. For instance, in the example of Figure 8.1, we assumed

$N = 20$, which entails that the central period is in the range $-10 \le n \le 9$. Therefore, $g(n) = g_p(n)$ when $-10 \le n \le 9$, but $g(n) \ne g_p(n)$ when n lies outside the
range $-10 \le n \le 9$. For example, $g(20) = 0$, while $g_p(20) = g_p(0) = g(0) \ne g(20)$.
But, inside the central period, the functions are identical, and so Equation (8.4)
is true.

According to Equation (8.4), it is possible to recover $g[n]$ from samples of the
DTFT, but we need to be careful because the sum returns a periodic version of
$g[n]$, and not $g[n]$ itself. So, now we know that we can discretize the DTFT, and
thus get a pair of transforms involving only discrete functions. All we need to do is
to define **the discrete function $G[k]$, which contains samples of the DTFT**
$G(v)$. Thus, defining:

$$G[k] = G\left(\frac{k}{N}\right) \tag{8.6}$$

where $G\left(\dfrac{k}{N}\right)$ is the DTFT $G(v)$ evaluated at the 'frequency' $v = k/N$. Since:

Equation (7.23)

$$G(v) = \sum_{n=-\infty}^{\infty} g[n]e^{-i2\pi vn}$$

Then it follows from Equation (8.6) that:

$$G[k] = \sum_{n=-\infty}^{\infty} g[n]e^{-i2\pi\frac{k}{N}n} \tag{8.7}$$

Equation (8.7) and Equation (8.4) (combined with Equation (8.6)) are a pair
of transforms involving only discrete functions. So, we kind of already have what
we set out to obtain in this chapter. However, to get a pair of equations that
are more like the DFT, we need to make a 'practical' consideration. Recall that,
though $g[n]$ is a function where n is allowed to range from $-\infty$ to $+\infty$, in practice
we cannot have an infinite range if we want to store the information of $g[n]$ in a
computer. Thus, we need to make a distinction between the mathematical object
$g[n]$, which is limitless in the sense that n is allowed to range from $-\infty$ to $+\infty$, and
the computational object storing $g[n]$, which is a finite vector, with a finite number
of points. So, suppose that we choose to store $g[n]$ in a vector with N_{pt} points. We
can, in principle, index the vector in many ways; but, as we will see in Section 8.2, it
is convenient to index it ranging from $-N_{pt}/2$ to $N_{pt}/2 - 1$ (notice that this is the
way I indexed it in the Worked Exercise: Example of a Matlab Routine to Calculate
the DTFT of Chapter 7). With this choice, the sum in Equation (8.7) is reduced to:

$$G[k] = \sum_{n=-\frac{N_{pt}}{2}}^{\frac{N_{pt}}{2}-1} g[n]e^{-i2\pi\frac{k}{N}n} \tag{8.8}$$

Now let us rewrite Equation (8.4) using the definition of Equation (8.6):
If n is in the central period of $g_p(t)$, then:

$$g[n] = \frac{1}{N} \sum_{k=-\frac{N}{2}}^{\frac{N}{2}-1} G[k]e^{i2\pi\frac{k}{N}n} \tag{8.9}$$

Equation (8.8) and Equation (8.9) form a Fourier transform pair involving only discrete functions, which is what we wanted to get in this section.

Let us think about this pair a bit. Recall that N is the number of points we used to discretize the central period of the DTFT (see Figure 8.1c). As such, N is the number of points of the vector $G[k]$. The number of points of the vector $g[n]$, on the other hand, is N_{pt}. Even though there is nothing wrong with Equation (8.8) and Equation (8.9), the fact that they involve vectors with different sizes is quite inconvenient from the computational point of view. So, instead of having these two parameters independent of each other, we choose, by convention and by convenience, to sample the DTFT with the same number of points of the vector $g[n]$. In other words, we impose that $N = N_{pt}$. I emphasize that this is a choice, not a mathematical requirement. But, besides convenience, this choice brings a material advantage. Recall that N_{pt} is the number of points of the vector which is storing the 'original' signal $g[n]$, and that Equation (8.9) turned $g[n]$ (which was originally nonperiodic) into a periodic function with period N. Thus, by imposing $N = N_{pt}$, we are imposing that the period of $g[n]$ has the same 'size' of the computational window. Consequently, the periods (other than the central period) of the sum in Equation (8.9) lie outside the computational window, so we do not even see them when we plot the vectors. That is not to say that these periods do not play a role: they are quite relevant, and we will later look at two major consequences of them: the circular time shift and the circular convolution. But, for now, I want you to notice that the choice of $N = N_{pt}$ ensures that only the central period lies within the computational window, as illustrated in Figure 8.2.

With the choice $N = N_{pt}$, Equation (8.8) and Equation (8.9) form a Discrete Fourier Transform pair, and as such deserve a box, as shown below (Equation (8.10)). Notice that the range covering the computational window, that is, the central period, is explicitly specified in the pair, as a reminder that the mathematical objects $g[n]$ and $G[k]$ are periodic, but that we are computing only the central periods (recall that $G[k]$ is periodic because $G[k]$ are samples of the DTFT, which is periodic; also recall that $g[n]$ is periodic because we discretized the inverse DTFT). Furthermore, I am calling this pair a CENTRALIZED DFT pair. I am using the term 'centralized' because this form involves samples of the central period, both in time and frequency domains. **Since the Fourier transform corresponds to the central period of the DTFT, the centralized DFT pair is the DFT we need to connect with the Fourier transform.** But this is not the form that is implemented in the algorithm *fft*. In Section 8.2, we will see the DFT implemented by the *fft*, learn how to convert it to the centralized DFT and, finally, how to obtain samples of the Fourier transform using the *fft*.

CENTRALIZED DFT PAIR

DIRECT CENTRALIZED DFT

$$G[k] = \sum_{n=-\frac{N}{2}}^{\frac{N}{2}-1} g[n]e^{-i2\pi\frac{k}{N}n} \; for -\frac{N}{2} \leq k \leq \frac{N}{2} - 1$$

INVERSE CENTRALIZED DFT

$$g[n] = \frac{1}{N} \sum_{k=-\frac{N}{2}}^{\frac{N}{2}-1} G[k]e^{i2\pi\frac{k}{N}n} \; for -\frac{N}{2} \leq n \leq \frac{N}{2} - 1 \qquad (8.10)$$

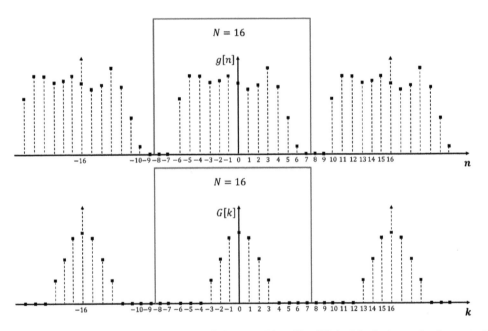

Figure 8.2 Illustration of the consequence of choosing $N = N_{pt}$. With this choice, only the central period of $g[n]$ lies inside the computational window. The illustration assumes $N = N_{pt} = 16$. The computational window ranges from -8 to $+7$. The red boxes highlight the computational windows under the assumption of $N = N_{pt} = 16$. The figure does not show an actual DFT pair: it is only an illustration.

8.2   The DFT and the Fast Fourier Transform (*fft*)

The most widely used algorithm to calculate the DFT is the fast Fourier transform (*fft*), precisely because it is fast. Mind you: here we are not interested in learning HOW the *fft* computes the DFT. Instead, we are interested in learning WHAT the *fft* computes, how to convert it to the centralized DFT, and then to the Fourier transform.

The *fft* computes a DFT, but not the centralized one (that is, not Equation (8.10)). I will first write the DFT computed by the *fft*, and then discuss how it is connected with the centralized DFT, and finally with the Fourier transform itself. Then we will see a couple of examples illustrating how to use the *fft*.

The DFT computed by the *fft* is shown in the box below:

DFT PAIR computed by the *fft*
DIRECT DFT

$$G[k] = \sum_{n=1}^{N} g[n] e^{-i2\pi \frac{(k-1)}{N}(n-1)} \text{ for } 1 \le k \le N$$

INVERSE DFT

$$g[n] = \frac{1}{N} \sum_{k=1}^{N} G[k] e^{i2\pi \frac{(k-1)}{N}(n-1)} \text{ for } 1 \le n \le N \qquad (8.11)$$

Compare Equation (8.11) and Equation (8.10). They both assume vectors with N points, and period N, but they use different parts of the periods to compute the direct DFT and the inverse DFT: while the centralized DFT (Equation (8.10)) takes samples across the central period (that is, n and k range from $-N/2$ to $N/2 - 1$), the DFT computed by the *fft* (Equation (8.11)) takes half of the samples from the central period, and the other half from the next period (that is, n and k range from 1 to N). This difference is illustrated in Figure 8.3 and Figure 8.4 (both figures assume the same example of Figure 8.2). Figure 8.3 shows the samples used in the centralized DFT, highlighted in red colour. This is the same computational window illustrated in Figure 8.2. The DFT calculated by the *fft*, however, takes a different set of samples, as illustrated in Figure 8.4.

Recall that our final goal is to obtain the Fourier transform. As we learned in Chapter 7, the Fourier transform corresponds to the central period of the DTFT. Since the DFT is the sampled DTFT, the central period of the DTFT is also the central period of the DFT. So, to get the Fourier transform, we need the central period of the DFT, that is, we need the centralized DFT, as illustrated in Figure 8.3. Thus, we need to find a way of getting the centralized DFT from the DFT calculated by the *fft*.

Fortunately, there is a neat way of transforming the DFT of the *fft* into the centralized DFT, and vice versa: by using the command *fftshift*. This command takes a vector, cuts it in the middle, and then swaps the first and second halves, as illustrated in Figure 8.5. Because the signal is periodic (see Figure 8.4), its second half (highlighted by the green box in Figure 8.5) is identical to the left part of the

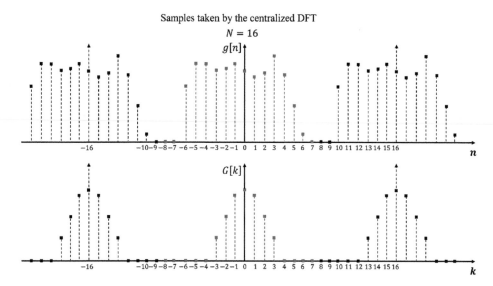

Figure 8.3 Illustration of the samples used in the centralized DFT (Equation (8.10)). The samples used in the centralized DFT are shown in red colour. The figure does not show an actual DFT pair: it is only an illustration.

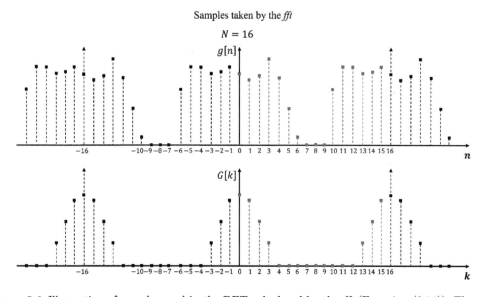

Figure 8.4 Illustration of samples used in the DFT calculated by the *fft* (Equation (8.11)). The samples used in the DFT calculated by the *fft* are shown in red colour. The figure does not show an actual DFT pair: it is only an illustration.

central period. Thus, the signal we get after the command *fftshift* is the central period, as shown in Figure 8.5c. Notice that we need to fix the indices n and k accordingly: after the *fftshift*, they need to range from $-N/2$ to $N/2 - 1$, which is the range of the central period.

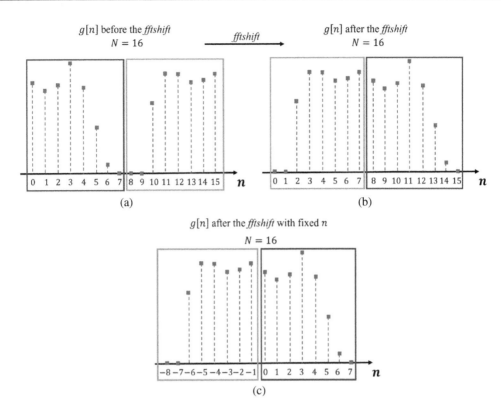

Figure 8.5 Illustration of the command *fftshift*. (a) An example signal with 16 samples. (b) The signal resulting from the *fftshift* consists of the original signal with the two halves swapped. (c) The resulting signal can be interpreted as being the central period – notice the new values of n.

In sum: the command *fftshift* transforms the DFT of the *fft* into the centralized DFT, and vice versa.

You may have noticed something funny: in all illustrations of the DFT calculated by the *fft* (Figure 8.4 and Figure 8.5a), I considered the first sample to be indexed by n (*or* k) $= 0$ instead of n (*or* k) $= 1$, even though the *fft* involves a sum from n (*or* k) $= 1$ to n (*or* k) $= N$ (see Equation (8.11)). Let me explain why this is the correct choice. The reason why the *fft* involves a sum beginning in n (*or* k) $= 1$ is that the first index of a vector in Matlab is 1 instead of 0. But notice that the arguments of the exponentials in Equation (8.11) involve $k - 1$ and $n - 1$. Consequently, the sample stored in the first slot of the vector corresponds to n (*or* k) $= 0$. For example, suppose we fix $k = 1$ in the direct DFT. In this case, we get:

$$G[1] = \sum_{n=1}^{N} g[n]e^{-i2\pi\frac{0}{N}(n-1)} = \text{for } 1 \leq k \leq N \tag{8.12}$$

So, what sample of the DTFT is being stored in $G[1]$? In other words, what is the value of v in the argument of the exponential? Well, it clearly is $v = 0$. Thus, $G[1]$ is the value of the DTFT sampled at $v = 0$. Likewise, $G[2]$ is the value of the DTFT sampled at $v = 1/N$, $G[3]$ is the value of the DTFT sampled at $v = 2/N$, and so on, until $G[N]$, which is the value of the DTFT sampled at $v = (N-1)/N = 1 - 1/N$. This is the reason why I took the first sample to lie at the origin. Furthermore, notice that the command *fftshift* takes the first sample (that is, $G[1]$), which is the sample for $v = 0$, and place it in the slot $N/2 + 1$. Thus, the centralized DFT must be defined in a such way that the origin lies at $N/2 + 1$. Accordingly, we indexed the sums of the centralized DFT (Equation (8.10)) to run from $-N/2$ to $N/2 - 1$. In this manner, the origin is at the position $N/2 + 1$. For example, if $N = 10$, then the index $k = 0$ is stored at the position $5 + 1 = 6$ (the vector k runs $-5 - 4 - 3 - 2 - 1\ 0\ 1\ 2\ 3\ 4$, so the 0 is in the sixth slot). An analogous reasoning holds for the inverse DFT.

We have seen that, with the help of the command *fftshift*, we can use the *fft* to obtain the centralized DFT, and from the centralized DFT we can get the Fourier transform. Recall that we need the centralized DFT because the Fourier transform corresponds to the central period of the DTFT.

Once we get a Fourier transform, we can then apply the theory of Chapter 3 (Representation of Systems) to analyze systems using a computer. We will do that in Chapter 9, but for now we focus attention on a few examples showing how to use the *fft* to obtain the centralized DFT, the Fourier transform, and its inverse.

Worked Exercise: Getting the Centralized DFT Using the Command *fft*

To exemplify the *fft*, we will use the same Gaussian function of the worked exercises of Chapter 7. Apart from the pedagogical benefit of following the same thread, this choice has the further advantage that we can check whether our results are correct because we know the analytical Fourier transform of the Gaussian function.

Thus, in this example, we will discretize the Gaussian function $g(t) = 5e^{-\left(\frac{t}{10}\right)^2}$ and then use the *fft* to obtain the DFT and the centralized DFT.

Solution:

The discretization of the Gaussian function follows the same procedure shown in the Worked Exercise: Example of a Matlab Routine to Calculate the DTFT, of Chapter 7. Thus:

```
clear all
close all
Npt = 2^10; %number of points
n = (-Npt/2:1:Npt/2-1); %definition of "centralized" vec-
tor n
```

```
ti = -300; %beginning of the time window
tf = -ti; %end of the time window

Ts = (tf-ti)/(Npt); %sampling period
fs = 1/Ts; %sampling frequency

t = n*Ts; % definition of time vector

A = 5;    %amplitude of the gaussian
sigma = 10; %width of the gaussian

g = A*exp(-(t./sigma).^2); %discretize the gaussian

%Plot centralized g
figure(1)
plot(n,g,'b','LineWidth',2)
xlabel('n')
ylabel('Centralized Gaussian')
```

But now we need an additional vector n to index the samples as in the DFT calculated by the *fft*. In the lines of code below, I call this additional vector n_fft. The vector resulting from the operation of the command *fft* is called g_fft. Compare the plot of the centralized Gaussian (Figure 8.6) and the plot of the Gaussian function to be used in the *fft* (Figure 8.7) and notice the difference in the indexes n.

```
g_fft = fftshift(g); %create vector for operation with fft
n_fft = (0:1:Npt-1); %definition of vector n accord-
ing to fft

%Plot g for fft
figure(2)
plot(n_fft,g_fft,'b','LineWidth',2)
xlabel('n')
ylabel('Gaussian for fft')
```

We also need vectors k and k_fft. Even though they are respectively identical to the vectors n and n_fft, I created the vectors k and k_fft explicitly for pedagogical reasons, so we can more easily differentiate between 'time' and 'frequency' domains. Notice in the lines of code below that the vector k_fft is associated (plotted) with

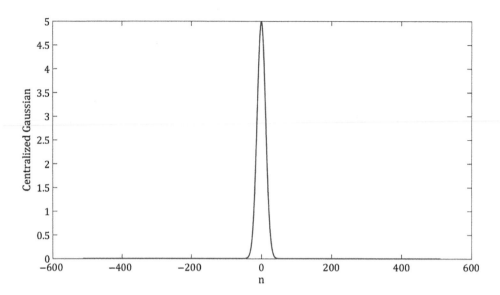

Figure 8.6 Plot of the centralized Gaussian, generated by the command above (figure 1)

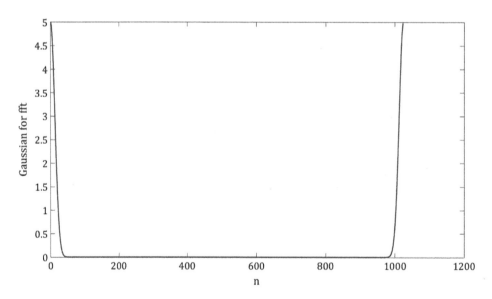

Figure 8.7 Plot of the Gaussian to be used in the *fft*, generated by the command above (figure 2).

the vector G_fft, which is the DFT of the vector g_fft. The vector k, however, is associated (plotted) with the vector G, which is the centralized DFT, obtained by applying the command *fftshift* to the vector G_fft.

```
k = n; %definition of "centralized" vector k

k_fft = n_fft; %definition of vector k according to fft

G_fft = fft(g_fft); %DFT of g_fft


%Plot DFT of g_fft
figure(3)
plot(k_fft,abs(G_fft),'r','LineWidth',2)
xlabel('k')
ylabel('magnitude of DFT')



G = fftshift(G_fft); %Centralized DFT

%Plot centralized G
figure(4)
plot(k,abs(G),'r','LineWidth',2)
xlabel('k')
ylabel('Centralized DFT')
```

Notice, once again, the difference in the indexes between the DFT generated by the *fft* (Figure 8.8) and the centralized DFT (Figure 8.9).

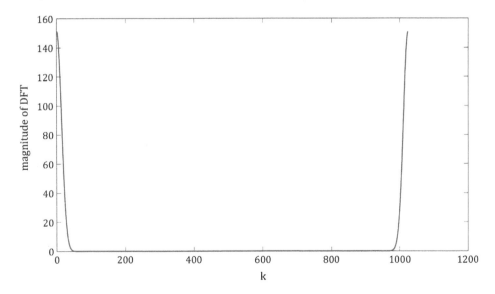

Figure 8.8 Plot of DFT calculated by the *fft*, generated by the command above (figure 3).

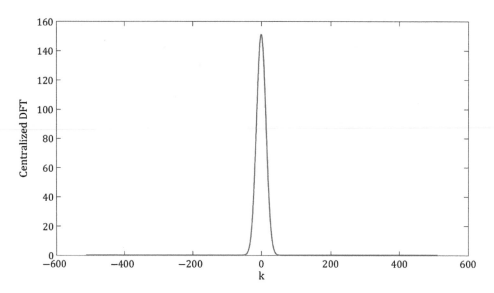

Figure 8.9 Plot of the centralized DFT, generated by the command above (figure 4).

Worked Exercise: Getting the Fourier Transform with the *fft*

Obtain the Fourier transform of the Gaussian function of the previous exercise using the *fft*.

Solution:

Now that we have learned how to get the centralized DFT using the *fft*, it is quite easy to obtain the Fourier transform. We know that the DFT are samples of the DTFT, and that the Fourier transform is a scaled version of the central period of the DTFT. So, all we need to do is to fix the amplitude of the centralized DFT – by dividing it by f_S or multiplying it by T_S – see Figure 7.4, and we get a vector with samples of the Fourier transform. But we need to be careful to keep track of what samples we have. In other words, we need a frequency vector that contains the values of the frequencies corresponding to the samples of the Fourier transform vector. For example, suppose G_FT is the vector storing the values of the Fourier transform, and f is the vector storing the values of the frequencies. If $G\_FT[k]$ is the value of the Fourier transform evaluated at the frequency f_k, then we need to generate f so that $f[k] = f_k$.

We have all the conceptual tools to generate the frequency vector f. The first step is to create the vector v covering the central period. We know that v is indexed by k as $v = k/N$ in the centralized DFT; thus, we need to get the vector k generated in the previous example and divide it by the number of points. Once we have v, we need to multiply it by the sampling frequency to get f (see Equation (7.24)). And that is it! The lines of code below show the implementation of the vectors G_FT and f (the lines of code below are a continuation of the code of the previous exercise: in particular, the centralized DFT vector G was generated in the previous exercise using the *fft* assisted by the *fftshift*). Notice that the Fourier transform obtained with the *fft* (as in the lines of code below) is the same Fourier transform we had got by direct computation of the DTFT, shown in Figure 7.6a. Importantly, notice that, since $f = v \cdot f_S$, and the highest value that the vector v takes is ± 0.5, that means the highest frequency sampled by the DFT is $f = \pm 0.5 \cdot f_S$.

```
G_FT = G*Ts; %obtain vector with samples of Fourier
transform

v = k./Npt; %definition of vector v

f = v*fs; %definition of vector f

%Plot Fourier Transform

figure(5)
plot(f,abs(G_FT),'k','LineWidth',2)
xlabel('f')
ylabel('Fourier Transform of g')
```

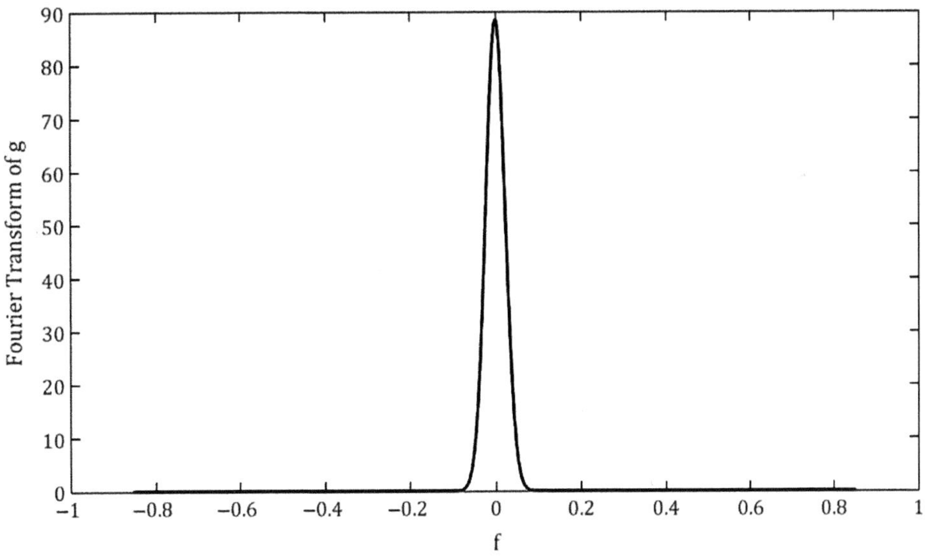

Figure 8.10 Plot of Fourier transform of the Gaussian function of the previous exercise, generated by the command above (figure 5). This is the same result of Figure 7.6a.

Worked Exercise: Obtaining the Inverse Fourier Transform Using the ifft

As a last example, let us start from the analytical Fourier transform of the Gaussian function, and then obtain the temporal Gaussian function by inverse Fourier transformation using the command *ifft*, which calculates the inverse DFT defined in Equation (8.11).

Solution:

The first step is to generate the vector f, from which we can discretize the analytical Fourier transform. But we have already generated this vector in the previous exercise, so we can discretize the analytical Fourier transform straight away:

```
%Analytical Fourier transform of gaussian function
G_FT_An = A*sigma*sqrt(pi)*exp(-(pi*sigma*f).^2);
```

Next, we need to turn this vector into the centralized DFT by fixing the amplitude:

```
  G_DFT_Centralized = G_FT_An.*fs;%Obtain central-
ized DFT from analytical FT
```

Now we turn the centralized DFT into the DFT to be used by the *ifft*.

```
  G_DFT = fftshift(G_DFT_Centralized); %Obtain DFT for ifft
```

Now the vector is ready to be used by the *ifft*:

```
  g_ifft = ifft(G_DFT); %Obtain inverse DFT
```

The vector g_ifft, however, is not centralized, so we still need one more step:

```
  g_centralized = fftshift(g_ifft); %Obtain the central-
ized inverse Fourier transform
```

The vector $g\_centralized$ contains samples of the gaussian $g(t) = 5e^{-\left(\frac{t}{10}\right)^2}$. The command below plots this vector against the time vector used to discretize $g(t)$ in the Worked Exercise: Getting the Centralized DFT Using the Command *fft*, of this chapter.

```
figure(6)
plot(t,g_centralized,'b','LineWidth',2)
xlabel('time')
ylabel('g(t)')
```

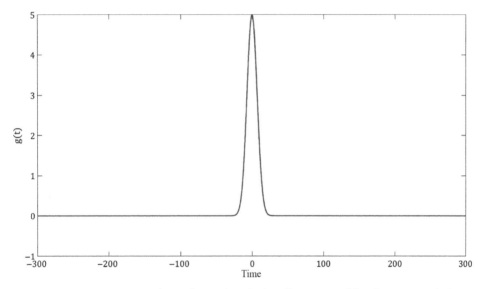

Figure 8.11 Inverse Fourier transform obtained with the *ifft*, generated by the command above (figure 6).

Now that we understand the DFT and its relation to the command *fft*, it is time to look at the DFT properties. Since we are ultimately interested in the Fourier transform, which is obtained from the centralized DFT, we will look at the properties of the centralized DFT instead of the DFT calculated by the *fft*. But recall that we can easily get the centralized DFT from the *fft* using the command *fftshift*.

Since the DFT is just the discretized DTFT, most properties of the former can be obtained from the properties of the latter by replacing v with k/N. However, there are two peculiarities related to the periodicity of the functions that deserve particular attention. These peculiarities are related to the properties of circular time shift and circular convolution, to which we turn our attention.

8.3 The Circular Time Shift

The property of circular time shift of the DFT is analogous to the property of temporal shift of the DTFT (Equation (7.32)). In Section 7.5.1 ('Time' Shifting), we deduced Equation (7.32) by evaluating the direct DTFT of $g_2[n] = g[n - n_0]$, which led to $G_2(v) = e^{-i2\pi v n_0} G(v)$.

This time, we will do it the other way around: we begin by assuming that the DFT of $g[n]$ is $G[k]$, and then we ask what the inverse DFT of $G_2[k] = e^{-i2\pi \frac{k}{N} n_0} G[k]$ is.

We proceed by applying the inverse DFT, as defined in Equation (8.10), to $G_2[k]$:

$$g_2[n] = \frac{1}{N} \sum_{k=-\frac{N}{2}}^{\frac{N}{2}-1} G_2[k] e^{i2\pi \frac{k}{N} n} = \frac{1}{N} \sum_{k=-\frac{N}{2}}^{\frac{N}{2}-1} e^{-i2\pi \frac{k}{N} n_0} G[k] e^{i2\pi \frac{k}{N} n} \qquad (8.13)$$

Rearranging:

$$g_2[n] = \frac{1}{N} \sum_{k=-\frac{N}{2}}^{\frac{N}{2}-1} G[k] e^{i2\pi \frac{k}{N} (n-n_0)} \qquad (8.14)$$

Notice that the sum in Equation (8.11) is the inverse DFT of $g[n]$, evaluated at the index $n - n_0$. Indeed:

$$g[n - n_0] = \frac{1}{N} \sum_{k=-\frac{N}{2}}^{\frac{N}{2}-1} G[k] e^{i2\pi \frac{k}{N} (n-n_0)} \qquad (8.15)$$

Thus, we conclude that:

if:

$$G_2[k] = e^{-i2\pi \frac{k}{N} n_0} G[k]$$

then:

$$g_2[n] = g[n - n_0] \qquad (8.16)$$

Equation (8.16) is the property of circular time shift. It is almost identical to the property of time shift of the DTFT (Equation (7.32)): in both cases, multiplying the frequency domain representation of a signal by an exponential with imaginary argument induces a 'time' shift. There is, however, an important difference: $g[n]$ in Equation (7.32) is not a periodic function, whereas $g[n]$ in Equation (8.16) is a periodic function. The fact that $g[n]$ in Equation (8.16) is a periodic function results in a funny feature, illustrated in Figure 8.12. Recall that the inverse DFT computes only the samples belonging to the computational window (highlighted in red in Figure 8.12). So, it can happen that some samples reach the end of the computational window, thus 'disappearing' from it. That is what is happening to samples $g[5]$ and $g[6]$ in Figure 8.12: when we shift $g[n]$ by three slots, thus obtaining $g[n-3]$, the samples $g[5]$ and $g[6]$ 'disappear', because they were shifted to the slots 8 and 9, which do not belong to the computational window of this example. However, it looks like these samples 'appeared' in the beginning of the computational window (in the slots -8 and -7 of $g[n-3]$), because the samples of the left period (originally at slots -11 and -10 of $g[n]$) 'invaded' the computational window. Hence the name 'circular' shift: it looks like we are rotating the samples around the computational window.

Unaware people seeing this happening are justified in concluding that a leprechaun is hiding inside the computer and playing with their mental sanity. You will be asked to check this property for yourself in the exercise list, hopefully with no such psychological hazard.

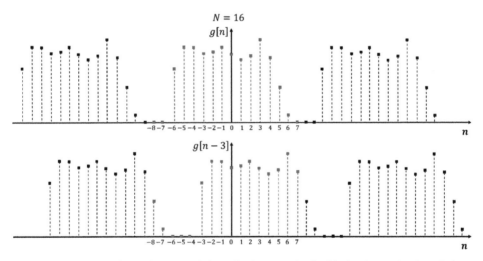

Figure 8.12 Illustration of circular time shift: only the samples highlighted in red colour belong to the computational window. When $g[n]$ is shifted by 3 units (resulting in $g[n-3]$), the samples $g[5]$ and $g[6]$ 'disappear' from the right-hand side of the computational window, and 'appear' on the left-hand side, giving the impression of a circular shift.

8.4 The Circular Convolution

The property of circular convolution is the counterpart of the DTFT property of convolution. Again, the significant difference between the DTFT and the DFT properties is that the signal in the time domain is not periodic in the DTFT, but it is periodic in the DFT.

Thus, on the one hand, when we take the product between two DTFTs, the resulting signal is the convolution in the time domain, and this convolution involves two nonperiodic signals (see Section 7.5.4 [Convolution in the 'time' domain]). On the other hand, as we will learn in this section, when we take the product between two DFTs, the resulting signal in the time domain is also a convolution, but now between two periodic signals.

As mentioned in Section 7.5.5 (Product in the Time Domain), it is customary to distinguish between the convolution involving nonperiodic signals and the convolution involving periodic signals using the terms 'linear' and 'circular'. Thus, the linear convolution is the convolution between nonperiodic signals, as defined in Equation (7.43). The circular convolution is the convolution between periodic signals. The relationship between linear and circular convolutions is a central feature of the theory of discrete systems, and we will discuss this relationship in Section 8.5. We begin, however, by obtaining the property of circular convolution.

To establish the property of circular convolution, we ask the question: what happens in the 'time' domain when we take the product between the DFTs of two signals? To answer this question, suppose that $a[n]$ and $b[n]$ are the inverse DFTs of $A[k]$ and $B[k]$. We want to evaluate the inverse DFT of the product $A[k]B[k]$ in terms of the signals $a[n]$ and $b[n]$. If $g[n]$ is the inverse DFT of $A[k]B[k]$, then:

$$g[n] = \frac{1}{N} \sum_{k=-\frac{N}{2}}^{\frac{N}{2}-1} A[k]B[k]e^{i2\pi \frac{k}{N}n} \tag{8.17}$$

The easiest way to obtain the convolution is to follow a procedure like in Section 7.5.4: express $A[k]$ in terms of the inverse DFT $\left(A[k] = \sum_{m=-\frac{N}{2}}^{\frac{N}{2}-1} a[m]e^{-i2\pi \frac{k}{N}m}\right)$, then combine the exponentials and identify that the sum over k is the inverse DFT of $B[k]$ evaluated at the point $n - m$, which as such returns $b[n - m]$. Here, however, I will show you a somewhat longer route to introduce a function that will be used later in Section 8.6, and to highlight the role of this function in the periodicity. If you wish, you can skip the derivation below and go straight to section 8.5.

Since our goal is to express $g[n]$ in terms of $a[n]$ and $b[n]$, it makes sense to express $A[k]$ and $B[k]$ in terms of $a[n]$ and $b[n]$ using the direct DFT. Thus:

$$g[n] = \frac{1}{N} \sum_{k=-\frac{N}{2}}^{\frac{N}{2}-1} \left[\sum_{m=-\frac{N}{2}}^{\frac{N}{2}-1} a[m]e^{-i2\pi \frac{k}{N}m} \right] \left[\sum_{q=-\frac{N}{2}}^{\frac{N}{2}-1} b[q]e^{-i2\pi \frac{k}{N}q} \right] e^{i2\pi \frac{k}{N}n} \tag{8.18}$$

Rearranging:

$$g[n] = \frac{1}{N} \sum_{m=-\frac{N}{2}}^{\frac{N}{2}-1} \sum_{q=-\frac{N}{2}}^{\frac{N}{2}-1} a[m]b[q] \sum_{k=-\frac{N}{2}}^{\frac{N}{2}-1} e^{i2\pi \frac{k}{N}(n-m-q)} \tag{8.19}$$

The sum over k does not depend on the signals $a[n]$ and $b[n]$, so we can evaluate it explicitly. To do it, consider the sum:

$$S = \sum_{k=-\frac{N}{2}}^{\frac{N}{2}-1} e^{i2\pi \frac{k}{N}p} \tag{8.20}$$

where p is an integer number. Notice that the sum depends on the value of p, so we can interpret it as function of p – this is the function we will also use in Section 8.6.

We can evaluate the sum in Equation (8.20), and then make the identification $p = n - m - q$ to obtain the sum over k in Equation (8.19).

To evaluate the sum in Equation (8.20), we begin by noticing that, if p is a multiple of N, that is, if $p = uN$, where u is an integer number, then:

$$for\ p = uN$$

$$S = \sum_{k=-\frac{N}{2}}^{\frac{N}{2}-1} e^{i2\pi \frac{k}{N}p} = \sum_{k=-\frac{N}{2}}^{\frac{N}{2}-1} e^{i2\pi \frac{k}{N}uN} = \sum_{k=-\frac{N}{2}}^{\frac{N}{2}-1} e^{i2\pi ku} = \sum_{k=-\frac{N}{2}}^{\frac{N}{2}-1} 1 = N \tag{8.21}$$

If you did not understand why $e^{i2\pi ku} = 1$, recall that ku is an integer number. If $p \neq uN$, we can evaluate the sum by shifting it by one unit:

$$e^{i2\pi \frac{1}{N}p} \cdot S = \sum_{k=-\frac{N}{2}}^{\frac{N}{2}-1} e^{i2\pi \frac{1}{N}p} e^{i2\pi \frac{k}{N}p} = \sum_{k=-\frac{N}{2}}^{\frac{N}{2}-1} e^{i2\pi \frac{(k+1)}{N}p} = \sum_{k=-\frac{N}{2}+1}^{\frac{N}{2}} e^{i2\pi \frac{k}{N}p} \tag{8.22}$$

And then subtract the unshifted sum from the shifted sum:

$$e^{i2\pi \frac{1}{N}p} \cdot S - S = \sum_{k=-\frac{N}{2}+1}^{\frac{N}{2}} e^{i2\pi \frac{k}{N}p} - \sum_{k=-\frac{N}{2}}^{\frac{N}{2}-1} e^{i2\pi \frac{k}{N}p} \tag{8.23}$$

Notice that only the terms $k = N/2$ in the first sum and $k = -N/2$ in the second sum survive the subtraction on the right-hand side of Equation (8.23). Thus:

$$for\ p \neq uN$$

$$\left[e^{i2\pi \frac{1}{N}p} - 1 \right] S = e^{i2\pi \frac{N}{2}\frac{1}{N}p} - e^{i2\pi \frac{-\frac{N}{2}}{N}p} = e^{i\pi p} - e^{-i\pi p} = 0 \tag{8.24}$$

From which we conclude that:

$$for\ p \neq uN$$

$$S = 0 \tag{8.25}$$

Combining Equation (8.25) with Equation (8.21), we conclude that S is a discrete periodic function of p, with period N, and which is zero everywhere expect at multiples of the period. This is an impulse train, but in the discrete domain, and multiplied by N.

To clean up the notation a bit, we define the impulse train in the discrete domain as the discrete function $\delta_{pN}[n]$:

$$\delta_{pN}[n] = 1 \quad for\, n = uN$$

$$\delta_{pN}[n] = 0 \quad for\, n \neq uN \tag{8.26}$$

Notice that the subscript 'pN' is a reminder that δ_{pN} is a periodic function with period N. Using the definition in Equation (8.26), with the results of Equation (8.25) and Equation (8.21), we conclude that:

$$S = \sum_{k=-\frac{N}{2}}^{\frac{N}{2}-1} e^{i2\pi\frac{k}{N}p} = N\delta_{pN}[p] \tag{8.27}$$

Substituting Equation (8.27) into Equation (8.19), we get:

$$g[n] = \frac{1}{N} \sum_{m=-\frac{N}{2}}^{\frac{N}{2}-1} \sum_{q=-\frac{N}{2}}^{\frac{N}{2}-1} a[m]b[q] \sum_{k=-\frac{N}{2}}^{\frac{N}{2}-1} e^{i2\pi\frac{k}{N}(n-m-q)}$$

$$= \frac{1}{N} \sum_{m=-\frac{N}{2}}^{\frac{N}{2}-1} \sum_{q=-\frac{N}{2}}^{\frac{N}{2}-1} a[m]b[q]N\delta_{pN}[n-m-q]$$

$$= \sum_{m=-\frac{N}{2}}^{\frac{N}{2}-1} \sum_{q=-\frac{N}{2}}^{\frac{N}{2}-1} a[m]b[q]\delta_{pN}[n-m-q] \tag{8.28}$$

The next step is to evaluate the sum over q. Before we do it, let us rearrange the argument of $\delta_{pN}[n-m-q]$ a little. First, notice that the definition of $\delta_{pN}[n]$ (Equation (8.26)) entails that $\delta_{pN}[n]$ is an even function (that is, $\delta_{pN}[n] = \delta_{pN}[-n]$). Thus, we can rewrite the argument of $\delta_{pN}[n-m-q]$ in Equation (8.28) as:

$$g[n] = \sum_{m=-\frac{N}{2}}^{\frac{N}{2}-1} \sum_{q=-\frac{N}{2}}^{\frac{N}{2}-1} a[m]b[q]\delta_{pN}[(n-m)-q] = \sum_{m=-\frac{N}{2}}^{\frac{N}{2}-1} \sum_{q=-\frac{N}{2}}^{\frac{N}{2}-1} a[m]b[q]\delta_{pN}[q-(n-m)] \tag{8.29}$$

According to Equation (8.26), $\delta_{pN}[q-(n-m)] = 1$ when $q = (n-m) + uN$ and $\delta_{pN}[q-(n-m)] = 0$ when $q \neq (n-m) + uN$.

Recall that we are evaluating $g[n]$ inside the computational window, which means that the smallest value of n is $-N/2$ and the largest value of n is $N/2-1$. Moreover, the sum over m in Equation (8.29) ranges from $-N/2$ to $N/2-1$. Consequently, $n-m$ ranges from $-N$ to $N-2$. Furthermore, since the sum over q ranges from $-N/2$ to $N/2 - 1$, for each combination $(n - m)$, there is only one value of u for which the condition $q = (n - m) + uN$ can be satisfied within this range of q. Indeed, if $-N \leq (n - m) \leq -N/2 - 1$, then the condition $q = (n - m) + uN$ can only be satisfied if $u = 1$. Moreover, if $-N/2 \leq (n - m) \leq N/2 - 1$, then the condition $q = (n - m) + uN$ can only be satisfied if $u = 0$. Finally, if $N/2 \leq (n - m) \leq N - 2$, then the condition $q = (n - m) + uN$ can only be satisfied if $u = -1$.

Consequently, for each combination $(n - m)$, there is only one value of q for which $\delta_{pN}[q - (n - m)] = 1$. In other words, for each combination $(n - m)$, all terms of the sum over q are zero, except for a single term, which satisfies $q = (n - m) + uN$, and at which $\delta_{pN}[q - (n - m)] = 1$. Therefore, the sum over q in Equation (8.29) reduces to:

$$g[n] = \sum_{m=-\frac{N}{2}}^{\frac{N}{2}-1} \sum_{q=-\frac{N}{2}}^{\frac{N}{2}-1} a[m]b[q]\delta_{pN}[q - (n - m)] = \sum_{m=-\frac{N}{2}}^{\frac{N}{2}-1} a[m]b[(n - m) + uN] \quad (8.30)$$

Finally, since $b[n]$ is the inverse DFT of $B[k]$, it follows that $b[n]$ is a periodic function, with period N. Consequently, $b[(n - m) + uN] = b[n - m]$, and Equation (8.30) reduces to*:

Circular convolution:

$$g[n] = \sum_{m=-\frac{N}{2}}^{\frac{N}{2}-1} a[m]b[n - m] \quad (8.31)$$

The sum in Equation (8.31) defines the circular convolution between the periodic signals $a[n]$ and $b[n]$.

We have concluded that:

if:

$$G[k] = A[k]B[k]$$

then:

$$g[n] = \sum_{m=-\frac{N}{2}}^{\frac{N}{2}-1} a[m]b[n - m] \quad (8.32)$$

Equation (8.32) is the DFT property of circular convolution.

---

* *notice that, even if $b[n]$ is nonperiodic, the last sum in Equation (8.30) still results in a circular convolution, because the term uN in the argument of $b[(n - m) + uN]$ maps points outside the computational window to points inside it (recall that the computational window begins in $-N/2$ and ends in $N/2 - 1$) . For example, we have seen that, if $-N \leq (n - m) \leq -N/2 - 1$, then $u = 1$; consequently $0 \leq (n - m) + N \leq N/2 - 1$. Therefore, the argument of $b[(n - m) + N]$ is a number between 0 and $N/2 - 1$, which is inside the computational window, even though $(n - m)$ in this example is outside it. The effect of such a mapping is equivalent to the convolution with a periodic $b[n]$. But it is easier to picture $b[n]$ as a periodic function straight away than to keep track of this mapping.*

The relationship between the circular and linear convolutions plays a pivotal role in the analysis of discrete systems, so we dedicate Section 8.5 to discuss how they are related.

8.5 Relationship Between Circular and Linear Convolutions

There are three significant differences between the linear convolution (Equation (7.43), which I repeat below for convenience) and the circular convolution (Equation (8.31)).

The first difference is that the signals in the linear convolution are nonperiodic, whereas the signals in the circular convolution are periodic.

The second difference is that, whereas the sum in the linear convolution ranges from $m = -\infty$ to $m = \infty$, the sum in the circular convolution ranges only over the central period (the computational window), that is, over $-N/2 \le m \le N/2 - 1$.

The third difference is that it is implied that $g[n]$ in the linear convolution must be evaluated for all integers n, whereas it is implied that $g[n]$ in the circular convolution must be evaluated only inside the computational window, that is, only for $-N/2 \le n \le N/2 - 1$.

Equation (7.43) (linear convolution)

$$g[n] = a[n] * b[n] = \sum_{m=-\infty}^{+\infty} a[m]b[n - m]$$

The mechanism of calculus of the circular convolution is essentially the same of the linear convolution: the index n is turned into the index m, the signal $b[m]$ is flipped (that is, $b[m]$ is turned into $b[-m]$), and then we shift the flipped signal according to the value of n. But, since the signals are periodic, adjacent periods of $b[-m]$ 'invade' the computational window when $b[-m]$ is shifted to $b[n - m]$.

This invasion of adjacent periods is relevant to the analysis of discrete systems. Recall from Chapter 3 (Representation of Systems) that the output of continuous LTI systems is the convolution between the input and the impulse response in the time domain, and product between the input and the frequency response in the frequency domain. Thus, the analysis of systems is heavily reliant on the property of convolution. As we will see in Chapter 9, unsurprisingly, the same relationships hold for discrete LTI systems: the output is related to the input through convolution in the 'time' domain and product in the 'frequency' domain. So, the analysis of LTI discrete systems involves taking the product of the DFT of the input signal with the DFT of the impulse response. As we have just seen, this operation results in the CIRCULAR, not linear, convolution in the 'time' domain. However, usually our goal is to analyze continuous time signals and systems through their discrete counterparts. Thus, in practice, we are interested in the linear convolution, not in the circular convolution. In other words, **the product between DFTs gives us the circular convolution, but what we usually want is the linear convolution**. So, we need to ask the question: can we get the linear convolution from the circular convolution?

Let us be a bit clearer about this question. Suppose that we have two signals of interest, $a[n]$ and $b[n]$, which are not periodic, and we want to calculate their LINEAR convolution in a computer. The easiest way to do that is calculate their DFT, take the product between their DFTs, and then calculate the inverse DFT from their product. We have just seen that this operation results in the CIRCULAR convolution. But is it possible that the circular convolution coincides with the linear convolution?

The answer is: yes, it is possible, if there is sufficient 'space' in the computational window. It is best to explain this by means of an example. Suppose that we pick the same two signals $a[n]$ and $b[n]$ of Exercise 11 of Chapter 7, defined as: $a[-1] = 1$, $a[0] = 2$, $a[1] = 3$, $a[n] = 0$ otherwise; $b[-1] = 4$, $b[0] = 0$, $b[1] = -1$, $b[n] = 0$ otherwise.

We need to distinguish between the size of the signal and the size of the vector storing the signal (the latter is the size of the computational window). Recall that the parameter N in the DFT has two meanings: (i) it is the period of the signal and (ii) it is also the size of the vector storing the signal, that is, the size of the computational window. The size of the signal, as opposed to the size of the vector, is the number of samples carrying information about the signal. In the example we just picked, both signals $a[n]$ and $b[n]$ have three samples carrying information (indexed by $n = -1, 0, 1$), so the size of each of these signals is 3. But we could store these samples in a vector with many more points. For example, as illustrated in Figure 8.13a and Figure 8.13b, we could store them in a vector with $N = 8$ points. Such a computational window is highlighted in red in Figure 8.13. Now suppose we proceed to calculate the circular convolution. Following the calculation recipe, we change the axis from n to m, and then flip the signal $b[m]$, as shown in Figure 8.13c and Figure 8.13d. Now, let us see what happens when we compute the circular convolution at the extremes of the computational window, that is, at $n = -4$ and at $n = 3$.

Let us begin with $n = -4$. Compare $b[-4 - m]$ (Figure 8.13e) with $b[-m]$ (Figure 8.13d). The signal $b[-4 - m]$ is obtained by shifting the signal $b[-m]$ to the left by four slots. Thus, the sample indexed by $m = 7$ of $b[-m]$ (Figure 8.13d), which is outside the computational window, ended up at the index $m = 7 - 4 = 3$ of $b[-4 - m]$, which is inside the computational window (Figure 8.13e). Since $m = 3$ is inside the computational window, this sample is an intruder: it belongs to an adjacent period, not to the central period. As such, it would not be there if we were computing the linear convolution, because the signals in the linear convolution are not periodic (so there are no adjacent periods).

However, this sample is a harmless intruder because it is multiplying $a[3] = 0$, so it does not affect the value of $g[-4]$. Consequently, the value of $g[-4]$ is the

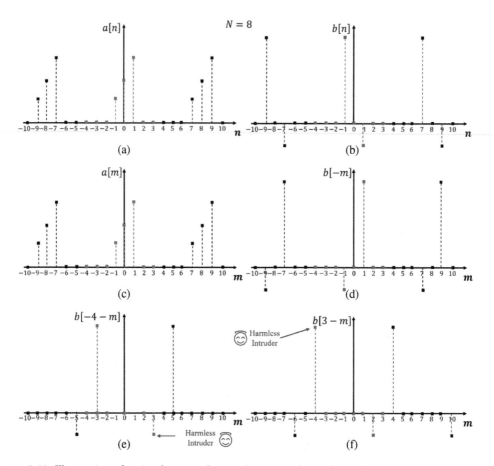

Figure 8.13 Illustration of a circular convolution that coincides with the linear convolution. The computational window is highlighted by the red colour: (a) example signal $a[n]$, (b) example signal $b[n]$, (c) signal $a[m]$, (d) signal $b[-m]$, (e) signal $b[-4 - m]$, and (f) signal $b[3 - m]$. The circular convolution coincides with the linear convolution when adjacent periods do not play a role, like in this example: every time a sample from an adjacent period of $b[-m]$ invaded the computational window, it was multiplied by a 0 of the signal $a[m]$. Consequently, the adjacent periods did not affect the calculation. The samples that invaded the computational window were the samples of $b[-m]$ (figure d) indexed by $m = 7$ and $m = -7$. The former ($m = 7$) invaded the computational window at the index $m = 3$ of figure (e), whereas the latter invaded the computational window at the index $m = -4$ of figure (f). For the circular convolution to coincide with the linear convolution, the computational window must be sufficiently large to fit the signal resulting from the linear convolution. In this example, the size of $a[n]$ and $b[n]$ is 3, which entails that their linear convolution has a size of $2 \times 3 - 1 = 5$. The computational window has a size of $N = 8$. Since $N > 5$, the circular convolution coincides with the linear convolution.

same for both linear and circular convolutions (by the way, you can easily check by inspection of Figure 8.13c and Figure 8.13e that $g[-4] = 0$).

Something similar happens in the other extreme of the computational window, that is, in the evaluation of $g[3]$. Figure 8.13f shows the signal $b[3 - m]$, which is obtained by shifting the signal $b[-m]$ to the right by three slots. As shown in Figure 8.13f, the sample indexed by $m = -7$ of $b[-m]$ (Figure 8.13d) ended up at the index $m = -7 + 3 = -4$ of $b[3 - m]$ (Figure 8.13f), thus invading the computational window of $b[3 - m]$. However, it did not affect the calculation of $g[3]$, because $a[-4] = 0$.

In our example, then the circular convolution between $a[n]$ and $b[n]$ coincided with the linear convolution because the computational window was larger than the signal resulting from the linear convolution. More precisely, in our example, both $a[n]$ and $b[n]$ have three samples, so the linear convolution between them results in a signal $g[n]$ with $2 \times 3 - 1 = 5$ samples. Since we used a computational window with size $N = 8$, it has enough slots to fit the five samples resulting from the linear convolution. Consequently, the circular convolution coincided with the linear convolution. In the exercise list, you will be asked to compute the circular convolution and double check that it indeed coincides with the linear convolution (the latter has already been calculated in Exercise 11 of Chapter 7).

We have seen a happy example, that is, one in which the computational window was sufficiently large to accommodate the linear convolution. Now, let us see a counterexample, assuming the same $a[n]$ and $b[n]$ as before, but now assuming $N = 4$. This time, $N < 5$, so the computational window is not sufficiently large. Consequently, there is interference of adjacent periods. As shown in Figure 8.14, the interference happens in the calculation of $g[-2]$: the sample indexed by $m = 3$ of $b[-m]$ (Figure 8.14d), which is outside the computational window, ended up at $m = 3 - 2 = 1$ of $b[-2 - m]$ (Figure 8.14e), which is inside the computational window. Since $m = 1$ belongs to the computational window, this sample is an intruder. Furthermore, it is a harmful intruder because $a[1] \neq 0$, so the sample of $b[-2 - m]$ indexed by $m = 1$ affects the calculation of $g[-2]$. Consequently, $g[-2]$ resulting from the circular convolution (which, by the way, is $g[-2] = 1 \times 4 + 2 \times 0 + 3 \times (-1) = 1$) is different from $g[-2]$ resulting from the linear convolution (which, by the way, is $g[-2] = 1 \times 4 + 2 \times 0 + 3 \times 0 = 4$).

Thus, if $N = 4$, then the circular convolution does NOT coincide with the linear convolution. If what we want is the linear convolution, then we get a 'wrong' result if we work it out using the product between the DFTs. In the exercise list, you will be asked to calculate the circular convolution assuming $N = 4$ and $N = 8$ by hand, and then to write a Matlab code to calculate them using the product between the DFTs, and then compare the results.

To summarize: usually we want to obtain the linear convolution using the product between DFTs, but this product results in the circular convolution. These two convolutions coincide if the computational window is sufficiently large to accommodate the signal resulting from the linear convolution. Thus, a good practice is to always leave plenty of 'space' in the computational window, that is, lots of zeros, to guarantee that the circular convolution coincides with the linear convolution. This process of adding zeros to a signal is usually called 'padding'.

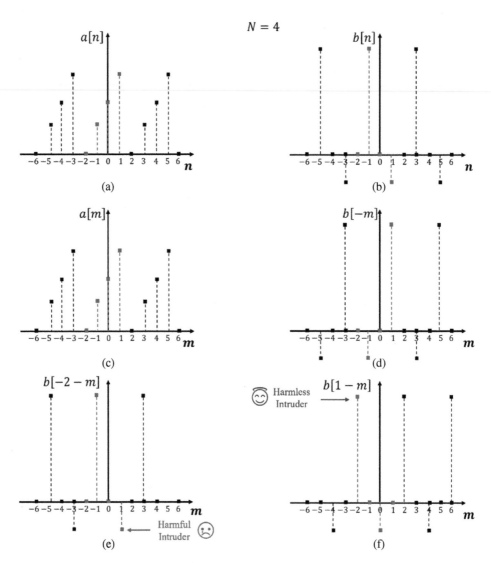

Figure 8.14 Example of a circular convolution that does not coincide with the linear convolution. This time, the computational window is smaller than the signal resulting from the linear convolution (that is, $N < 5$), so the circular convolution does not coincide with the linear convolution. The problem happened in the calculation of $g[-2]$: as shown in figure (e), there is a harmful intruder. This intruder came from the index $m = 3$ of $b[-m]$ (figure d), which is outside the computational window, but invaded the computational window of $b[-2-m]$ at the index $m = 1$. Since $a[1] \neq 0$, this intruder affected the calculation of $g[-2]$. Consequently, $g[-2]$ of the circular convolution does not coincide with $g[-2]$ of the linear convolution. You can check by yourself that the circular convolution results in $g[-2] = 1 \times 4 + 2 \times 0 + 3 \times (-1) = 1$, whereas the linear convolution results in $g[-2] = 1 \times 4 + 2 \times 0 + 3 \times 0 = 4$.

8.6 Parseval's Theorem for the DFT

In this section, we will obtain an expression for the energy of a signal in terms of its DFT. This expression is Parseval's theorem for DFTs.

Recall that, by definition, the energy of a discrete signal is given by Equation (7.49), whose sum ranges from $n = -\infty$ to $n = +\infty$. Assuming that the signal has been stored in a vector of size N, we can rewrite the sum in Equation (7.49) as:

$$E_d = \sum_{n=-\frac{N}{2}}^{\frac{N}{2}-1} |g[n]|^2 \tag{8.33}$$

To express this sum in terms of the DFT of $g[n]$, we proceed in the usual manner by writing $g[n]$ and its complex conjugate explicitly in terms of the inverse DFT. Thus:

$$g[n] = \frac{1}{N} \sum_{k=-\frac{N}{2}}^{\frac{N}{2}-1} G[k] e^{i2\pi \frac{k}{N} n}$$

$$g^*[n] = \frac{1}{N} \sum_{k=-\frac{N}{2}}^{\frac{N}{2}-1} G^*[k] e^{-i2\pi \frac{k}{N} n}$$

Consequently:

$$E_d = \sum_{n=-\frac{N}{2}}^{\frac{N}{2}-1} |g[n]|^2 = \sum_{n=-\frac{N}{2}}^{\frac{N}{2}-1} \frac{1}{N^2} \sum_{k=-\frac{N}{2}}^{\frac{N}{2}-1} \sum_{q=-\frac{N}{2}}^{\frac{N}{2}-1} G[k] G^*[q] e^{i2\pi \frac{k}{N} n} e^{-i2\pi \frac{q}{N} n}$$

Rearranging:

$$E_d = \frac{1}{N^2} \sum_{k=-\frac{N}{2}}^{\frac{N}{2}-1} \sum_{q=-\frac{N}{2}}^{\frac{N}{2}-1} G[k] G^*[q] \sum_{n=-\frac{N}{2}}^{\frac{N}{2}-1} e^{i2\pi \frac{n}{N} (k-q)} \tag{8.34}$$

The sum over n in Equation (8.34) has the same form of the sum in Equation (8.27), with the dummy index n in Equation (8.34) playing the role of the dummy index k of Equation (8.27), and with $k - q$ in Equation (8.34) playing the role of p in Equation (8.27). Therefore, we already know that the result of this sum is $N\delta_{pN}[k - q]$, which entails that:

$$E_d = \frac{1}{N^2} \sum_{k=-\frac{N}{2}}^{\frac{N}{2}-1} \sum_{q=-\frac{N}{2}}^{\frac{N}{2}-1} G[k] G^*[q] N\delta_{pN}[k - q]$$

Using the fact that $\delta_{pN}[k - q]$ is an even function:

$$E_d = \frac{1}{N} \sum_{k=-\frac{N}{2}}^{\frac{N}{2}-1} \sum_{q=-\frac{N}{2}}^{\frac{N}{2}-1} G[k]G^*[q]\delta_{pN}[q - k] \tag{8.35}$$

Now we evaluate the sum over q by taking advantage of the definition of $\delta_{pN}[n]$ (Equation (8.26)). Notice that, since the sums of Equation (8.35) range only over the central period of the computational window, it follows that the minimum value $q - k$ can obtain is $q - k = -N/2 - (N/2 - 1) = -N + 1$, whereas its maximum value is $q - k = N/2 - 1 - (-N/2) = N - 1$. Consequently, the only multiple of N accessible in the double sum in Equation (8.35) is $u = 0$, which entails that $\delta_{pN}[q - k] = 1$ only when $q = k$. Hence, the sum over q in Equation (8.35) reduces to:

$$E_d = \frac{1}{N} \sum_{k=-\frac{N}{2}}^{\frac{N}{2}-1} \sum_{q=-\frac{N}{2}}^{\frac{N}{2}-1} G[k]G^*[q]\delta_{pN}[q - k] = \frac{1}{N} \sum_{k=-\frac{N}{2}}^{\frac{N}{2}-1} G[k]G^*[k]$$

Thus, we have proven that:

$$E_d = \sum_{n=-\frac{N}{2}}^{\frac{N}{2}-1} |g[n]|^2 = \frac{1}{N} \sum_{k=-\frac{N}{2}}^{\frac{N}{2}-1} |G[k]|^2 \tag{8.36}$$

Equation (8.36) is Parseval's theorem for DFT's. Recall that we can connect this result with the energy of the continuous signal through Equation (7.59).

In Chapter 9, we take a closer look at the representation of discrete systems, again with a focus on LTI systems. We will see that they are closely analogous to their continuous time part, so we will be navigating charted waters.

8.7 Exercises

Exercise 1

Consider the signal $g[n]$, which is zero everywhere, except at $g[-1] = 5$, $g[0] = 1.1$, and $g[1] = 2.5$. Now consider the signal $x[n]$, defined as:

$$x[n] = \frac{1}{16} \sum_{k=-8}^{7} \sum_{m=-8}^{7} g[m]e^{-i2\pi\frac{k}{16}m}e^{i2\pi\frac{k}{16}n}$$

Find $x[-32]$, $x[-10]$, $x[-17]$, and $x[17]$.

Exercise 2

Write a Matlab code to check if your answer to Exercise 1 of Chapter 8 is correct.

Exercise 3

Write a Matlab routine to calculate the centralized DFT of the Gaussian used in the Worked Exercise: Getting the Centralized DFT Using the Command *fft*, of this chapter. Your routine cannot use the command *fft*. Compare your result with the centralized DFT obtained in that worked exercise.

Exercise 4

Suppose we define the silly DFT pair, as shown in the blue box below.

Using the Gaussian function discretized in the Worked Exercise: Getting the Centralized DFT Using the Command *fft*, write a Matlab code to calculate the direct silly centralized DFT $G[k]$, and to calculate the inverse silly centralized DFT of $G[k]$. Compare the inverse silly centralized DFT with the inverse centralized DFT. Explain your results. Notice that the silly centralized DFT assumes that the vector $g[n]$ has size N and that the vector $G[k]$ has size $N/4$.

SILLY CENTRALIZED DFT PAIR
DIRECT SILLY CENTRALIZED DFT

$$G[k] = \sum_{n=-\frac{N}{2}}^{\frac{N}{2}-1} g[n]e^{-i2\pi \frac{k}{\frac{N}{4}}n} \quad for -\frac{N}{8} \le k \le \frac{N}{8} - 1$$

INVERSE SILLY CENTRALIZED DFT

$$g[n] = \frac{1}{\frac{N}{4}} \sum_{k=-\frac{N}{8}}^{\frac{N}{8}-1} G[k]e^{i2\pi \frac{k}{\frac{N}{4}}n} \quad for -\frac{N}{2} \le n \le \frac{N}{2} - 1$$

Exercise 5

Implement a Matlab code to discretize the Gaussian as in the Worked Exercise: Getting the Centralized DFT Using the Command *fft*. Then, using the property of circular time shift, implement a code to shift the Gaussian by $n0\_1 = 100$, $n0\_2 = 511$, and $n0\_3 = 850$. Notice that half of the Gaussian for $n0\_2 = 511$ has gone to the beginning of the computational window, and that the Gaussian shifted by $n0\_3 = 850$ appears on the left of the unshifted Gaussian, even though the shift is to the right-hand side. Check whether the peaks of the shifted Gaussians lie at the expected values of n.

Exercise 6

When performing time shift, it is usually more convenient to express the argument of the exponential in terms of the time vector. Thus, prove that:

$$e^{-i2\pi \frac{k}{N} n_0} G[k] = e^{-i2\pi f t_0} G[k]$$

where f is the frequency vector and $t_0 = n_0 T_S$.

Then, implement a Matlab code to time shift the Gaussian with $t_{0-1} = 100 \cdot T_S$, $t_{0-2} = 511 \cdot T_S$, and $t_{0-3} = 850 \cdot T_S$ using the frequency vector in the argument of the exponential. Notice that the shifted vectors obtained in this exercise are identical to the shifted vectors obtained in Exercise 5 of Chapter 8. The form used in the current exercise is the most common form used in practice.

Exercise 7

Consider the following two commands, where g is the vector obtained by discretizing the Gaussian as in the Worked Exercise: Getting the Centralized DFT Using the Command *fft*:

$$G - 1 = fftshift(fft(fftshift(g)));$$

$$G - 2 = fftshift(fft(g));$$

Theoretically, both G_1 and G_2 are real vectors, but sometimes there is numerical noise in their imaginary part, which can be removed with the function real().
Plot both $real(G\_1)$ and $real(G\_2)$ together and explain their differences.
Hint: use the property of time shifting to explain their difference.

Exercise 8

Calculate, 'by hand', the circular convolution between the signals $a[n]$ and $b[n]$ defined in Exercise 11 of Chapter 7 ($a[-1] = 1$, $a[0] = 2$, $a[1] = 3$, $a[n] = 0$ otherwise; $b[-1] = 4$, $b[0] = 0$, $b[1] = -1$, $b[n] = 0$ otherwise). Do the calculation twice: first, assuming $N = 8$, and then assuming $N = 4$. Recall that the calculations should be done only for the central period. Compare your results with the linear convolution worked out in Exercise 11 of Chapter 7.

Exercise 9

Considering the same signals $a[n]$ and $b[n]$ of Exercise 8 of Chapter 8, write two Matlab codes to calculate their circular convolution by taking the product between their DFTs. In one code assume $N = 8$ and in the other code assume $N = 4$. Check if the results obtained with Matlab agree with the results obtained 'by hand' in Exercise 8 of Chapter 8.

Exercise 10

Write a Matlab code to discretize the frequency response of a harmonic oscillator (Equation (3.69)) assuming $m = k = 1$ and $\gamma = 0.1$. Then, use the command *ifft*, assisted by the command *fftshift*, to obtain the impulse response. Use $N = 2^{10}$ and $f_S = 2$. Plot the magnitude of the frequency response, and the impulse response (there may be numerical noise in the imaginary part of the impulse response, which can be removed using the command real()).

After obtaining the impulse response using the *ifft*, discretize the analytical expression of the impulse response (Equation (6.29) – recall that $u(t)$ is the step function – see Equation (6.22)) and compare it with the result obtained using the *ifft* by plotting them together.

Exercise 11

Repeat Exercise 10 of Chapter 8, but now assuming $m = 7$. Notice that the impulse response obtained from the *ifft* is no longer zero for negative time. According to the theory developed in Chapter 3 (Representation of Systems), the impulse response must be zero for negative time, because the harmonic oscillator is a causal system. So, what is happening? Why can we see nonzero values of the impulse response?

Exercise 12

Functions with discontinuities often feature ripples around the discontinuity when they are calculated through an inverse DFT (or direct DFT if the discontinuity is in the frequency domain). These ripples arise by virtue of their boundless Fourier transforms: to 'accommodate' the discontinuity, there is no upper bound f_0 in their Fourier transforms. But a numerical calculation, such as the DFT, requires a cut-off frequency (and we have learned that the highest frequency involved in the DFT is $f = 0.5 \cdot f_S$, so this is the cut-off frequency). It is precisely the exclusion of the frequencies $f > 0.5 \cdot f_S$ from the DFT that causes the ripples. These ripples are known as the 'Gibbs phenomenon'.

The most well-known manifestation of the Gibbs phenomenon is in the rectangular function (Equation (2.7)). Recall that the Fourier transform of the rectangular function is a *sinc* function (Equation (2.86)). To get acquainted with the Gibbs phenomenon, discretize the *sinc* function assuming c = 10 and N = 2^{10}. Do the calculation for three different sampling frequencies: $f_S = 0.5$, $f_S = 2$, and $f_S = 10$. Then, use the *ifft* to calculate the inverse Fourier transform of the *sinc* function using these three different sampling frequencies. Notice that the amplitude of the ripples in the centre of the rectangular function (around $t = 0$) reduces as f_S increases, because the higher f_S is, the more frequencies are included in the DFT. However, these amplitudes do not decrease around the discontinuity: they are only crammed into the discontinuity as the sampling frequency increases.

Exercise 13

Starting from the frequency response of an RC low-pass filter (Equation (4.7)) use the *ifft* to calculate the impulse response of the filter and compare it with the analytical equation (see Exercise 9 of Chapter 5 for the analytical equation). Assume $RC = 10$ and choose a sensible sampling frequency to discretize the frequency response. A sensible sampling frequency should be sufficiently high so as not to leave a significant part of the spectrum out, but not too high so as to result in a too short time window. Notice the Gibbs phenomenon around the origin ($t = 0$) of the impulse response, which features a discontinuity.

Discrete Systems

Learning Objectives

In Chapters 7 and 8, we developed a 'frequency' domain representation of discrete signals. Recall from Chapter 3 that a major motivation for a frequency domain representation is that it diagonalizes the operator of LTI systems, which leads to the concept of frequency response. In this chapter, we will learn that, like their continuous time counterparts, the frequency domain representation of discrete systems is also simplified. Indeed, again like their continuous time counterpart, the output is related to the input via convolution in the 'time' domain and product in the 'frequency' domain. Thus, the theory developed in Chapter 3 (Representation of Systems) can be translated to discrete systems. Consequently, we can use the theory of Chapter 3, supplemented by the theory of Chapters 7 and 8, to analyze continuous time systems through computer implementation of their discrete versions. As an example, we will look at the basic ideas of digital filters.

9.1 Introduction and Properties

In Chapter 3 (Representation of Systems), we learned that a system is anything that modifies a signal. Thus, a discrete system is anything that modifies a discrete signal; in other words, a discrete system is anything that turns a sequence of numbers into another sequence of numbers.

We again represent the system by a capital letter followed by curly brackets. Thus, the representation of the action of the discrete system $T\{\ \}$ on the input signal $x[n]$, resulting in the output signal $y[n]$ reads:

$$y[n] = T\{x[n]\} \tag{9.1}$$

Essentials of Signals and Systems, First Edition. Emiliano R. Martins.
© 2023 John Wiley & Sons Ltd. Published 2023 by John Wiley & Sons Ltd.
Companion website: www.wiley.com/go/martins/essentialsofsignalsandsystems

Discrete systems can be classified into many types according to their properties, and we again focus on their four main properties: linearity, 'time invariance',
'causality', and 'stability'.

9.1.1 Linearity

The property of linearity of discrete systems is closely analogous to their continuous
time counterpart (Equation (3.2)). Thus, a system $T\{\ \}$ is linear if and only if:

$$T\{ax_1[n] + bx_2[n]\} = aT\{x_1[n]\} + bT\{x_2[n]\} \tag{9.2}$$

where a and b are constants.

9.1.2 'Time' invariance

Again, in close analogy to their continuous time counterpart (Equation (3.3)), the
property of 'time' invariance of a discrete system reads:

if:

$$y[n] = T\{x[n]\}$$

then $T\{\ \}$ is time invariant if and only if:

$$T\{x[n - n_0]\} = y[n - n_0] \tag{9.3}$$

9.1.3 Causality

In Section 3.1.3 (Causality), we learned that a continuous time causal system is one
in which the output at time t_0 depends on the input only for times $t \le t_0$ – that is,
it depends on $x(t \le t_0)$. The definition of causality of discrete systems follows the
same idea: a discrete system is causal when the output at 'time' n_0 depends on the
input $x[n]$ only for 'times' $n \le n_0$.

Recall that, since a physical system must be causal, so most of continuous time
systems are causal. In the discrete domain, however, it is quite common to find
noncausal systems. After all, a discrete system can be any operation that turns
a sequence of numbers into another sequence of numbers. For example, a system
that takes the average between $x[n]$ and its 'future' $x[n + 1]$ (that is, $T\{x[n]\} =
(x[n] + x[n + 1])/2$) is a noncausal system.

9.1.4 Stability

Like its continuous time counterpart (Section 6.1 [Motivation: Stability in LTI Sys-
tems]), the criterion of stability is connected with the criterion of 'boundness'. A

discrete signal is said to be bounded if it is never infinite. Formally, a signal $g[n]$ is bounded if there exists a finite number A such that $|g[n]| \leq A$ for all n.

The definition of stability in discrete systems is identical to its definition in continuous time systems: a system is stable if any bounded input results in a bounded output. In Chapter 10, we will see that the criterion of stability of LTI systems requires the DTFT of its impulse response to exist. Thus, an unstable LTI system has no DTFT, which motivates the introduction of the z-transform, to be treated in Chapter 10.

9.2   Linear and Time Invariant Discrete Systems

Following the same philosophy of Chapter 3 (Representation of Systems), now we focus attention on LTI systems. In this section, we will show that the output of an LTI system is given by a convolution in the 'time' domain, and product in the 'frequency' domain, just like their continuous time counterpart. The proof is analogous to the continuous time counterpart, developed in Section 3.8 (The Physical Meaning of Eigenvalues: The Impulse and Frequency Responses).

We begin by defining the discrete impulse function $\delta[n]$ as:

$$\delta[0] = 1$$

$$\delta[n] = 0 \ for \ n \neq 0 \tag{9.4}$$

Notice that $\delta[n]$ is a quite simple signal: it is 0 'everywhere', except at the 'origin', $n = 0$, where it has the value 1. We can also define a 'time' shifted function $\delta[n - n_0]$: according to Equation (9.4), $\delta[n - n_0]$ is 0 'everywhere', except at $n = n_0$, where it has the value 1.

In close analogy with its continuous time counterpart, the set of all functions of the type $\delta[n - n_0]$, consisting of all possible integer values n_0, forms a basis for the 'time' domain representation of discrete signals. Indeed, any signal $x[n]$ can be expressed as a linear combination of functions of the type $\delta[n - n_0]$:

$$x[n] = \sum_{n_0=-\infty}^{+\infty} x[n_0]\delta[n - n_0] \tag{9.5}$$

To check that Equation (9.5) is true, pick a value of n, say $n = 3$, and evaluate the sum:

$$x[3] = \sum_{n_0=-\infty}^{+\infty} x[n_0]\delta[3 - n_0]$$

Since $\delta[3 - n_0] = 0$ if $n_0 \neq 3$, it follows that all terms of the sum are zero, except the term $n_0 = 3$. Thus:

$$x[3] = \sum_{n_0=-\infty}^{+\infty} x[n_0]\delta[3 - n_0] = x[3]\delta[3 - 3] = x[3]\delta[0] = x[3]$$

The interpretation of Equation (9.5) follows the same philosophy of its continuous time counterpart (Equation (2.24)): the signal $x[n]$ is being expressed as a linear combination of signals $\delta[n - n_0]$, where each signal $\delta[n - n_0]$ has the coordinate $x[n_0]$. Notice that Equation (9.5) is the convolution between $x[n]$ and $\delta[n]$ (the convolution in the discrete domain was defined in Equation (7.43)). Thus:

$$x[n] = \sum_{n_0=-\infty}^{+\infty} x[n_0]\delta[n - n_0] = x[n] * \delta[n] \tag{9.6}$$

We can take advantage of the linearity of an operator $T\{\ \}$ by describing its action on a signal $x[n]$ in terms of its action on the basis functions $\delta[n - n_0]$. We begin by expressing $x[n]$ as linear combination of basis functions:

$$T\{x[n]\} = T\left\{ \sum_{n_0=-\infty}^{+\infty} x[n_0]\delta[n - n_0] \right\}$$

Recall that the signals in the sum are the terms $\delta[n - n_0]$, whereas $x[n_0]$ plays the role of a coefficient (a coordinate). Thus, if $T\{\ \}$ is linear, it follows that:

$$T\{x[n]\} = T\left\{ \sum_{n_0=-\infty}^{+\infty} x[n_0]\delta[n - n_0] \right\} = \sum_{n_0=-\infty}^{+\infty} x[n_0]T\{\delta[n - n_0]\} \tag{9.7}$$

Furthermore, if $T\{\ \}$ is also time invariant, it follows that:

$$T\{\delta[n - n_0]\} = h[n - n_0] \tag{9.8}$$

where $h[n]$ is the impulse response of the system:

$$h[n] = T\{\delta[n]\} \tag{9.9}$$

Using Equation (9.8) in Equation (9.7), we find that:

$$T\{x[n]\} = \sum_{n_0=-\infty}^{+\infty} x[n_0]h[n - n_0] = x[n] * h[n] \tag{9.10}$$

Thus, we have concluded that the output $y[n]$ of an LTI system is given by the convolution between the input and the impulse response:

$$y[n] = h[n] * x[n] \tag{9.11}$$

Defining the frequency response $H(v)$ as the DTFT of $h[n]$:

$$H(v) = \sum_{n=-\infty}^{\infty} h[n]e^{-i2\pi vn} \tag{9.12}$$

And using the property of convolution in the 'time' domain of DTFTs (Equation (7.46)) in Equation (9.11), we find:

$$Y(v) = H(v)X(v) \tag{9.13}$$

where $Y(v)$ and $X(v)$ are the DTFTs of the output and input, respectively.

Equation (9.11) and Equation (9.13) are, respectively, the 'time' and 'frequency' domains representation of LTI systems. They are closely analogous to their continuous time counterparts (Equation (3.60) and Equation (3.61), respectively).

There is, however, a catch! In practice, we obtain the output of LTI systems by taking the product between the DFTs instead of the DTFTs. Thus, in practice, instead of using Equation (9.13), we use:

$$Y[k] = H[k]X[k] \tag{9.14}$$

And then obtain $y[n]$ with the inverse DFT. So, what we calculate in practice is the circular convolution, but what we want is the linear convolution (Equation (9.11) is the linear convolution). Therefore, as discussed in Section 8.5: Relationship Between Circular and Linear Convolutions, we need to be careful to leave enough 'space' in the vectors, so that the circular convolution coincides with the linear convolution.

Worked Exercise: Further Advantages of Frequency Domain

In addition to its simplicity, a formulation in the frequency domain often offers more accurate results than a formulation in the time domain. One example is the implementation of an LTI system performing a time derivative. Thus, consider the continuous time system $T\{\ \}$ whose action on $x(t)$ results in its time derivative:

$$y(t) = T\{x(t)\} = \frac{dx}{dt} \tag{9.15}$$

Now suppose that we want to investigate this system using a computer. There are two ways we could do that: one way is to calculate the numerical derivative in the 'time' domain, by setting:

$$y[n] = \frac{x[n] - x[n-1]}{T_S} \tag{9.16}$$

Equation (9.16) is an example of a time domain formulation: the output is obtained directly in the time domain.

Another way is to use a frequency domain formulation, which means that we obtain the output in the frequency domain by means of Equation (9.14), where $H[k]$ and $X[k]$ are vectors containing samples of the Fourier transform. In other words, $H[k]$ and $X[k]$ are the centralized DFT, with the amplitudes multiplied by T_S (recall that we need to multiply by T_S, or divide by f_S, because both DTFT and DFT have a factor of f_S in the amplitude when compared to the Fourier transform – compare Figure 7.4a with Figure 7.4c).

And how can we obtain these vectors $H[k]$ and $X[k]$? In this example (and sometimes in real life), we already know the frequency response: it is $H(f) = i2\pi f$ (see Equation (2.132)). Thus, $H[k] = i2\pi f$, where $f = v \cdot f_S$ is the frequency vector obtained from the DTFT vector v (see Worked Exercise: Getting the Fourier Transform with the *fft*, of Chapter 8).

What about $X[k]$? Well, that obviously depends on the input, and we have learned how to obtain samples of the Fourier transform of the input using the *fft* function in Chapter 8. So, we already have all the basic theoretical tools to deal with discrete systems.

In this example, let us use our old friend – the Gaussian function $x(t) = 5e^{-\left(\frac{t}{10}\right)^2}$ – as the input of the system. Our job is to obtain the output using both methods: the time domain (Equation (9.16)) and the frequency domain (Equation (9.14) – with the amplitude correction factors). Then, we will compare the results we obtained with the analytical derivative, and check whether they are equally accurate, or if one of them is more accurate than the other. Of course, we also need to explain the result.

Solution:

A Matlab code implementing the time and frequency domain formulations is shown below:

```
clear all
close all

N = 2^10; %number of points


n = (-N/2:1:N/2-1); %definition of vector n


ti = -300; %beginning of the time window
tf = -ti; %end of the time window

Ts = (tf-ti)/(N); %sampling period
fs = 1/Ts; %sampling frequency
t = n*Ts; % definition of time vector

k = n; %definition of vector k
```

```
v = k./N; %definition of vector v
f = v*fs; %definition of vector f

A = 5;    %amplitude of the gaussian
sigma = 10; %width of the gaussian


x = A*exp(-(t./sigma).^2); %discretize the gaussian

y_analytical = -2*(t./(sigma.^2)).*x; %discretize ana-
lytical derivative

y_numerical_time_domain = x*0; %initialize vector for
numerical derivative in the %time domain

%------------------------------
%The time domain formulation

%calculate derivative in the time domain

for u = 2: length(x)

    y_numerical_time_domain(u) = (x(u) - x(u-1))./Ts;
end
%------------------------------

%calculate error of numerical derivative obtained
in the time domain
error_time_domain = sum(abs(y_analytical-y_numerical_
time_domain))

%------------------------------
%The frequency domain formulation

%calculate derivative in the frequency domain

H = 1i*2*pi*f; %frequency response of system perform-
ing time derivative;

X_CDFT = fftshift(fft(fftshift(x)));  % obtain central-
ized DFT of x[n]
X_FT = X_CDFT*Ts;  %obtain samples of Fourier Trans-
form of x(t)
```

```
%-----------------------------------------
% Y(f) = H(f)X(f)
Y_FT = H.*X_FT; %obtain samples of Y(f)
%-----------------------------------------
Y_CDFT = Y_FT*fs;   %obtain centralized DFT of y[n]
%-----------------------------------------
y_ifft = fftshift(ifft(fftshift(Y_CDFT)));   %obtain
inverse DFT of output
%-----------------------------------------



%calculate error of numerical derivative obtained in the
time domain
error_frequency_domain = sum(abs(y_analytical-y_ifft))


%plot analytical derivative together with numerical
derivatives, calculated in the %time and frequency domains
figure(1)
plot(t,y_analytical,'k',t,y_numerical_time_domain,'ro',
t,y_ifft,'b+')
xlabel('time')
ylabel('y(t) = dx/dt')
legend('analytical derivative','derivative calculated
in the time domain','derivative calculated in the
frequency domain')
axis([-40 40 -0.45 0.45])
```

The results are plotted in Figure 9.1. The solid black line is the analytical derivative, the red circles are the results obtained with the time domain formulation, and the blue crosses are the results obtained with the frequency domain formulation. Plain visual inspection already shows that the frequency domain formulation is more accurate. The vectors 'error_time_domain' and 'error_frequency_domain' quantify the accuracy by taking the magnitude of the difference with respect to the analytical equation, point by point, and adding them up. In my machine, I get error_time_domain $= 0.857$, and error_frequency_domain $= 3.2 \times 10^{-13}$. A huge difference!

But why? Why is it more accurate to calculate the derivative in the frequency domain? The answer is in Equation (7.35): when we use the formulation in the time domain, we are essentially multiplying $X(f)$ by $f_s(1 - e^{-i2\pi v})$. But we know that the correct factor is $i2\pi f$, not $f_s(1 - e^{-i2\pi v})$. And that is what we do in the frequency domain: we multiply by the exact function $i2\pi f$, so we get a more accurate result.

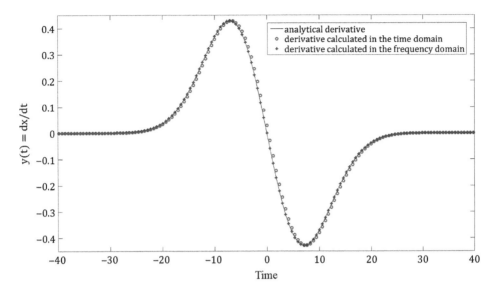

Figure 9.1 Plot obtained using the command above (figure 1).

Another way of thinking about this business is like this: even though we do not know dx/dt analytically (in this exercise, for the sake of illustration, we picked a function of which we do know the analytical derivative, but in practice we seldom know it), we do know that the Fourier transform of dx/dt is $i2\pi f X(f)$. Thus, in the frequency domain formulation, we are getting samples of the EXACT (as opposed to numerically approximated) Fourier transform of dx/dt. Consequently, we get a much more accurate result with the frequency domain formulation.

If you are wondering whether this conclusion is general, or only valid for this specific system chosen in this example, recall that any LTI system boils down to a bunch a differentiations. So, this result is, in fact, quite general.

9.3   Digital Filters

Earlier, I mentioned that a major motivation for a theory of discrete signals and systems is the possibility of analyzing continuous time systems using a computer. This is still true for digital filters, but they have the additional benefit of allowing the designer to build mathematical filters, which could be quite hard, not to say impossible, to obtain with a physical implementation. In this approach, the designer samples the signal of interest, then digitally process it (in other words, make the discrete signal pass through a digital filter), and then use a digital-to-analog converter to generate the filtered physical signal.

Unsurprisingly, digital filters play an important role in modern technologies, and there is a vast literature about them (see References with Comments for suggestions). With the theory we learned in this book, we already have the essential concepts to embark on a study of digital filters, but a thorough exposition of them is outside our scope. So, here, we discuss some basic ideas pertaining to digital filters.

If the digital filter is a representation of a physical filter, then its implementation follows the same procedure shown in the frequency domain formulation of the Worked Exercise: Further Advantages of Frequency Domain (Chapter 9): one discretizes the frequency response of the continuous time filter and multiplies it by the Fourier transform of the input, usually obtained with the *fft* algorithm. If the time domain representation of the output is of interest, it can be obtained using the *ifft* algorithm.

But often a digital filter is not a representation of a physical system. For example, during the pandemic of Covid-19 (which I hope is a thing of the past by the time you are reading this book), many of us got obsessed with checking the moving average of the cases of Covid contamination and Covid-related deaths. These moving averages are an example of a low-pass digital filter: they 'smooth' out the curves, thus removing the fast (high-frequency) oscillations. And these moving average filters do not need to be causal. For example, we could very well implement a filter $T\{\ \}$ such that:

$$y[n] = T\{x[n]\} = \frac{x[n-1] + x[n] + x[n+1]}{3} \qquad (9.17)$$

This is an example of a noncausal low-pass filter: it is noncausal because the output $y[n]$ depends on $x[n+1]$, which is in the 'future'.

Digital filters are often classified into 'recursive' and 'nonrecursive'. The filter of Equation (9.17) is an example of a nonrecursive filter: it is nonrecursive because the output appears only as $y[n]$ in the equation. In contrast, in a recursive filter, the output at other 'times', like $y[n-1]$, also appears.

Filters described by difference equations are often recursive. For example, let us suppose we want to discretize the time domain representation (that is, the differential equation) of the RC filter, whose equation I repeat below for convenience:
Equation (4.4)

$$RC\frac{dy(t)}{dt} + y(t) = x(t)$$

Recall that $x(t)$ is the input and $y(t)$ is the output. Turning this differential equation into a difference equation, we get:

$$\frac{RC}{T_S}y[n] - \frac{RC}{T_S}y[n-1] + y[n] = x[n] \qquad (9.18)$$

The filter described by Equation (9.18) is an example of a recursive filter, since it involves not only $y[n]$, but also $y[n-1]$.

As mentioned in Chapter 5 (Filters), filter design is a science on its own, and we cannot delve into it in this book. But there is a vast literature in digital signal processing (see References with Comments) dealing with digital filter designs. With the concepts learned in this book, you are ready to embark on advanced studies of digital filter designs and digital signal processing.

9.4 Exercises

Exercise 1

Obtain the frequency response of the moving average digital filter described by Equation (9.17).

Exercise 2

Find the frequency response (the DTFT of the impulse response) of the digital RC filter (Equation (9.18)), and compare it with the frequency response (the Fourier transform of the impulse response) of the continuous time RC filter (Equation (4.7)). Considering the discussion in the Worked Exercise: Further Advantages of Frequency Domain, of this chapter, and of Section 7.5.2 (Difference), discuss whether it is worth implementing the frequency response from the difference equation (Equation (9.18)), or if it is better to discretize the frequency response of the continuous time filter (Equation (4.7)) straight away.

Exercise 3

Consider a digital filter defined by the following difference equation:

$$2y[n] - y[n-1] = x[n]$$

where $y[n]$ and $x[n]$ are the output and input, respectively.

(a) Is this a low-pass, band-pass, or high-pass filter?

Comment: recall that the DFTF is a periodic function, and it is the behaviour of the frequency response within the central period $(-0.5 \leq v < 0.5)$ that determines the classification of the filter.

(b) Find the cut-off frequency of this filter.

Comment: the cut-off frequency is the value of v at which $|H(v)| = \dfrac{max|H(v)|}{\sqrt{2}}$.
Since $H(v)$ is a periodic function of v, there are many solutions. It suffices to find the solution within the central period.

(c) If this digital filter is meant to process physical signals (in other words, if the filtered discrete signal is to be turned back into a physical signal), then what sampling frequency is required to set the cut-off frequency at $f_c = 24\,kHz$?

Exercise 4

Repeat parts (a), (b), and (c) of Exercise 3 of Chapter 9, but now considering the following difference equation:

$$2y[n] + y[n-1] = x[n]$$

Exercise 5

The 'window method' of digital filter design consists of obtaining an impulse function $h[n]$ by multiplying the rectangular function $w[n]$ by a desired impulse response $h_d[n]$ (that is, $h[n] = h_d[n]w[n]$). The rectangular function is defined as (see Exercise 6 of Chapter 7):

$$w[n] = 1 \quad for - n_0 \leq n \leq n_0$$

$$w[n] = 0 \quad for \ n \ outside \ the \ range - n_0 \leq n \leq n_0$$

Discuss the relationship between the DTFT of $h_d[n]$ and the DTFT of $h[n]$. In particular, discuss what happens to their DTFTs as n_0 increases.
  Hint: see Exercise 6 of Chapter 7

Exercise 6
Write a Matlab code to obtain the output of a harmonic oscillator (whose frequency response is given by Equation (3.69)), when the input is a sinusoidal wave of frequency $f_0 = 0.4$ Hz, which is chopped at $t = 100$ and $t = -100$ (that means the input is zero outside the interval $-100 < t < 100$). Set the parameters of the harmonic oscillator to be $m = \gamma = k = 1$. You must discretize the frequency response (Equation (3.69)) straight away, but the input must be discretized in the time domain, and then the *fft* must be used to obtain the Fourier transform of the input. Once you got the Fourier transform of the output, use the *ifft* function to obtain the output in the time domain. Then plot the input and output in the time domain and compare your result with Figure 3.5.

Exercise 7
Using the same code of Exercise 6 of Chapter 9, obtain the output in the time domain when the sinusoidal wave frequency is set to $f_0 = 0.2$ Hz. Explain why the amplitude of the output in the stationary state is higher for $f_0 = 0.2$ Hz than for $f_0 = 0.4$ Hz.

Exercise 8
Write a Matlab code to obtain the output of an RC low-pass filter when the input is a rectangular function with width $c = 10$ ms (see Equation (2.7)). You must discretize the frequency response of the RC low-pass filter (Equation (4.7)), but the input must be discretized in the time domain, and the *fft* must be used to obtain its Fourier transform. Once the Fourier transform of the output is obtained, the *ifft* must be used to get the output in the time domain. Use $RC = 2$ ms and compare your result with Figure 3.7. Finally, compare the exponential decay part of the plot with the analytical equation for the discharge of a capacitor through a resistor.

Introduction to the z-transform

Learning Objectives

The z-transform is the discrete time counterpart of the Laplace transform. As such, it is a generalization of the DTFT, motivated by the need of finding a frequency domain representation of signals that have no DTFT. In this chapter, we will learn the essential concepts of the z-transform. The main learning objectives are to understand how the z-transform relates to the DTFT and to the Laplace transform.

10.1   Motivation: Stability of LTI Systems

In Chapter 6 (Introduction to the Laplace Transform), we motivated the generalization of the Fourier transform to the Laplace transform by calling attention to the property of stability of LTI systems: the impulse response of an unstable LTI system has no Fourier transform, so the latter needs to be generalized to establish a frequency domain formulation that includes unstable systems. Here, we follow the same line of reasoning: we first show that the impulse response of an unstable discrete LTI system has no DTFT, which motivates the generalization of the DTFT to a transform that may converge, even when the DTFT sum diverges. The approach followed here is similar to the logic followed in Section 6.1 (Motivation: Stability in LTI Systems).

In Section 9.1.4 (Stability), we learned that a discrete system is stable if any bounded input results in a bounded output. If the system is LTI, then the output

Essentials of Signals and Systems, First Edition. Emiliano R. Martins.
© 2023 John Wiley & Sons Ltd. Published 2023 by John Wiley & Sons Ltd.
Companion website: www.wiley.com/go/martins/essentialsofsignalsandsystems

is related to the input through a convolution with the impulse response. Thus:

$$y[n] = \sum_{n_0=-\infty}^{+\infty} x[n-n_0]h[n_0] \tag{10.1}$$

Now we inspect the magnitude of $|y[n]|$ by noticing that:

$$|y[n]| = \left| \sum_{n_0=-\infty}^{+\infty} x[n-n_0]h[n_0] \right| \leq \sum_{n_0=-\infty}^{+\infty} |x[n-n_0]||h[n_0]| \tag{10.2}$$

If the input is bounded, then there exists a finite number A such that $|x[n]| \leq A$ for all n. Thus, if the input is bounded, it follows that:

$$\sum_{n_0=-\infty}^{+\infty} |x[n-n_0]||h[n_0]| \leq A \sum_{n_0=-\infty}^{+\infty} |h[n_0]| \tag{10.3}$$

Equation (10.3) and Equation (10.2) entails that an LTI system is stable if the sum:

$$\sum_{n=-\infty}^{+\infty} |h[n]|$$

converges.

As it turns out, the convergence of this sum coincides with the existence of the DTFT of $h[n]$. Indeed:

$$|H(v)| \leq \sum_{n=-\infty}^{+\infty} |h[n]|\,|e^{-i2\pi vn}| \tag{10.4}$$

And, since $|e^{-i2\pi vn}| = 1$, the condition of convergence of $H(v)$ reduces to:

$$|H(v)| \leq \sum_{n=-\infty}^{+\infty} |h[n]| \tag{10.5}$$

Translating Equation (10.5) into English: if $\sum_{n=-\infty}^{+\infty} |h[n]|$ converges, then it is guaranteed that $|H(v)|$ is not infinite; in other words, that the DTFT of $h[n]$ exists.

To summarize: it is only guaranteed that the DTFT of the impulse response $h[n]$ converges if $h[n]$ is the impulse response of a stable system. Consequently, to obtain a 'frequency' domain representation of an unstable system, we need to generalize the DTFT to a transform that may converge even when the sum $\sum_{n=-\infty}^{+\infty} |h[n]|$ does not converge. This generalization leads to the z-transform.

10.2   The z-transform as a Generalization of the DTFT

Following the same logic leading to the Laplace transform, we need to generalize the representation in the basis of exponential functions with unit magnitude to a representation of exponentials with magnitude different than one. In Section 6.2 (The Laplace Transform as a Generalization of the Fourier Transform), we generalized $e^{i2\pi ft}$ to $e^{\sigma t}e^{i2\pi ft}$. Therefore, in the Laplace transform, the parameter σ defines the magnitude of the basis function. Now we generalize the basis function $e^{i2\pi vn}$ to $r^n e^{i2\pi vn}$, where r is a real and positive number. Obviously, now it is the parameter r that defines the magnitude of the basis function $r^n e^{i2\pi vn}$. Do not worry at this moment about why the Laplace transform uses $e^{\sigma t}$, but now we are using r instead of something like $e^{\sigma n}$: we will discuss that in Section 10.3.

With the help of the new parameter r, we generalize the DTFT to a transformation of two variables. Thus:

$$H(v,r) = \sum_{n=-\infty}^{+\infty} h[n]r^{-n}e^{-i2\pi vn} \tag{10.6}$$

Notice that, in the particular case $r = 1$, the transform of Equation (10.6) reduces to the DTFT. Furthermore, depending on the value of r, $H(v,r)$ may or may not converge. The range of r wherein $H(v,r)$ converges defines the ROC of the transform.

Equation (10.6) defines the z-transform of $h[n]$. This is not, however, the most common notation used for the transform. The most common notation, of course, is in terms of the complex number z, defined as:

$$z = re^{i2\pi v} \tag{10.7}$$

Using the definition of Equation (10.7), we can express the transform of Equation (10.6) as:

$$H(z) = \sum_{n=-\infty}^{+\infty} h[n]z^{-n} \tag{10.8}$$

Equation (10.8) is the 'official' direct z-transform. To find the inverse z-transform, we notice that Equation (10.6) is the DTFT of $h[n]r^{-n}$. Consequently, $h[n]r^{-n}$ can be obtained by taking the inverse DTFT of $H(v,r)$. Thus:

$$h[n]r^{-n} = \int_{-0.5}^{+0.5} H(v,r)e^{i2\pi vn}dv \tag{10.9}$$

Expressing $H(v,r)$ as $H(z)$ and rearranging, we get:

$$h[n] = r^n \int_{-0.5}^{+0.5} H(z)e^{i2\pi vn}dv$$

Since the integral is over v, we can bring r inside the integral. Thus:

$$h[n] = \int_{-0.5}^{+0.5} H(z) r^n e^{i2\pi vn} dv = \int_{-0.5}^{+0.5} H(z) z^n dv \qquad (10.10)$$

The last equality defines the inverse z-transform. As usual, we collect the direct and inverse transforms in a box:

Z-TRANSFORM PAIR
DIRECT Z- TRANSFORM

$$X(z) = \sum_{n=-\infty}^{+\infty} x[n] z^{-n}$$

INVERSE Z-TRANSFORM

$$x[n] = \int_{-0.5}^{+0.5} X(z) z^n dv \qquad (10.11)$$

As in the Laplace transform, it is convenient to distinguish between right-sided and left-sided functions. The definition is basically the same as in the Laplace transform: a right-sided function is one that has a 'beginning', but that may never 'end', whereas a left-sided function is one that may never 'begin', but that has an 'end'. Thus, a right-sided function may extend towards positive n, and a left-sided function may extend towards negative n.

We can again use the step function to define right- and left-sided functions. The definition of the discrete step function $u[n]$ is:

$$u[n] = 1 \; for \; n \geq 0$$

$$u[n] = 0 \; for \; n < 0 \qquad (10.12)$$

Unsurprisingly, and again in analogy with the Laplace transform, the ROC of the z-transform of right- and left-sided functions are different. In the following worked exercise, we will find the z-transform and ROC of a right-sided function. In the exercise list, you will be asked to find the z-transform and ROC of a left-sided function.

Worked Exercise: Example of z-transform

Find the z-transform and ROC of the right-sided function $x[n] = a^n u[n]$.

Solution:

According to Equation (10.11), the direct z-transform is given by:

$$X(z) = \sum_{n=-\infty}^{+\infty} x[n]z^{-n} = \sum_{n=-\infty}^{+\infty} a^n u[n]z^{-n}$$

The step function $u[n]$ ensures that all terms of the sum with negative n are zero. Therefore:

$$X(z) = \sum_{n=-\infty}^{+\infty} a^n u[n]z^{-n} = \sum_{n=0}^{+\infty} a^n z^{-n} = \sum_{n=0}^{+\infty} \left(\frac{a}{z}\right)^n$$

This sum only converges if:

$$\frac{|a|}{|z|} < 1$$

Since $|z| = r$ (see Equation (10.7)), the ROC reduces to:

$$r > |a|$$

Within the ROC, $X(z)$ can be found using the trick of shifting the sum by one term. Thus, noticing that:

$$\left(\frac{a}{z}\right) X(z) = \left(\frac{a}{z}\right) \sum_{n=0}^{+\infty} \left(\frac{a}{z}\right)^n = \sum_{n=0}^{+\infty} \left(\frac{a}{z}\right)^{(n+1)} = \sum_{n=1}^{+\infty} \left(\frac{a}{z}\right)^n$$

We find:

$$X(z) - \left(\frac{a}{z}\right) X(z) = \sum_{n=0}^{+\infty} \left(\frac{a}{z}\right)^n - \sum_{n=1}^{+\infty} \left(\frac{a}{z}\right)^n = \left(\frac{a}{z}\right)^0 = 1$$

Therefore:

if $x[n] = a^n u[n]$, then:

$$X(z) = \frac{1}{1 - \dfrac{a}{z}}, \quad ROC : r > |a| \tag{10.13}$$

Equation (10.13) is the z-transform counterpart of Equation (6.23). Notice that their ROC are different: whereas the ROC of Equation (6.23) is a plane delimited by a vertical line crossing the horizontal axis at $s = -\sigma$ (see Figure 6.2a), the ROC

z plane

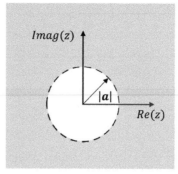

Figure 10.1 Illustration of region of convergence of a right-sided function: the region of convergence is the exterior of a circle.

of Equation (10.13) is the external region delimited by a circle of radius $r = |a|$, as illustrated in Figure 10.1.

The difference between the ROC of Equation (10.13) and the ROC of Equation (6.23) is consequent on the different definitions of s (Equation (6.12)) and z (Equation (10.7)). The motivation for these different definitions is the subject of Section 10.3.

10.3   Relationship Between the z-transform and the Laplace Transform

In Section 6.2 (The Laplace Transform as a Generalization of the Fourier Transform), we defined $s = \sigma + i2\pi f$ (Equation (6.12)). This is a 'Cartesian' definition, in the sense that the significant parameters (σ and f) are the real and imaginary parts of s. In Section 10.2, however, we defined $z = re^{i2\pi v}$ (Equation (10.7)). This is a 'polar' definition, in the sense that the significant parameters (r and v) are the magnitude and phase of z. In this section, we discuss the motivation for a polar definition, instead of a Cartesian one.

The motivation for the polar definition is intimately connected with the periodicity of the DTFT. First, notice that the z-transform, being a generalization of the DTFT, is also periodic in v, with unit period. Indeed:

$$X(v+1, r) = \sum_{n=-\infty}^{+\infty} x[n] r^{-n} e^{-i2\pi(v+1)n} = \sum_{n=-\infty}^{+\infty} x[n] r^{-n} e^{-i2\pi v n} e^{-i2\pi n}$$

$$= \sum_{n=-\infty}^{+\infty} x[n] r^{-n} e^{-i2\pi v n} = X(v, r) \tag{10.14}$$

So, put yourself in the shoes of the person developing this theory. Suppose that, following the footsteps of the great Pierre–Simon Laplace, you decide to generalize $e^{i2\pi v n}$ to $e^{\sigma n} e^{i2\pi v n}$ and to define the complex number $z = \sigma + i2\pi v$, thus leading to

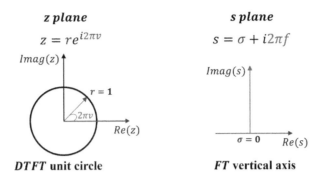

z plane

$z = re^{i2\pi v}$

$Imag(z)$

$r = 1$

$2\pi v$

$Re(z)$

DTFT unit circle

s plane

$s = \sigma + i2\pi f$

$Imag(s)$

$\sigma = 0$ $Re(s)$

FT vertical axis

Figure 10.2 Illustration of the difference between the 'location' of the DFTFT in the z plane and the 'location' of the Fourier Transform (FT) in the s plane. The DTFT is 'located' on the unit circle in the z plane, and the Fourier Transform is located on the vertical axis crossing the origin in the s plane.

the transform $X(v, \sigma) = X(z) = \sum\limits_{n=-\infty}^{+\infty} x[n]e^{-\sigma n}e^{-i2\pi v n} = \sum\limits_{n=-\infty}^{+\infty} x[n]e^{-zn}$. That looks like Laplace's transform, and you are happy that proper tribute has been paid to the French scholar. But then you notice that $X(v + 1, \sigma) = X(v, \sigma)$. That means that the information in the z plane is confined to the region between $-0.5 \leq v < 0.5$. Following the usual convention of ascribing the vertical axis to the imaginary part, that means we are only 'using' a tiny bit of the vertical axis, because the information outside the region $-0.5 \leq v < 0.5$ is redundant: it is just a repetition of the information inside this region. Being the parsimonious person I am sure you are, you ponder if you could find a definition of the variable z that would not waste space in the z plane with the repetitions. So, you notice that if v was the phase of z (as opposed to its imaginary part), then every time you 'walked a distance' of $\Delta v = 1$, you would go full circle and end up at the same point in the z plane. That would be neat, because then two points $X(v, \sigma)$ and $X(v + 1, \sigma)$, which have the same transform, would also be located at the same point in the z plane. That being a clever and elegant solution, you decide to define z in polar coordinates, placing v in the argument, that is, you define $z = re^{i2\pi v}$. Now, two numbers z_1 and z_2, with the same r, but with v separated by a 'distance' $\Delta v = 1$, lie at the same point in the z plane. Indeed, if $v_2 = v_1 + 1$, then:

$$z_2 = re^{i2\pi v_2} = re^{i2\pi(v_1+1)} = re^{i2\pi v_1}e^{i2\pi} = re^{i2\pi v_1} = z_1 \qquad (10.15)$$

That is why the z-transform is defined in polar coordinates. Notice that, whereas the Fourier transform is mapped onto the vertical axis of the s plane (in other words, the Fourier transform lies along the axis $\sigma = 0$, that is, $s = i2\pi f$), the z-transform is mapped onto the unit circle $r = 1$, with the angle defining the value of v. An illustrative comparison between the two planes is shown in Figure 10.2.

10.4 Properties of the z-transform

The properties of z-transforms are closely analogous to the properties of the DTFT. Since their proofs are also similar to the proofs of the DTFT, they are left as exercises. Here, we provide a short list of some of the most important properties.

10.4.1 'Time' shifting

If:

$$g_2[n] = g[n - n_0]$$

then:

$$G_2(z) = z^{-n_0} G(z), \quad ROC \; excludes \; r = 0 \; if \; n_0 > 0 \tag{10.16}$$

10.4.2 Difference

If:

$$g_2[n] = g[n] - g[n - 1]$$

then:

$$G_2(z) = G(z) - z^{-1} G(z) = (1 - z^{-1}) G(z), ROC \; excludes \; r = 0 \tag{10.17}$$

10.4.3 Sum

If:

$$g_2[n] = \sum_{m=-\infty}^{n} g[m]$$

then:

$$G_2(z) = \frac{G(z)}{(1 - z^{-1})}, ROC \; exlcudes \; r = 1 \tag{10.18}$$

10.4.4 Convolution in the Time Domain

If:

$$g[n] = a[n] * b[n]$$

then:

$$G(z) = A(z) B(z) \tag{10.19}$$

10.5   The Transfer Function of Discrete LTI Systems

The transfer function of a discrete LTI system is the z-transform of the impulse response. We learned in Chapter 9 that the output $y[n]$ of a discrete LTI system is related to the input $x[n]$ through a convolution with the impulse response $h[n]$ (Equation (9.11)). Thus, according to the property of convolution in the time domain (Equation (10.19)), the z-transform of the output $Y(z)$ is related to the z-transform of the input $X(z)$ through a product with the transfer function $H(z)$:

$$Y(z) = H(z)X(z) \qquad\qquad (10.20)$$

In Section 6.6 (Zeros and Poles), we learned that the transfer function of a continuous time LTI system can be obtained from the differential equation describing the system by straightforward application of the property of differentiation in the time domain. Analogously, the transfer function of a discrete LTI system can be obtained from the difference equation describing the system using the property of 'time' shifting (Equation (10.16)). You will be asked to do that in the exercise list.

In Section 6.2 (The Laplace Transform as a Generalization of the Fourier Transform), we noticed that the ROC of the Laplace transform of a right-sided function extends to the right part of vertical line in the s plane. Since, due to causality, the impulse response of a physical system is a right-sided function, then the ROC of the transfer function must also extend to the right part of a vertical line in the s plane. In the worked exercise of Section 10.2 (The z-transform as a Generalization of the DTFT), we learned that the ROC of the z-transform of a right-sided function is exterior to a circle in the z-plane (see Figure 10.1). Thus, if a discrete LTI system is causal, then the ROC of its transfer function must also be exterior to a circle in the z-plane. Recall, however, that the criterion of causality is relaxed in discrete systems: they do not necessarily need to be causal.

Finally, recall that the ROC of the transfer function of a stable continuous time LTI system must include the Fourier Transform, that is, it must include the vertical axis crossing the origin ($\sigma = 0$) of the s plane (see Section 6.2 [The Laplace Transform as a Generalization of the Fourier Transform], and Figure 10.2). Analogously, the ROC of the transfer function of a stable discrete LTI system must include the DTFT, that is, it must include the unit circle ($r = 1$) in the z plane (see Figure 10.2).

10.6   The Unilateral z-transform

Recall, from Section 6.7 (The Unilateral Laplace Transform), that the unilateral Laplace transform was introduced to deal with signals of the form $g(t)u(t)$, that is, with signals that 'begin' at time $t = 0$. We also learned in the same section that this transform is quite useful to deal with initial conditions.

The counterpart of the unilateral Laplace transform is the unilateral z-transform. Thus, we define the unilateral z-transform of $g[n]$ as the z-transform of the signal $g[n]u[n]$:

$$G(z) = \sum_{n=-\infty}^{+\infty} g[n]u[n]z^{-n} = \sum_{n=0}^{+\infty} g[n]z^{-n} \tag{10.21}$$

The last sum in Equation (10.21) defines the unilateral z-transform.

The properties of the unilateral z-transform are basically identical to the properties of the z-transform, except for the 'time' shifting property and, consequently, for the difference property.

To find the property of 'time' shifting of unilateral z-transforms, we need to express the unilateral z-transform of $g[n-1]$ in terms of the unilateral z-transform of $g[n]$. Then, we use this result to find the property for an arbitrary $g[n - n_0]$.

To find the unilateral z-transform of $g[n-1]$, we first note that this is equivalent to the z-transform of $g[n-1]u[n]$. We also note that the property of 'time' shifting (Equation (10.16)) entails that, if $G(z)$ is the z-transform of $g[n]u[n]$, then the z-transform of $g[n-1]u[n-1]$ must be $z^{-1}G(z)$. Now, note that the only difference between $u[n]$ and $u[n-1]$ is at the origin $(n = 0)$: the former is one at the origin, whereas the latter is zero at the origin. Consequently, we can express $u[n-1]$ in terms of $u[n]$ by subtracting a function that is zero everywhere, expect at the origin, from the function $u[n]$. The impulse function (Equation (9.4)) is such a function. Thus:

$$u[n-1] = u[n] - \delta[n] \tag{10.22}$$

Now we can use the fact that the z-transform of $g[n-1]u[n-1]$ is $z^{-1}G(z)$ to find the z-transform of $g[n-1]u[n]$. Thus, from the time-shifting property (Equation (10.16)), we know that:

$$\sum_{n=-\infty}^{+\infty} g[n-1]u[n-1]z^{-n} = z^{-1}G(z) \tag{10.23}$$

Using Equation (10.22) on the left-hand side of Equation (10.23):

$$\sum_{n=-\infty}^{+\infty} g[n-1]u[n-1]z^{-n} = \sum_{n=-\infty}^{+\infty} g[n-1](u[n] - \delta[n])z^{-n}$$

Consequently:

$$\sum_{n=-\infty}^{+\infty} g[n-1]u[n]z^{-n} - \sum_{n=-\infty}^{+\infty} g[n-1]\delta[n]z^{-n} = z^{-1}G(z)$$

Since the impulse function is nonzero only at the origin (see Equation (9.4)), the second sum reduces to the term $n = 0$:

$$\sum_{n=-\infty}^{+\infty} g[n-1]\delta[n]z^{-n} = g[0-1]\delta[0]z^{-0} = g[-1]$$

Consequently:

$$\sum_{n=-\infty}^{+\infty} g[n-1]u[n]z^{-n} = z^{-1}G(z) + g[-1]$$

We have thus proved that:

if:

   $G(z)$ is the unilateral z-transform of $g[n]$
   Then, the unilateral z-transform of $g_1[n] = g[n-1]$ is:

$$G_1(z) = z^{-1}G(z) + g[-1] \tag{10.24}$$

Equation (10.24) can be used to find the property of higher order time shifting. Thus, for example, if $G_1(z)$ is the unilateral z-transform of $g_1[n] = g[n-1]$, then the unilateral z-transform of $g_2[n] = g_1[n-1] = g[n-2]$ must be:

$$G_2(z) = z^{-1}G_1(z) + g_1[-1] = z^{-1}[z^{-1}G(z) + g[-1]] + g[-2] \tag{10.25}$$

Notice that $g[-1]$ and $g[-2]$ are initial conditions of the signal. In the exercise list, you will be asked to apply the property of time shifting of unilateral z-transforms to solve a difference equation involving initial conditions.

These are the essential concepts of z-transforms.

And with that we part ways, at least for today. I hope you have enjoyed learning the essential concepts of signals and systems and I wish you good luck in your further studies.

10.7 Exercises

Exercise 1

Find the z-transform and ROC of the left-sided function $x[n] = b^n u[-n]$.

Exercise 2

Obtain, by inspection, the inverse z-transform of:

$$H(z) = 5\frac{1}{1-\dfrac{z}{9}} + 4\frac{1}{1-\dfrac{3}{z}}, \quad ROC: 3 < r < 9$$

Exercise 3

If $H(z)$ of Exercise 2 of Chapter 10 is the transfer function of an LTI system, is this a stable or unstable system? And is the system causal?

Exercise 4
Suppose that the impulse response of a discrete LTI system has been stored in a computational vector with N points. Further, suppose that you use the *fft* to obtain the DFT of the impulse response. Where are these N DFT samples 'located' in the z plane?

Exercise 5
Prove the time-shifting property of the z-transform (Equation (10.16)).

Exercise 6
Prove the property of convolution in the time domain (Equation (10.19)).

Exercise 7
Consider an LTI discrete system described by the following difference equation:

$$\sum_{m=0}^{M} b_m y[n-m] = \sum_{p=0}^{P} a_p x[n-p]$$

where $y[n]$ is the output and $x[n]$ is the input. Find the transfer function of the system.

Exercise 8
If the z-transform of $g[n]u[n]$ is $G(z)$, find the z-transform of $(g[n] - g[n-1])u[n]$ and the z-transform of $g[n]u[n] - g[n-1]u[n-1]$ in terms of $G(z)$.

Exercise 9
Consider a digital filter defined by the following difference equation:

$$2y[n] - y[n-1] = x[n]$$

where $y[n]$ and $x[n]$ are the output and input, respectively.
(a) Assuming $x[n] = \delta[n]$ and $y[-1] = 5$, apply the unilateral z-transform to both sides of the difference equation to obtain $Y(z)$. Then, find $y[n]$ from $Y(z)$ by inspection.

(b) Check whether your solution indeed satisfies the difference equation by evaluating $y[0]$, $y[1]$, and $y[2]$ explicitly using the difference equation and comparing the result with the result predicted by the solution obtained in part (a).

Exercise 10
Consider a signal $g_2[n] = g[n]u[n]$. The initial value theorem asserts that:

$$\lim_{z \to \infty} G_2(z) = g[0]$$

where $G_2(z)$ is the z-transform of $g_2[n]$. Prove the initial value theorem.

References with Comments

1. **Theory of representation**
   The emphasis on the algebraic meaning of the theory of representation follows the same philosophy of Cohen-Tannoudji '*Quantum Mechanics*', Volume 1, Chapter 2. The main difference is that Cohen focuses on spatial frequencies, but the theory is the same.

2. **Impulse function and impulse response**
   A more advanced and formal treatment of impulse functions can be found in G. Barton '*Elements of Green's Functions and Propagations Potentials, Diffusion, and Waves*'. Barton's book is a good introduction to Green's functions, which are a generalization of the description of systems through the impulse response.

3. **Fourier Transforms**
   A more advanced and formal treatment of Fourier Transforms can be found in Ronald N. Bracewell '*The Fourier Transform and its Applications*'.

4. **Signals and system**
   Two references for further studies in signals and systems are Allan Oppenheim '*Signals and Systems*' and Simon Haykin '*Signals and Systems*'.

5. **Digital signal processing and digital filters**
   Two world reference in digital signal processing are Allan Oppenheim '*Discrete-time Signal Processing*' and John G. Proakis '*Digital Signal Processing*'.

Essentials of Signals and Systems, First Edition. Emiliano R. Martins.
© 2023 John Wiley & Sons Ltd. Published 2023 by John Wiley & Sons Ltd.
Companion website: www.wiley.com/go/martins/essentialsofsignalsandsystems

Appendix A: Laplace Transform Property of Product in the Time Domain

According to Equation (2.152) of Section 2.12.9 (Product in the Time Domain): if:

$$g(t) = a(t)b(t)$$

then:

$$G(f) = A(f) * B(f)$$

That is, product in the time domain corresponds to convolution in the frequency domain. This property looks a bit different in the Laplace domain. We begin by writing the Laplace transform of $g(t)$ explicitly:

$$G(s) = \int_{-\infty}^{+\infty} g(t)e^{-st}dt = \int_{-\infty}^{+\infty} a(t)b(t)e^{-st}dt$$

Now we express $a(t)$ in terms of their inverse Laplace transform. For the sake of clarity, I will use the form of Equation (6.16). Thus:

$$a(t) = \int_{-\infty}^{+\infty} A(\sigma, f)e^{\sigma t}e^{i2\pi ft}df$$

$$b(t) = \int_{-\infty}^{+\infty} B(\sigma, f)e^{\sigma t}e^{i2\pi ft}df$$

Therefore:

$$G(s) = \int_{-\infty}^{+\infty} \left[\int_{-\infty}^{+\infty} A(\sigma, f)e^{\sigma t}e^{i2\pi ft}df\right]\left[\int_{-\infty}^{+\infty} B(\sigma, f)e^{\sigma t}e^{i2\pi ft}df\right]e^{-st}dt$$

Rearranging:

$$G(s) = \int_{-\infty}^{+\infty} \int_{-\infty}^{+\infty} \int_{-\infty}^{+\infty} A(\sigma_1, f_1)B(\sigma_2, f_2)e^{(\sigma_1+\sigma_2-\sigma)t}e^{i2\pi(f_1+f_2-f)t}dtdf_2df_1$$

Essentials of Signals and Systems, First Edition. Emiliano R. Martins.
© 2023 John Wiley & Sons Ltd. Published 2023 by John Wiley & Sons Ltd.
Companion website: www.wiley.com/go/martins/essentialsofsignalsandsystems

But, according to Equation (2.77):

$$\int_{-\infty}^{+\infty} e^{i2\pi(f_1+f_2-f)t}dt = \delta(f_1 + f_2 - f)$$

Therefore:

$$G(s) = \int_{-\infty}^{+\infty} \int_{-\infty}^{+\infty} A(\sigma_1, f_1)B(\sigma_2, f_2)e^{(\sigma_1+\sigma_2-\sigma)t}\delta(f_1 + f_2 - f)df_2df_1$$

Rearranging the argument of the impulse function:

$$G(s) = \int_{-\infty}^{+\infty} \int_{-\infty}^{+\infty} A(\sigma_1, f_1)B(\sigma_2, f_2)e^{(\sigma_1+\sigma_2-\sigma)t}\delta(f_2 - (f - f_1))df_2df_1$$

Integrating over df_2 and using the sifting property of the impulse function:

$$G(s) = \int_{-\infty}^{+\infty} A(\sigma_1, f_1)B(\sigma_2, f - f_1)e^{(\sigma_1+\sigma_2-\sigma)t}df_1$$

Thus, we have concluded that:

if:

$$g(t) = a(t)b(t)$$

then:

$$G(\sigma, f) = \int_{-\infty}^{+\infty} A(\sigma_1, f_1)B(\sigma_2, f - f_1)e^{(\sigma_1+\sigma_2-\sigma)t}df_1 \qquad (A.1)$$

Equation (A.1) is the Laplace transform property of product in the time domain. Notice that it reduces to Equation (2.152) in the particular case of $\sigma = \sigma_1 = \sigma_2 = 0$, as expected.

Appendix B: List of Properties of Laplace Transforms

A list of useful properties of Laplace transforms is provided in the table.

Time domain	Laplace domain	ROC
$g_2(t) = g(t - t_0)$	$G_2(s) = e^{-st_0}G(s)$	The ROC of $G_2(s)$ coincides with the ROC of $G(s)$
$g_2(t) = e^{s_0 t}g(t)$	$G_2(s) = G(s - s_0)$	If s lies within the ROC of $G(s)$, then $s + Re\{s_0\}$ lies within the ROC of $G_2(s)$
$g_2(t) = \dfrac{dg(t)}{dt}$	$G_2(s) = sG(s)$	The ROC of $G_2(s)$ includes the ROC of $G(s)$
$g_2(t) = \int_{-\infty}^{t} g(\tau)d\tau$	$G_2(s) = \dfrac{G(s)}{s}$	Same as the right part of the ROC of $G(s)$ (the right part is $Re\{s\} > 0$; notice that it excludes the origin).
$g_2(t) = a(t) * b(t)$	$G_2(s) = A(s)B(s)$	The ROC of $G_2(s)$ includes the intersection of the ROC of $A(s)$ with the ROC of $B(s)$
$g_2(t) = -tg(t)$	$G_2(s) = \dfrac{dG(s)}{ds}$	The ROC of $G_2(s)$ coincides with the ROC of $G(s)$
$g_2(t) = g(\alpha t)$	$G_2(s) = \dfrac{1}{|\alpha|}G\left(\dfrac{s}{\alpha}\right)$	If s lies within the ROC of $G(s)$, then αs lies within the ROC of $G_2(s)$

Essentials of Signals and Systems, First Edition. Emiliano R. Martins.
© 2023 John Wiley & Sons Ltd. Published 2023 by John Wiley & Sons Ltd.
Companion website: www.wiley.com/go/martins/essentialsofsignalsandsystems

Index

Essentials of Signals and Systems, First Edition. Emiliano R. Martins.
© 2023 John Wiley & Sons Ltd. Published 2023 by John Wiley & Sons Ltd.
Companion website: www.wiley.com/go/martins/essentialsofsignalsandsystems